Record Label Marketing

Record Label Marketing, Third Edition is the essential resource to help you understand how recorded music is professionally marketed. Fully updated to reflect current trends in the industry, this edition is designed to benefit marketing professionals, music business students, and independent artists alike.

As with previous editions, the third edition is accessible for readers new to marketing or to the music business. The book addresses classic marketing concepts while providing examples that are grounded in industry practice. Armed with this book, you'll master the jargon, concepts, and language you need to understand how music companies brand and market artists in the digital era.

Features new to this edition include:

■ Social media strategies, with coverage of step-by-step tactics used by major and independent labels, presented in a new section contributed by Ariel Hyatt, owner of CYBER PR®

■ An in-depth look at SoundScan and other big data matrices used as tools by all entities in the music business

■ An exploration of the varieties of branding, with particular attention paid to the impact of branding on the artist and the music business, in a new chapter contributed by Tammy Donham, former Vice President of the Country Music Association

■ A robust companion website (**focalpress.com/cw/macy**) with web links, exercises, suggestions for further reading and instructor resources such as PowerPoint lecture outlines, a test bank, and lesson plans

Amy Macy is an Associate Professor and recent Music Business Internship Coordinator who received both her undergraduate degree in Music Education and her Master's degree in Business Administration from Belmont University. She has taught Marketing of Recordings, Record Retail Operations, Survey of the Recording Industry, The Nashville Music Business, and the Lecture Series, developed and delivered online courses in all these subjects, and headed up an Old Time String Band Music Ensemble. Amy is co-author with Tom Hutchison and Paul Allen of Record Label Marketing, editions 1 and 2.

Clyde Philip Rolston is Professor of Music Business in the Mike Curb College of Entertainment and Music Business at Belmont University. Prior to joining the faculty at Belmont University, he was a Vice President of Marketing at Centaur Records, Inc. Dr. Rolston received a PhD in Marketing from Temple University and has taught marketing to music business students for over 12 years.

Record Label Marketing

How Music Companies Brand and Market Artists in the Digital Era

Third Edition

Amy Macy, Clyde Rolston, Paul Allen, and Tom Hutchison

Focal Press
Taylor & Francis Group

NEW YORK AND LONDON

This edition published 2016
by Focal Press
711 Third Avenue, New York, NY 10017

and by Focal Press
2 Park Square, Milton Park, Abingdon, Oxon OX14 4RN

Focal Press is an imprint of the Taylor & Francis Group, an informa business

Notices
Knowledge and best practice in this field are constantly changing. As new research and experience broaden our understanding, changes in research methods, professional practices, or medical treatment may become necessary.

Practitioners and researchers must always rely on their own experience and knowledge in evaluating and using any information, methods, compounds, or experiments described herein. In using such information or methods they should be mindful of their own safety and the safety of others, including parties for whom they have a professional responsibility.

Product or corporate names may be trademarks or registered trademarks, and are used only for identification and explanation without intent to infringe.

First edition published by Focal Press, 2006
Second edition published by Focal Press, 2010

Library of Congress Cataloging-in-Publication Data
Hutchison, Thomas W. (Thomas William)
   Record label marketing : how music companies brand and market artists in the digital
era / Tom Hutchinson, Amy Macy, Paul Allen. — Third edition.
      pages cm
  1. Music trade. 2. Sound recordings—Marketing. 3. Sound recording industry. I. Macy,
Amy. II. Allen, Paul, 1946– III. Title.
   ML3790.H985 2016
   780.68′8—dc23
   2015012801

ISBN: 978-0-415-71515-7 (hbk)
ISBN: 978-0-415-71514-0 (pbk)
ISBN: 978-1-315-88205-5 (ebk)

Typeset in Kuenstler
by Apex CoVantage, LLC

Printed and bound in the United States of America by Sheridan Books, Inc. (a Sheridan Group Company).

The Third Edition of Record Label Marketing is dedicated to our friend, colleague, and coauthor, Tom Hutchison.

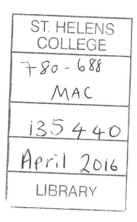

Contents

Acknowledgments

Amy Macy

I especially want to thank my coauthor, Clyde Rolston, who put up with my bad eating habits and poor grammar to make this 3rd edition happen—you are a superb writing partner! Also to Tammy Donham and Ariel Hyatt whose chapter contributions are invaluable. I'm grateful for my friendships, old and new, with those clever and talented folks who "are" the music business. This 3rd edition is built on the work of colleagues Tom Hutchison (who I think of daily on matters bookish and curricular) and Paul Allen, who has been an immeasurable support during this last writing. Thanks to new Leadership Music friends Tim Roberts, Operations/Program Director, CBS Radio, and Rob Simbeck, writer, Bob Kingsley's American Country Countdown; who have lent their expertise, advice, and support on many subjects throughout this project. Thanks to "old" friends Danny Bess, VP of Finance, UMG Nashville, Debbie Linn, Executive Director of Leadership Music, and John Conway, Senior Director, National Sales and Label Relations at Sony Music Entertainment, for their never-ending connection to the "real" world and for their on-going friendship. How 'bout Jim Lidestri and Chris Muratore at Border City Media BuzzAngle Music for blazing new trails in "big data" for music—you guys are thinking! Thanks for sharing your insights. And a BIG thank you for making "it" happen with our Nielsen friends to Josh Bennett and David Bakula.

I stay motivated by my colleagues at Middle Tennessee State University and the Department of Recording Industry who maintain an educational integrity that I so admire. Thank you to the students of this outstanding program who keep me inspired and let me be a part of their dreams. Mucho Gracias to the dance moms who helped to keep my lil' dancers going when I couldn't be there, especially Alice Charron. And lastly, to my family: Doug, Millie Mae, Emma Jo, Grammy, and Choo Choo—who went on without me, endured my all-nighters, listened to my anxieties, and gave me hope. Thank you all.

Clyde Rolston

I would like to thank my coauthor, Amy Macy, for her perseverance, guidance and positive attitude throughout the process of writing this edition. I am also grateful to Tom Hutchison for the opportunity to be a part of this book; to Tom and Paul Allen for their contributions to earlier editions that reverberate through the current edition. Many thanks to the colleagues who reviewed early manuscripts and whose thoughtful comments helped make this a better book. A special thanks to Ariel Hyatt, Tammy Donham, Jim Donio, Dave Pomery, Don Passman, and Fred Buc for their contributions to the book. I owe a big debt of gratitude to my students, department chair Rush Hicks, and my dean, Wesley Bulla, for allowing me the time and distraction of being a textbook writer. Thank you to my colleagues for their support, especially Larry Wacholtz, Sarita Stewart, Cheryl Slay-Carr, Dan Keen, Richard Churchman, and Barry Padgett. And most of all, thank you to my wife, Allison, and my children, Elliott, Jessica, and Kristen, whose faith, love, and support make all of my days better.

About the Authors

Amy Sue Macy, Professor in the Department of Recording Industry at Middle Tennessee State University, received both her undergraduate degree in Music Education and her Masters degree in Business Administration from Belmont University. For fifteen years, she worked for various labels including MTM, MCA, Sparrow Records, and the RCA Label Group. Amy created strategic marketing plans for launching new releases into the marketplace, working closely with artists and their managers including Martina McBride, Kenny Chesney, Clint Black, Alabama, and Lonestar. While maintaining a marketing focus, Amy was responsible for all sales at the national retail level with clients including Walmart, Kmart, Target, Best Buy, and Musicland, to name a few, all the while communicating key marketing strategies coast to coast with RLGs national distributor Bertelsmann Music Group (BMG).

Since securing her teaching gig at MTSU, she has served as the Music Business Internship Coordinator for eight years and has taught Marketing of Recordings, Record Retail Operations, Survey of the Recording Industry, the Lecture Series, and an Old Time String Band Music Ensemble and the student-run record label MATCH Records. She also teaches online utilizing the Desire2Learn Online System integrating Web conferencing software, collaborative study through wikis, virtual tests, and links to various websites and live "spreadsheet" activities. She has been the recipient of various teaching awards including the 2009 Distinguished Educator in Distance Learning and the 2009–2010 Outstanding Achievement in Instructional Technology.

Amy's love of music extends beyond the business world. She initially moved to Nashville to be a musician and has toured professionally with several artists internationally. She is a vocalist and is accomplished on guitar, fiddle, and banjo. Amy has performed in musicals at Nashville's famous Ryman Auditorium as well as the Tennessee Performing Arts Center, has been an artist in residence at the Country Music Hall of Fame and has recently created a concert series focusing on the "story behind the song" which highlights Civil War and Irish music that has migrated to America.

LinkedIn: linkedin.com/in/amymacy

Clyde Philip Rolston is Professor of Music Business in the Mike Curb College of Entertainment and Music Business at Belmont University. Prior to joining the faculty at Belmont University he was a Vice President of Marketing at Centaur Records, Inc. While with Centaur Records, Dr. Rolston engineered and produced many projects, including recordings by the Philadelphia Trio and the London Symphony Orchestra. He is an active member of the Music and Entertainment Industry Educators Association. Dr. Rolston received his Bachelors and Masters degrees from Louisiana State University and a PhD in Marketing from Temple University. He has taught merchandising and marketing to music business students for over 20 years. His research interests include music marketing, music acquisition and consumption, international music business, and consumer behavior.

For the past three years he has co-produced with Dan Keen a show for Music City Roots featuring alumni and current students from Belmont University. He has published articles on the music industry in the MEIEA Journal, the Journal of Arts Management, Law and Society, the NARAS Journal, and numerous conference proceedings.

Paul Allen is Associate Professor in the Department of Recording Industry at Middle Tennessee State University and coauthor of Record Label Marketing, also published by Focal Press. He is also a frequent lecturer at other universities on artist management and other music business subjects. His career work has included radio, TV, political management, and the music business.

The late **Tom Hutchison** was a professor of marketing in the Department of Recording Industry at MTSU, but was on leave to serve as the director of the School of Business and Management at Husson University. He worked with a wide range of popular artists including Faith Hill, The Dixie Chicks, The Roots, and Beck. Tom also conducted market research projects for Sony, MCA/Universal, DreamWorks, and Warner Music Group.

Contributors

Tammy Donham received her undergraduate degree in Marketing from Western Kentucky University and her Master's degree in Business Administration from Middle Tennessee State University. Donham worked for Fruit of the Loom, Inc. in several marketing capacities before moving to Nashville in 1996 where she worked for nearly 17 years for the Country Music Association (CMA).

She held various marketing-related positions within CMA ultimately rising to Vice-President of Marketing. While in her post as VP, she oversaw all marketing, creative services, and research efforts for the CMA Awards, CMA Music Festival and CMA Country Christmas events and television specials, including broadcast, digital, radio, out-of-home and print initiatives. She was CMA's lead liaison with ABC Television Marketing, Synergy and Affiliate teams and worked closely with these and other event partners to maximize promotional and brand-building opportunities for CMA properties across all platforms. Donham is a graduate of Leadership Music, as well as a member of the Academy of Television Arts & Sciences and the Country Music Association.

Donham began teaching Marketing of Recordings and Digital Strategies for the Music Business in the Recording Industry Department at Middle Tennessee State University in the fall of 2013.

LinkedIn: linkedin.com/in/tammydonham

Ariel Hyatt is the founder of a successful PR firm Cyber PR®, an international speaker, and author of four books (three out and one coming in 2015). Her trademarked, award winning Cyber PR® process marks the intersection of social media, PR, and online marketing. Her PR method is taught at several universities. However she is best known in her industry for her ability to simplify and explain things that creative minds don't necessarily love: PR, marketing, and social media.

She is so good at un-confusing, that she has been invited to present at over 70 conferences in 12 countries, including SXSW (where she has appeared 15 times), CMJ, Vivid Sydney, Hubspot Ignite, Campus Party London, The 140 Conference, You Are In Control (Reykjavik), and Social Media Week New York. Ariel's work has been lauded by established media outlets garnering her press in Oprah, CNN, *Wired*, *Billboard*, *Forbes*, and *The Washington Post*.

An entrepreneur who embraces challenge, Ariel taught herself crowdfunding by raising $61,000 in her own successful crowdfunding campaign. "I see crowdfunding as the new 'advance' book publishers are not giving them, record labels aren't giving them and small business loans from banks are next to impossible hard to secure for the creative class. Crowdfunding must be mastered."

In order to help people gain mastery she has written her latest book *Crowdstart, on Fears, Fans and Funding*. It will be released officially in the spring of 2015. It's a step-by-step guide on exactly what you need to do to complete a successful crowdfunding campaign extracted not only from Ariel's own experience, but also from many of the successful crowdfunding campaigns she has coached her clients through.

Ariel is also the author of three successful books on PR and new media, *Music Success in 9 Weeks* (now out in its 3rd edition), *Musician's Roadmap to Facebook & Twitter*, and *Cyber PR For Musicians*. These offer artists step-by-step plans to create profitable and sustainable businesses. Her newsletter and YouTube series, "Sound Advice," has over 20,000 subscribers, and she has written over 200 articles to date.

Ariel proudly serves on the advisory boards of Sweet Relief Musicians Fund, SXSW Accelerator, and on the education committee of The Moth.

LinkedIn: linkedin.com/in/arielhyatt

CHAPTER 1

Introduction

SUCCESS AND WHERE IS THIS GOING?

What is success in the music business? In today's environment, the answer to this question is a moving target:

- Is it a platinum-selling artist who has achieved household name status (i.e., Beyoncé, Bruno Mars, Katy Perry)?
- Is it an independent, do-it-yourself singer/songwriter like David Wilcox or Derek Webb who travels from venue to venue and sells 50,000 albums out of the back of their car, performs 90 live shows in year, and sells thousands of t-shirts and other cool merchandise?
- Or is it someone in between?

Most in the industry would answer: "It depends!" Who are the stakeholders in the equation? Does a record label sell enough records to cover expenses but to allow the company to invest in the next developing artist while paying its employees and keeping the lights on? Or does the independent artist make the music they desire while feeling in control of their artistic destiny and making a living? If both entities are satisfied with the answer—then ALL win.

In the current marketplace, the artist and their sound recording have certainly moved way past the sale of a CD in a brick-and-mortar retail store. The exchange for music has burgeoned into a multitude of opportunities as well as challenges. The adoption of the digital mp3 file at the consumer level continues to shift from "ownership" of music towards an

experience of service such as streaming and the **interaction** of a live concert in front of the artist.

This transference from ownership to mere **experience and interaction** has transformed the bottom line of the entire music industry. Look at where the business has moved. Overall music sales slipped –.3% between 2012 and 2013, physical sales dropped –12.3%. The trend has commenced to digital and not just in digital sales, but in revenues generated through streaming services, both free and subscription based. Digital sources grew 7.6% making digital sales $4.3 billion of the total $6.996 billion of 2013. That is 64% of the total revenue generated through music sales. Out of the $4.3 billion that is digital, $2.8 billion belongs to digital downloads with the remaining revenues generated from streaming and this sector is projected to only grow.

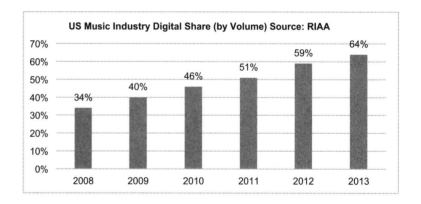

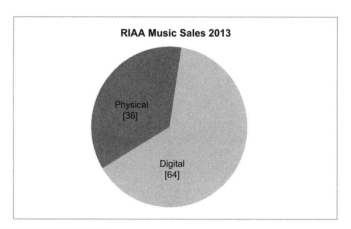

FIGURES 1.1, 1.2, 1.3, 1.4 *SoundExchange Revenue*

Source: RIAA

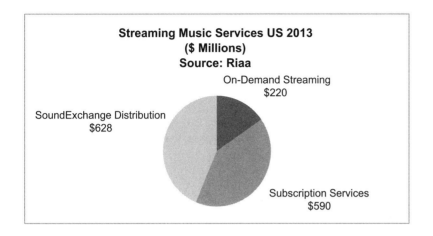

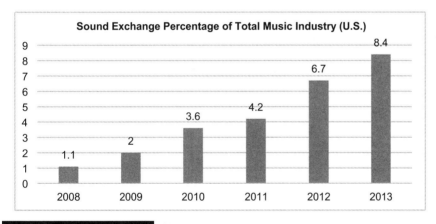

FIGURES 1.1, 1.2, 1.3, 1.4 *(Continued)*

The last graphic in this line-up shows that in 2013, 8.4% of the $7 billion generated in record sales were produced through SoundExchange, a service that collects royalties from the streaming of the Sound Recordings (not songs) and pays the label and artists. This number was only 1.1% in 2008, and this shift confirms the significance of streaming in today's value chain that comes from both subscription-based and ad-supported sites (SoundExchange, 2014).

The 2014 SoundScan data shows a similar trend. On-demand music streaming which includes data from AOL, Beats, Cricket, Google Play, Medianet, Rdio, Rhapsody, Slacker, Spotify, Xbox Music, and YouTube/Vevo reflects a year-over-year 2013 to 2014 increase +54.5%. The data below has been converted to album sales with downloaded songs reflecting 10 downloaded songs = 1 album and 1,500 streams (video and/or audio) = 1 album. And yet, all this streaming did not make up for the loss in overall sales, which declined –2.0% from 2013 to 2014. Good news: consumers are engaging music more than ever before with transactions in the form of sales or streams up by 54%. Bad news: these transactions are not translating into monetary gains to all in the record company revenue stream.

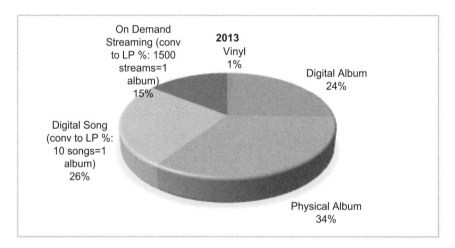

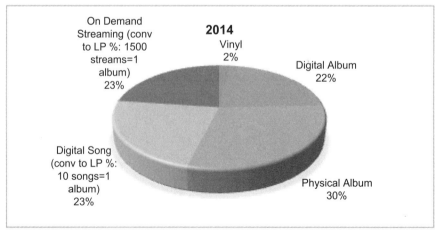

FIGURE 1.5   *2014 Sales by Configuration—Source Nielsen Year End Report 2014*

To stay viable and relevant, the music company needs to show value to not only the consumer, but to the artist and their music. Read on to appreciate a myriad of successful marketing efforts mounted by the industry, both major label effort and independents—highlighting that there is more than one pathway to success.

WHAT CREATES SUCCESS?

Princeton researcher Matthew J. Salganik has studied the notion that artistic success is *not* based on best performance or the quality of the music, but on chance or dumb luck. Dr. Salganik and his colleagues created nine identical online worlds that each contained 48 of the same pop/rock songs. In these parallel worlds, 30,000 students were evenly split up and asked to choose and download songs that they liked—thus creating a chart of best-liked songs in each of the nine worlds (Spiegel, 2014).

What Dr. Salganik discovered is that above a certain quality threshold, each "virtual" world created a very different "popularity" chart with none of the same songs downloaded as the top 10 tunes in any domain. So without the power of radio, charts, social media, word of mouth, or other influences, each song was left to its own quality, appeal, and chance as to whether it would be chosen for download.

Conversely, with the power of influence—radio airplay, social media, imaging, word of mouth, etc.—a quality recording has a greater chance to succeed. The record label can deliver this influence. Ideally, a label should be the branding agent, the megaphone, and the central intelligence agency on behalf of the artist and recording to create awareness of not only the music but that artist who created it.

MUSIC WITHIN THE ENTERTAINMENT INDUSTRY

Sound Recordings are an integral part of something much larger: the entertainment industry. Practically every segment of the entertainment economy is tethered by a sound recording, save the book and magazine segment. If analyzed alone, music sales of sound recordings are ranked tenth in revenue generation as a silo of entertainment products. However, imagine the other entertainment items without music as a part of their equation: movies, gaming, radio, theme parks, etc. None of these entities would be nearly as dynamic as stand-alone products without the enhancement of music as icing for their cake.

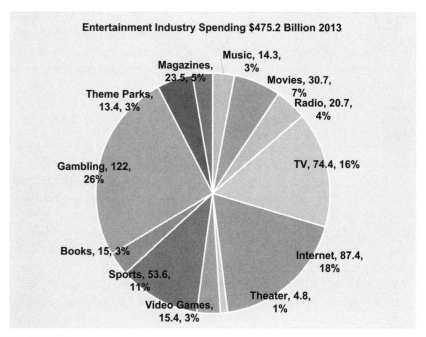

FIGURE 1.6 *Entertainment Industry Spending 2013*

Sources Listed: (Hendrix, 2013; PwC Price Waterhouse Coopers, 2013; Palmeri, 2013; Eichelberger, 2013; Revenue Releases/RAB, 2014; "When are Mobile Owners Using Apps?" 2014)

Branding

Creating a "household name" means that your act has broken through the clutter and found a footing in the daily vernacular of the culture. Branding of an act, be it through the music, an image, a strategic entertainment alliance such as a movie soundtrack, a partnership with a corporation or non-music item, has everything to do with the success of the artist.

> *Branding: The process involved in creating a unique name and image for a product in the consumers' mind. Branding aims to establish a significant and differentiated presence in the market that attracts and retains loyal customers.*

("Branding," 2014)

Music and its artists are innately unique, which shifts the branding and marketing challenge from product quality differences such as sonic superiority and pricing to that of artist individuality, song meaning, timing in the marketplace, and brand partnerships.

SO WHO ZOOMS WHOM?

With the introduction and adoption of the internet by consumers, the traditional business model of the music industry has become challenged and some would even say that it is broken or in "entropy"—being total disorder. But if you follow the money, it's (somewhat) easy to see a framework of players who have a stake in the success of the music and its artists.

REVENUE STREAMS OF THE INDUSTRY

Let's follow the money. Consumers start the flow of money in each of the revenue streams and although they look like silos, they are co-dependent on one another to exist. Without the song, there isn't a sound recording. Without the sound recording, there isn't exposure via radio or internet play. Without exposure, there isn't an artist in demand to go on tour. Without a tour, there aren't live venues hosting bands and selling merchandise. And who sits at the top of each

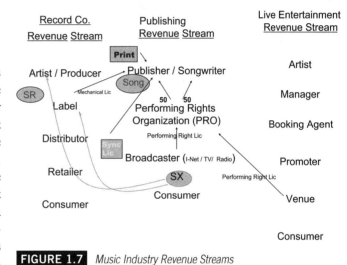

FIGURE 1.7 *Music Industry Revenue Streams*

of these revenue streams? It's the artist. From a probability standpoint, the artist has a strong chance of making the money, if the act manages his or her business smartly and makes him- or herself a brand.

RECORD LABEL SURVIVAL TODAY

Today's business model looks to take the artist and make them a brand. Label revenues no longer can rely on sales of CDs and downloads to succeed but must leverage the asset of the sound recording as well as the artist's brand to maximize revenue potential. The development of the 360 Deal advanced as the evolution of the digital download took hold with consumers, especially free downloading, which produced the long slide in album sales and the even longer drop of overall music sales since its all-time high in 2000 of $14 billion to something near $7 billion—and sliding.

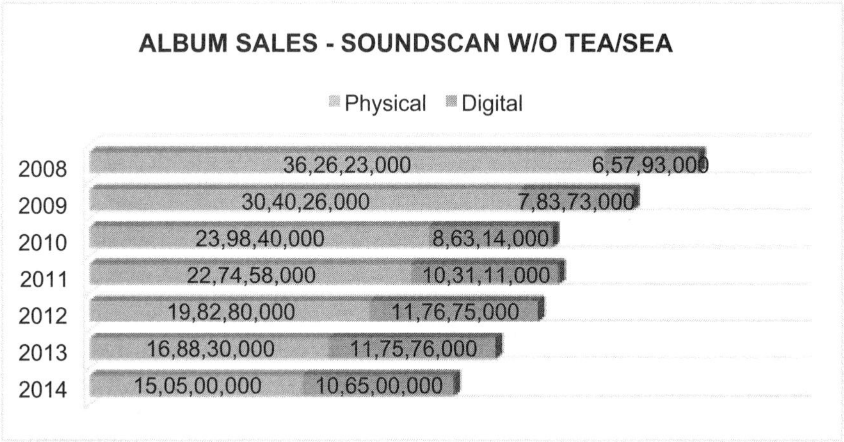

ALBUM SALES - SOUNDSCAN W/O TEA/SEA

■Physical ■Digital

Year	Physical	Digital
2008	36,26,23,000	6,57,93,000
2009	30,40,26,000	7,83,73,000
2010	23,98,40,000	8,63,14,000
2011	22,74,58,000	10,31,11,000
2012	19,82,80,000	11,76,75,000
2013	16,88,30,000	11,75,76,000
2014	15,05,00,000	10,65,00,000

FIGURE 1.8 *Album Sales without TEA and SEA*

Source: SoundScan

And no—digital single sales do not make up for the loss of album sales, which was a great hope. And streaming does not look to be the answer for labels either (as earlier revealed)—hence the inception of the 360 Deal. Artists have questioned the need for labels and many try the DIY (do-it-yourself) model. But managing all the pieces can take away from the creative juice and it is expensive to make yourself a house-hold name. The "other" answer would be if artists can bank roll their own artistic endeavors, including that of the A&R process, recording costs, imaging, publicity, radio airplay, social media, retail marketing, tour support, and endorsement strategies, then a label wouldn't have a purpose—or a job. But to hit critical mass and become a star, an artist still needs the deep pockets and expertise of a label or music company to get this done.

THE 360 DEAL

A 360 Deal is where the artist shares profits from other revenue sources in exchange for the upfront costs and marketing efforts of the record label. These "other" sources could include tour and merchandise revenue, corporate sponsorships, and publishing. More will be discussed on this model later, but its value to the artist and label can be seen in a multitude of ways.

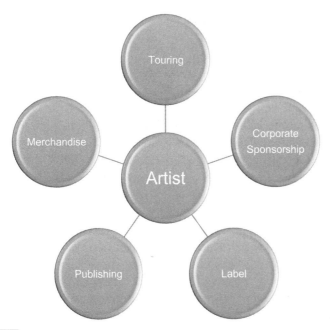

FIGURE 1.9 *360 Deal Model*

THE ADVANTAGE OF MULTIPLE REVENUE STREAMS

Artist have learned to tap the various revenue streams to maximize their financial gains. Look at how the song "Happy" placed Pharrell Williams on a new stratosphere. Although Pharrell Williams has worked for over 20 years as a songwriter, vocalist, and prolific producer (Producer Credits: Justin Timberlake, Kanye West, Gwen Stefani, Frank Ocean, Robin Thicke, Alicia Keyes, Ed Sheeran), just recently he has enjoyed the spotlight as a live solo performer. To correspond with the release of his hit "Happy," the "world's first 24-hour music video release" occurred featuring cameos from Whit Hertford, Kelly Osbourne, Magic Johnson, Urijah Faber, Sérgio Mendes, Jimmy Kimmel, Odd Future, Steve Carell, Jamie Foxx, Ana Ortiz, Miranda Cosgrove, Gavin DeGraw, and JoJo. "Happy" was also in the feature film *Despicable Me 2* with the minions from the film making several appearances throughout the music video as well. The song gained worldwide radio airplay, and "Happy" found itself on top of the U.S. Hot 100 charts for 10 weeks making it the biggest song of 2014. Pharrell's subsequent album release "G I R L" featuring "Happy" debuted #2 on the top 200 album chart.

"Happy" could be heard beyond that of the radio airwaves and was used for commercial use by "Beats by Dre" and Fiat featuring Sean Combs with "Happy" as the soundtrack. Remakes of the song were featured on several television shows including *The X Factor*, *The Voice*, and *Glee*—adding to the attractiveness of each of these TV offerings while generating lucrative publishing revenues.

With the recent artist success of Pharrell Williams, his demand as a soloist increased. He was booked to perform live at one of the nation's largest festivals Coachella, performed private concerts for Sprint's HTC One (M8) phone series, plugged into the Bruno Mars Moonshine Jungle tour, and headlined arenas across Europe. He also became the 14th featured American Express "Unstaged" artist to be streamed live in partnership with Vevo and YouTube.

All this action has led to several endorsement deals including an Adidas footwear line which Pharrell helped with the design of the product. His apparel lines Billionaire Boys Club and Ice Cream have flagship stores in both New York City and Tokyo including joint ventures with Jay-Z where his high-end sweat shirt lines have flourished. He has also enjoyed a partnership with Smirnoff Vodka celebrating "great mixes," featured in Australia with both a nationwide television and online campaign (Kellman, 2014).

Branding doesn't limit itself to just artists. Walt Disney's corporate profits rose 27% on the *Frozen* movie franchise, and the movie soundtrack has been an integral part of the success equation. In 2014, Disney's mega-hit *Frozen* out-generated revenues of any Disney animated feature film previously released, quadrupling its profit to $475 million. The movie sustained a #1 soundtrack recording on the *Billboard* 200 charts for over 25 weeks with its karaoke version debuting in the top 10 and spawned a live Broadway theater production. Movie merchandise representing lead characters Anna and Elsa sold out immediately, and the rarer still phenomena known as "cosplayers" was made popular where everyday fans dressed up like their favorite animated character. Countless remakes of the music were posted on YouTube, some professional—most novice. All generating revenue to the copyright holders with the advent of advertising on streaming sites. And Disney hopes to extend the popularity of the movie throughout its businesses by adding new features in its theme parks built around the film (Rothman, 2014).

MUSIC AS A TOOL

The outlook for sales on recorded music looks cloudy, and yet the use of the sound recording as a tool in the tool kit is essential for the success of an artist. Sound recordings and the elements that surround the success of an artist need to be managed by a team of experts. How this team makes revenue will be critical to the success of the organization, as well as to the success of the artist and their music. Record sales have been diminishing as the revenue shifts from one silo of the industry to another. Yet innovation and re-evaluation of how the business gets "done" needs to happen.

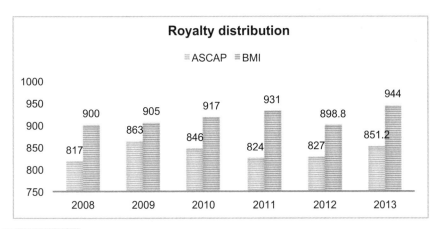

FIGURE 1.10 *ASCAP and BMI Royalty Distribution*

Source: www.ascap.com and www.bmi.com

PUBLISHING

A gap in industry data reveals that the National Music Publishers Association (NMPA) does not publish year-over-year data that would reveal a comprehensive look at how the U.S. Music Publishing Industry is fairing. The year-end 2013 report stated that the U.S. music publishing industry generated totals of $2.2 billion. However, NMPA also estimated that roughly $2.3 billion was lost due to outdated copyright law and government regulations.

"We are finally able to capture what the industry is worth and, more importantly, what our industry is losing," said David Israelite, NMPA President and CEO. "The new digital marketplace is changing how songwriters and their music publishing partners can thrive. As the marketplace evolves, it is essential our industry no longer be hamstrung by outdated laws and government regulation."

The collecting and dispersing of royalties through the Great Recession shows that although there were some issues in gathering monies, both

ASCAP and BMI acknowledge the use of new technologies has been successful in gaining access to new sources of revenue. As an example, ASCAP employed the use of a website called PLAY MUSIC that allowed for easy distribution of licenses for smaller entities, thus increasing the number of licenses issued. BMI entered more aggressive agreements with several digital outlets including Netflix and Hulu which increased their bottom line by $57 million year after year. Look at the shift of royalty distribution in just the past three years, with mechanical (record sales) losing ground and performance increasing dramatically as a result from these positions that the performing rights organizations are taking.

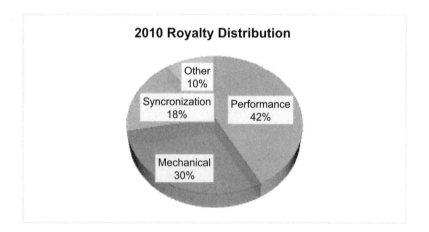

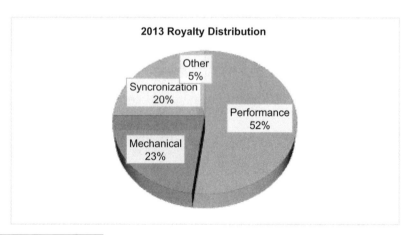

FIGURES 1.11, 1.12 *Publishing Royalty Distribution*

Source: NMPA

2013 Royalty Breakdown:

- Performance License—52%: Public performance royalties represent the most important income stream for songwriters and music publishers. While the performance right is not explicitly regulated by law, the Department of Justice imposed consent decrees on the performance rights organizations ASCAP and BMI in 1941. Incredibly, those consent decrees still are in effect today and do not include sunset provisions.
- Mechanical License—23%: Section 115 of the Copyright Act imposes a compulsory license that dates back to 1909. As a result of this World War I-era law, songwriters and music publishers are denied the right to negotiate the value of their intellectual property in a free market. For every song downloaded on iTunes, songwriters receive only 9.1 cents—the current rate set by the Copyright Royalty Board.
- Sync License—20%: The use of music synchronized with audiovisual content represents the third significant source of revenue for songwriters and publishers. Traditionally this has included using music in movies, television shows, and commercials. Newer forms of this right include music videos produced by record labels as well as user-generated content such as videos on YouTube. For songwriters and music publishers, this is a free market right not regulated by law or consent decrees.
- Other—5%: Songwriters and music publishers also receive income for the use of their content in other categories, such as sheet music and lyric websites (Sellmeyer).

TICKET SALES AND CONCERT REVENUES

Just as with Publishing, Live Entertainment has been nearly recession-proof with music lovers converting their discretionary dollars to the "experience" of music more so than to ownership, as noted in U.S. ticket revenues. In the last ten years, live music ticket sales have increased nearly 100% with over 20% growth during the Great Recession (2008–2013). As streaming continues to gain momentum and buying of sound recordings wanes, how artists monetize their assets has become the $64,000 question—and the answer is shifting.

Of the top 40 touring artists in 2012, concerts make up 69.9% of artist revenues. And if Taylor Swift and Adele were removed from the list, the average would increase to 72.5%. Madonna's concert revenues comprised 92.5% of her totals for the year; Bruce Springsteen's revenues made a close 92%. Interestingly, big artists who are past their record-selling prime usually have larger concert touring revenue than that of hot radio chart action artists with large streaming revenue.

In 2012, top revenue-generating artists who made less than $1 million touring would make more from the sale of recorded music, reflecting a negative correlation between concert revenue and music sales. Older artists realize a similar negative correlation between concert revenue and streaming since many of their fans are over 50 and don't know how to use computers and mobile phones (or just don't have the time or care to learn) but certainly have deep pockets to purchase pricey tickets to experience their aging favorite artist. Top touring acts like Rush, Neil Diamond, Elton John, and Rod Steward generated very little streaming revenue but were the leading earners on the road. Maroon 5 earned the most from streaming in 2012, comprising 3.5% of total revenue, with Drake earning 3.3% from streaming and One Direction taking in 2.5% from streaming (Peoples, 2013).

In recent years, it's the festival scene that has driven the calendar. Many artist hold open dates until the festival season has been booked and then back fill dates to complete routing between festival locations. For North American in 2014, individual ticket prices increased by $1.92 for an average of $71.44 per ticket while selling less tickets at 38.19 million, down by 4.7%. The top 100 tour revenue contracted slightly from 2013 by 2.2%. But overall revenue for the year was up by 21%. And the years to come look promising as the overall economy appears to be improving and music consumption is at an all-time high.

The take-away: as consumers shift from purchasing of sound recordings to streaming, and as artists become iconic and age with their fans, the income stream for that artist shifts. The challenge is to stay relevant for a long period of time to reap the benefit of live touring.

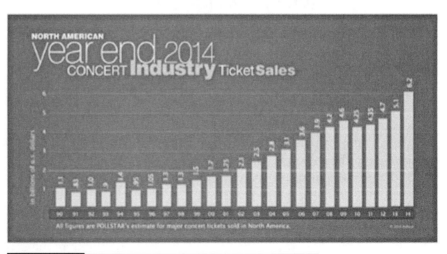

FIGURE 1.13 *North American Year End 2014 Concert Industry Ticket Sales*

Source: Pollstar

So of the $14 billion dollar business known as the Music Industry, in 2015 looks like this: $7 billion in music sales with revenues on the negative slide. Publishing has hit a plateau at $2.2 billion, but admittedly $2.3 billion more has been lost in old copyright laws that are in need of updating. And live entertainment ticket sales show a healthy climb at $5 billion—and increase of over 20% since the Great Recession of 2008. The wild card that is unquantifiable is artist merchandising that could be just as large as ticket sales—if not larger. But it's near impossible to estimate this value since there is no association tracking this data.

The music industry continues to evolve, with revenues shifting from traditional music sales to the publishing realm and live entertainment world. The "experience" of the artist and the use of the sound recording is a viable sales item and continues to dominate the industry in revenue volume, but the business has recognized that streaming looks to be the "new" normal as this format continues its adoption process with consumers. Revenues will only continue to shift from the record sales cut of the pie to the innovative ways that the sound recording will become a marketing tool for the licensing elements in publishing and live entertainment. It will be up to the record labels and their teams to figure out novel marketing strategies that will help them and their artists continue to remain sustainable in a fluid marketplace. (www.soundscan.com)

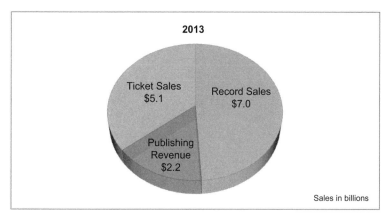

FIGURE 1.14 *2013 Music Industry Revenue*

REFERENCES

"Branding." BusinessDictionary.com. 2014. N.p. Web. June 8, 2015. http://www. businessdictionary.com/definition/branding.html#ixzz2bmpk2QfO.

Eichelberger, C. "Sports Revenue to Reach $67.7 Billion by 2017, PwC Report Says." *Bloomberg News*. Ed. Michael Sillup. N.p., November 13, 2013. Web. July 14, 2014.

Hendrix, J. "10 Key Takeaways from PwC's Global Entertainment and Media 2013." *NYC Media Lab*, June 5, 2013. Web. June 8, 2015.

Kellman, A. www.AllMusic.com, July 15, 2014. Web.

Palmeri, C. "U.S. Theme Park Revenue Is On Its Best Ride In Years." *Bloomberg Businessweek*, May 30, 2013. Web.

Peoples, G. "Madonna, Bruce Springsteen Lead Billboard's 2013 Top 40 Money Makers." *Billboard*.Biz. February 22, 2013. July 17, 2014.

PwC Price Waterhouse Coopers. "Global Entertainment Media Outlets—Segment Insights." Web. November 2013. http://www.pwc.com/gx/en/global-entertainment-media-outlook/segment-insights/internet-access.jhtml.

"Revenue Releases/RAB.com." *Revenue Releases | Radio Advertising Bureau*. N.p., n.d. Web. November 7, 2014.

Rothman, L. "Ice, Ice, Baby. *Frozen* Inspires a Totally Chilled-Out Cult Following." *Time*, February 24, 2014. July 14, 2014.

Sellmeyer, C. "U.S. Music Publishing Industry Valued at $2.2 Billion." *NMPA*. National Music Publishers Association, June 11, 2014. Web. July 15, 2014.

SoundExchange. www.soundexchange.com, March 18, 2014. Web.

Spiegel, A. "Good Art Is Popular Because It's Good, Right?" NPR, n.d. Web. February 27, 2014. www.npr.org.

"When Are Mobile Owners Using Apps?" *Marketing Charts*. November 4, 2014. Web. June 8, 2015.

www.ascap.com

www.bmi.com

www.SoundScan.com

CHAPTER 2

Marketing Concepts and Definitions

If you have previously taken an introductory marketing or consumer behavior course, much of this chapter and the next will be review, but don't use that as an excuse to skip the chapters because we will tie those concepts specifically to the music business. If you have not taken marketing before, this chapter will give you an introduction to some basic marketing concepts that you need to understand. We also look at how the Internet has impacted some of those practices.

SELLING RECORDED MUSIC

The concept of selling recorded music has been around for more than a century. While the actual storage medium for music has evolved—from cylinders to vinyl discs, to magnetic tape, digital discs, and now downloads and streaming services—the basic notion has remained the same: a musical performance is captured to be played back at a later time, at the convenience of the consumer. Music fans continue to enjoy the ability to develop music collections, whether in the physical compact disc format, or collections of digital files on their computer hard drives. Consumers also enjoy the portability afforded by contemporary music listening devices, allowing the convenience of determining the time and place for listening to music. The ways in which consumers select to access music have been undergoing changes in the past few years, as physical sales have diminished and the industry scrambles to find new business models for underwriting the cost of developing new creative products.

CONTENTS

Music consumption should not be confused with music purchases. The consumption of music by consumers has increased (Hefflinger, 2008), despite the fact that the sales of recorded music albums in the U.S. have decreased over the past five years, from 373.9 million units in 2009 to 289.4 million units in 2013, according to Nielsen Soundscan. In response to the decline in sales of recorded music, record labels are experimenting with new ways to monetize the consumption of their music. For example, new services like SoundCloud, LastFM, Pandora, and Spotify offer music fans the opportunity to listen to music without actually purchasing it. According to IFPI research, over 28 million people worldwide paid for music subscription services in 2013 (IFPI Facts and Figures). Record labels are compensated for licensed streams through advertising revenue sharing. These revenues are collected by SoundExchange and accounted for 8% of record sales revenue in 2013.

Thus recorded music is finding ways to make money much the same as television programming has done for over 50 years. For much of this time, the television programming industry relied solely upon advertising revenue to fund some of the most popular television shows in history. Other, more recent forms of revenue have come from premium (fee-based) programs and physical sales of shows (DVDs). Fee-based programming did not occur until the premium channels such as HBO and Showtime started developing their own proprietary shows. The physical sale of this commodity did not occur until consumers started collections of videotapes and DVDs. Even with these other forms of income, the bulk of revenue for producing television content still comes from advertising. Perhaps the recording industry can look to the television industry for ideas in developing new models to create revenue from creative products.

As the paradigm shifts from physical sale of recordings to a more complex model of generating revenue, marketing efforts must also evolve to respond to the plethora of income possibilities.

WHAT IS MARKETING?

In today's marketplace, the consumer is showered with an array of entertainment products from which to choose, making the process of marketing more important than ever. Competition is fierce for the consumers' entertainment budget. But, before explaining how recorded music is marketed to consumers, it is first necessary to gain a basic understanding of marketing. Even the concept of a record or album has undergone changes recently. Record is short for recorded music—that much has not changed. An album is still a collection of songs released as a unit, whether it's in the CD format or otherwise.

Kotler and Armstrong (2014) define marketing as "the process by which companies create value for customers and build strong customer relationships in order to capture value from customers in return." Marketing involves satisfying customer needs or desires. To study marketing, one must first understand the notions of product and consumer (or market). The first questions a marketer should answer are, "What markets are we trying to serve?" and, "What are their needs?" Here we are using the term "need" not in the way that Maslow defined needs (see Chapter 3), but in the generic sense that includes wants and desires, not just needs. Marketers must understand these consumer needs and develop products to satisfy those needs. Then, they must price the products effectively, make the products available in the marketplace, and inform, motivate and remind the customer of their availability. In the music business, this involves supplying consumers with the recorded music they desire where they want it and at a price they are willing to pay.

The market is defined as consumers who want or need your product and who have the willingness and ability to buy. This definition emphasizes that the consumer wants or needs something. A product is defined as something that will satisfy the customer's want or need. You may want a candy bar, but not necessarily need one. You may need surgery, but not necessarily want it. Markets used to be physical places where buyers and sellers met but today they are just as likely to be a website on the Internet.

THE MARKETING MIX

The marketing mix, often called the Four P's, refers to a blend of product, distribution (place), promotion and pricing strategies designed to produce mutually satisfying exchanges with the target market. Let's take a closer look at each of the components of the marketing mix.

Product

The marketing mix begins with the product. It would be difficult to create a detailed strategy for the other components without a clear understanding of the product to be marketed. Record label marketers are, however, often asked to create a basic marketing mix and marketing plan knowing very little about the final project because the music has not yet been recorded.

An array of products may be considered to supply a particular market. Then the field of potential products is narrowed to those most likely to perform well in the marketplace. In most industries this function is performed by the research and development (R&D) arm of the company, but in the music business, the artist and repertoire (A&R) department performs this task by searching for new talent and helping decide which songs will have the broadest consumer appeal.

New products introduced into the marketplace must somehow identify themselves as different from those that currently exist. Marketers go to great lengths to position their products to ensure that their customers perceive it as more suitable for them than the competitor's product. Product positioning is defined as the customer's perception of a product in comparison with the competition. This is achieved primarily through advertising and publicity.

Consumer tastes change over time. As a result, new products must constantly be introduced into the marketplace. New technologies render old products obsolete and encourage growth in the marketplace. For example, the introduction of the compact disc (CD) in 1983 created opportunities for the record industry to sell older catalog product to customers who were converting their music collections from the vinyl LPs to CDs. Similarly, when a recording artist releases a new recording, marketing efforts are geared toward selling the new release, rather than selling older recordings (although the new release may create some consumer interest in earlier works and they may be featured alongside the newer release at retail).

The Product Life Cycle

The product life cycle (PLC) is a concept used to describe the course that a product's sales and profits take over what is referred to as the lifetime of the product, from its inception to its demise.

It is characterized by four distinct stages: introduction, growth, maturity, and decline. Preceding this is the product development stage, before the product is introduced into the marketplace. The introduction stage is typically a period of slow growth as the product is launched into the marketplace. Profits are nonexistent because of heavy marketing expenses. The growth stage is a period of rapid acceptance into the marketplace during which profits increase. Maturity is a period of leveling in sales mainly because the market is saturated—most consumers have already purchased the product. Marketing is more expensive (to the point of diminishing returns) as efforts are made to reach resistant customers and to stave off competition. Decline is the period when sales fall off and profits are reduced. At this point, prices are cut to maintain market share (Kotler and Armstrong, 1996).

The PLC can apply to a variety of situations such as products (a particular album), product forms (artists and music genres), and even product classes (cassettes, CDs, and vinyl) that are referred to as formats in the music business. Product formats have the longest life cycle—the compact disc has been around since the early 1980s (and is currently in the decline phase). However, the life cycle of an average album release is 12–18 months. It is at this point that the label will generally terminate most marketing efforts and rely on catalog sales to deplete remaining inventory. When a product reaches the decline stage, the company may withdraw from the market or, as in the case of vinyl records, efforts or circumstances may lead to a revival of the product that will sustain a smaller number of competitors for an extended maturity stage.

Classic PLC and Product Adoption Model Overlay

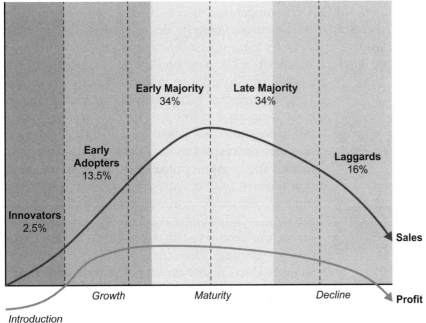

FIGURE 2.1

Not all products will have the nice bell shaped curve. A hit song may have a very steep, short introduction and growth stage and just as steep and quick decline, looking more like an inverted V than a bell. Other products, like vinyl records, may never be withdrawn from the market but, instead, achieve a mature stage where sales level off and maintain a steady, albeit lower, level of sales for many years.

Diffusion of Innovations

When a product is introduced into the marketplace, its consumption is expected to follow a pattern of diffusion. Diffusion of innovations is defined as the process by which the use of an innovation is spread within a market group, over time and over various categories of adopters (American Marketing Association, 2004). The concept of diffusion of innovations describes how a product typically is adopted by the marketplace and what factors can influence the rate (how fast) or level (how widespread) of adoption. The rate of adoption is dependent on consumer traits, the product, and the company's marketing efforts.

Consumers are considered adopters if they have purchased and used the product. Potential adopters go through distinct stages when deciding whether to adopt (purchase) or reject a new product. These stages are referred to as AIDA, which under one model (affective or driven by emotion) is represented as attention, interest, desire, and action. Another model (cognitive or driven by logic) uses awareness, information, decision, and action. These stages describe the psychological progress a buyer must go through in order to get to the actual purchase. First, a consumer becomes aware that they need to make a purchase in this product category. Perhaps the music consumer has grown tired of his or her collection and directs their attention toward buying more music. The consumer then seeks out information on new releases and begins to gain an interest in something in particular, perhaps after hearing a song on the radio or attending a concert. The consumer then makes the decision (and desires) to purchase a particular recording. The action is the actual purchase. Some products (like a car, or music to study by) have elements of both the emotional and the logical models present in the decision-making process.

This process may take only seconds or may take weeks, depending on the importance of the decision and the risk involved in making the wrong decision. Purchase involvement refers to the amount of time and effort a buyer invests in the search, evaluation, and decision processes of consumer behavior. If the consumer is not discriminating or the consequences of a poor decision is not a major financial or social risk, then the process may take only moments

The AIDA Model

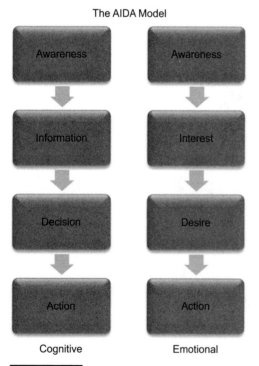

Cognitive   Emotional

FIGURE 2.2 *The AIDA decision-making process*

and is thus a low-involvement decision. On the other hand, if the item is expensive, such as a new car, or highly visible, like clothing, and if consequences of a wrong decision are severe, then the process may take much longer. Several factors influence the consumers' level of involvement in the purchase process, including previous experience, ease of purchase, product involvement, and perceived risk of negative consequences. As relative cost increases, so does the level of involvement.

PRODUCT INVOLVEMENT: AKA THE COLLECTOR

"Product involved" consumers tend to be early adopters of new products. This is the guy in your high school class who spent all his money from his part-time job at the record store on music and was the first to hear of new groups before they became famous. Product involvement is product specific and product involved consumers often become *opinion leaders* for their peer group. You might ask your friend who works in the record store his advice on music but not on fashion, since his wardrobe consists entirely of jeans and rock-related t-shirts. Product involvement is different from purchase involvement. Product involvement tends to be more central to one's life or self image—he's the car guy or the jazz guy. Purchase involvement is temporary or short-term and tends to be highest when there is social or economic risk to the decision. You may have done a lot of research and gathered a lot of information before you bought your car to be sure it was reliable and economical, but now that you have made your choice your identity and leisure time don't revolve around cars. You don't read about and talk about cars all the time. It's just a tool, a way to get to where you need to go. If you are "the car guy" and subscribe to several car magazines, watch car shows and basically eat, drink, and sleep cars, then you are product involved. In the entertainment world product-involved consumers are the collectors, the people who have ten recordings of the same song because there are (minute) differences in every one.

One major advantage of being product involved is that it may decrease purchase involvement—the time spent making the decision—because you routinely spend the time and effort gathering all the information that is needed to make your selection.

Consumers who adopt a new product are segmented into five categories:

Innovators
Early adopters
Early majority
Late majority
Laggards

Innovators are the first 2.5% of the market (Rodgers, 1995) and are eager to try new products. Innovators are above average in income and thus the cost of the product is not of much concern. Early adopters are the next 13.5% of the market, and adopt once the innovators have demonstrated that the new product is viable. Early adopters are more socially involved and are considered opinion leaders. Their enthusiasm for the new product will do much to assist its diffusion to the majority. The early majority is the next 34% and will weigh the merits before deciding to adopt. They rely on the opinions of the early adopters. The late majority represents the next 34%, and these consumers adopt when most of their friends have. The laggards are the last 16% of the market and generally adopt only when they feel they have no choice. Laggards adopt a product when it has reached the maturity stage and is being "deep discounted" or is widely available at discount stores. When introducing a new product, marketers target the innovators and early adopters. They will help promote the product through word of mouth. How the adopter categories drive the product life cycle is shown in Figure 2.1.

Products (or innovations) also possess characteristics that influence the rate and level of adoption. Those include:

> Relative advantage: the degree to which an innovation is perceived as better than what it supersedes. Cassette tapes were perceived to be superior to vinyl because they could be played on portable devices.
> Compatibility: the degree to which an innovation is perceived as consistent with existing values and experiences. Using a 4K TV is no different than using an HD TV, it just looks better.
> Complexity: the degree to which an innovation is perceived as difficult to understand and use. If your VCR is still flashing 12:00 then you are not an early adopter and complexity may be an issue that keeps you from adopting new technology.
> Trialability: the degree to which an innovation may be experimented with on a limited basis. This is why artists should give away free songs and allow sampling of every song on their new albums.
> Observability or communicability: the degree to which the results of an innovation are visible to or can be communicated to others." The more easily you can see or hear the difference, the faster consumers can make a decision to adopt, or not. (Rogers, 1995)

One way marketers can increase the potential for success is by allowing customers to "try before they buy." Listening stations in retail stores and online music samples have increased the level of trialability of new music. Marketers can improve sales numbers by ensuring the product has a relative advantage, that it is compatible, that it is not complex, and that consumers can observe and try it before they purchase.

Hedonic Responses to Music

Researchers in the fields of psychology and marketing strive to understand the hedonic responses to music. Hedonic is defined as "of, relating to, or marked by pleasure." Hedonic products are those whose consumption is primarily characterized by an affective or emotional experience. It is a study of why people enjoy listening to music and what motivates them to seek out music for this emotion-altering experience. Recorded music is considered a tangible hedonic product, compared to viewing a movie or attending a concert, which is an intangible hedonic product. "The purchase of a tangible hedonic portfolio product, such as a CD, gives the consumer something to take home and experience at her convenience, possibly repeatedly" (Moe and Fader, 1999). This convenience factor has fueled the increase in consumption of music, despite the fact that sales of recorded music have been falling this century.

Consumers like music for a variety of reasons, mostly connected to emotions or emotional responses to social situations involving music. A brief glance through articles and studies offers the following reasons:

To evoke an emotional feeling or regulate a mood
To evoke a memory or reminisce
As a distraction or escape from reality
To create a mood in an environmental setting or a cultural/sporting event
To combat loneliness and provide companionship
To foster social interaction with peers
To calm and relax
To stimulate (such as to stay awake while driving)
For dancing or other aerobic performances
To enhance/reinforce religious or cultural experiences
For therapeutic purposes (such as pain reduction)
To pass the time while working or waiting

By understanding the situations that drive consumers to purchase or consume music, marketers can be more effective in providing the right music to the right customers at the right time. Even retail placement of music can benefit from understanding the context in which the music will be consumed. For example, music designed to inspire sports fans may be made available in locations that fans are likely to visit on their way to or while attending a sporting event.

FIGURE 2.3

Altissimo! Recordings is dedicated to the continuing production and distribution of the vast array of music from the wonderful musicians of the United States military bands, orchestras, ensembles, and choruses. The album *A Patriotic Salute to Military Families* is a collection of marches, bugle calls, ceremonial, and Americana music and features performances from bands of ALL branches of the U.S. Military. The songs are used for many military and patriotic events, and the album charted at number seven in the *Billboard* Top Classical Albums Chart in July 2002.

Price

Pricing is more complex than how much it costs to make the product. Pricing structure must consider not only economic costs but market influences and business practices as well. Once the wholesale and retail pricing is determined, price-based incentives must then be considered. For example, should the product be put on sale and if so, when? Should coupons be issued? Should the retailers receive wholesale price breaks for quantity orders or other considerations?

There are generally three methods for deciding the retail price of a product: cost-based pricing, competition-based pricing, and consumer-based (value-based) pricing.

Cost-Based Pricing

Cost-based pricing is achieved by determining the cost of product development and manufacturing, marketing and distribution, and company overhead. Then an amount is added to cover the company's profit goal. The weakness with this method is that it does not take into account competition and consumer demand.

$$\text{Break-even point (units)} = \frac{\text{Total fixed costs}}{\text{Price} - \text{Variable costs per unit}}$$

When determining cost-based pricing, consideration must be given to the fixed costs of running a business, the variable costs of manufacturing products, and the costs related to marketing and product development. Fixed costs include items such as overhead, salaries, utilities, mortgages, and other costs that are not related to the quantity of goods produced and sold. In most business situations, variable costs are usually associated with manufacturing and vary depending on the number of units produced. For recorded music, that would include discs, CD booklets, jewel cases, and other packaging. However, within the marketing budget of a record label, the costs of manufacturing and mechanical royalties are considered fixed once the number of units to be manufactured is determined and the recording costs have been computed. Then there are other (semi) variable costs, which can vary widely, but are not directly related to the quantity of the product manufactured. They would include recording costs (A&R) and marketing costs. For purposes of the formula below, the costs of marketing are considered variable costs, and the recording and manufacturing costs, having already been determined for the project, are considered fixed at this point.

$$\text{Break-even point (units)} = \frac{\text{Total fixed costs}}{\text{Price} - (\text{total variable costs/units})}$$

Table 2.1 Standard Business Model vs. Record Label Business Model

	Variable costs	Fixed costs	Semi-variable
Standard business model	Manufacturing cost per unit, royalties, licensing, or patent fees per unit, packaging per unit, shipping and handling per unit	Overhead, salaries, utilities, mortgages, insurance	Marketing, advertising, research and development (R&D)
Record label business model	All marketing aspects, including advertising, discounts, promotion, publicity expenses, sales promotions, etc.	Overhead, salaries, utilities, mortgages, insurance, manufacturing costs, production costs (A&R)	

Competition-Based Pricing

Competition-based pricing attempts to set prices based on those charged by the company's competitors—rather than demand or cost considerations. The company may charge more, the same, or less than its competitors, depending on its customers, product image, consumer loyalty, and other factors.

Companies may use an attractive price on one product to lure consumers away from the competition to stimulate demand for their other products. This is known as leader pricing. The item is priced below the optimum price in hopes that sales of other products with higher margins will make up the difference or that purchase will create brand loyalty. The most extreme case of this is loss leader pricing. Under loss leader pricing, the company actually loses money on one product in an attempt to bring in customers who will purchase other products that are more profitable. Compact discs were once used by retailers like Best Buy and Circuit City (who are now out of business) as loss leaders because they believed it would generate traffic to the store and increase sales of other, more profitable items. Record labels responded to the cries of independent record stores (who had no high margin appliances to sell) by instituting a minimum advertised price policy (MAP). Lawsuits were filed to overturn MAP, but before a decision could be reached, the practice ended because the stores realized that CDs were not an effective loss leader. Consumers did not simply return to the store where they bought CDs to buy refrigerators. They went online and found the best deal they could and spent the money they saved on cheap CDs at the appliance store!

Companies are prohibited from coordinating pricing policies to maximize profits and reduce competitive pricing. Such price fixing is illegal under the Sherman Act (1890), which prevents businesses from conspiring to set prices (Finch, 1996). The Robinson-Patman Act (1936) prohibits any form of price discrimination, making it illegal for manufacturers to sell products to competing wholesale or retail buyers at different prices unless those price differentials can be justified. But neither of these laws keeps competitors from shopping the competition and adjusting their prices accordingly, nor does it keep retailers like Wal-Mart or Apple from dictating prices because of their power in the distribution channel.

Value-Based (Consumer-Based) Pricing

Consumer-based pricing uses the buyer's perceptions of value, a reversal from the cost-plus approach. The music business has at times sought to maintain a high-perceived value for compact discs by tweaking and repackaging the product. Lady Gaga, U2, Green Day, Eric Clapton and others have released special editions or boxed sets of albums in addition to the standard versions. Consumers are willing to pay a premium price well beyond the economic value of these special editions because they are limited releases or have exclusive content. This same concept is applied in the concert side of the industry which is selling deluxe packages that may include special items, meals, and a meet and greet with the artist, priced well beyond the actual costs and any normal profit, just because fans are willing to pay for the experience.

In an effort to maintain the perceived value of recorded music, record labels are now focusing on how to add value to music downloads, which provide a limited amount of value-added materials compared to a physical CD. Electronic "digital booklets" are available to consumers who download albums. These digital booklets, usually in PDF file form, contain the artwork, liner notes, and lyrics normally found in a CD booklet.

Crowdfunding

A recent trend in the arts and entertainment industries is the use of crowdfunding to finance new projects. Crowdfunding is the collecting of money from multiple sources to fund the production or venture rather than getting the funding from a single source. Crowdfunding replaces the label and gives the artist greater control over the project. Several companies offer crowdfunding services to would-be artists, the most popular being Kickstarter. For their services Kickstarter takes 5% of the money from successfully funded projects. Their financial partners, PayPal in the U.S., take another 5%. Kickstarter only collects money from project supporters if the fundraising goal is met; however competitors like Indiegogo have options that allow the artist to keep whatever is raised, less the fees, of course. To attract supporters, artists typically offer incentives ranging from a free download for a $1 donation to a house party for a donation in the thousands.

Crowdfunding has not been without its controversies. Because the company gets a percentage of the money raised, there is no incentive for them to cut off the fund-raising when the artist reaches the goal. In the case of former Roadrunner Records artist Amanda Palmer, more than 10 times the original goal was raised. This raises ethical issues for both the artist and for Kickstarter (Clover, 2012).

CROWDFUNDING: THE CASE OF AMANDA PALMER AND THE GRAND THEFT ORCHESTRA

On April 30, 2012, recording-artist Amanda Palmer launched a Kickstarter campaign to raise $100,000 to release a new album. Palmer offered incentives to contributors based on the amounts of their gifts, from a download for $1, to a CD and art book for $125, to $5000 for a show at your house, to $10,000 for dinner with her and she would paint your portrait. Palmer raised over $379,000 from more than 6,600 supporters in just two days. When the campaign ended on May 31, 2012, Palmer had raised more than $1.19 million and had over 24,000 backers donating from $1 to $10,000. The album debuted at #10 on the *Billboard* charts. In response to a question on Twitter, Palmer posted the following on her blog on May 13, 2012:

first i'll pay off the lovely debt—stacks of bills and loans and the like—associated with readying all of the stuff that had to happen BEFORE i brought this project to kickstarter. for the past 8 months or so, i wasn't touring—and therefore wasn't making much income—but every step of the way, there were expenses. so, during that time, i borrowed from various friends and family who i'd built up trust with over the years. (Palmer, 2012)

Palmer goes on to lay out the estimated expenses for the entire project, including incentives that were part of the fundraising effort. In the end she claimed she would do well if she had $100,000 left over after the entire project was completed.

Palmer then set out on a tour to promote the album, but not before putting out a request for "professional-ish" musicians to join the band and play for free. "We will feed you beer, hug/high-five you up and down, give you merch, and thank you mightily." She claimed that she did not have the resources to pay them. Her request was met with a huge backlash, including a petition posted at change.org and protests from musician unions (change.org, 2012).

The crowd-funding platform Kickstarter.com also faced criticism for allowing Palmer to raise more funds than requested. Some critics argued that contributions should be limited to the amount requested to fund the project. Kickstarter posts the running total so that contributors can know how much money has already been raised before deciding to donate funds. Others argued that Kickstarter.com should not permit "established artists" to post fund requests, because the platform was created to allow aspiring artists to find support for their projects. Eventually the company responded with a press release defending its position. One industry analyst said, "the Kickstarter team tried to shoot down one of the biggest complaints about celebrities using Kickstarter: that it takes away attention and funding from other worthy projects from lesser-known people on the website . . . the Veronica Mars and Zach Braff projects have brought tens of thousands of new people to Kickstarter," the founders wrote. But the analyst adds, "Even if it's the case that famous people like Braff end up attracting new donors to the website, there's still a more fundamental question about whether the presence of these celebrities end up encouraging or discouraging more people to take a chance on launching a campaign of their own" (Fiegerman, 2013). Of course, Kickstarter receives 5% of the money from funded projects, so they have a lot to gain from allowing larger sums to be raised. Only 43.94% of all project offerings have been successfully funded (www.kickstarter.com).

Amanda Palmer defended her actions as "the future of music." According to Palmer, "we're moving to a new era where the audience is taking more responsibility for supporting artists at whatever level" (Peoples, 2013). Palmer argues that social media allows for an unprecedented connection with fans, that now the artist has opportunities to connect with fans on a personal basis, which is more important than the traditional music-industry based formula for artist control.

—From Padgett, Barry L. and Rolston, Clyde Philip, "Crowd Funding: A Case Study at the Intersection of Social Media and Business Ethics." Journal of the International Academy for Case Studies, Allied Academies, 2014.

Name Your Own Price

Radiohead was one of the first to use "name your own pricing" methods to sell their albums. After a small fee of about $.25 the fan could pay whatever price they wanted to download the entire *In Rainbows* album. The idea was great publicity and it probably made the band more money than they would have made with a record label; still, most people paid less than three dollars for the download. Both Palmer and Radiohead have the advantage of having once been on a major label and had household-name recognition. For a new, or independent artist, these types of funding are difficult to achieve.

FIGURE 2.4

Promotion

Promotion includes the activities of advertising, personal selling, sales promotion, and public relations. It involves informing, motivating, and reminding the consumer to purchase the product. In the recording industry, the four traditional methods of promotion include: radio promotion (getting airplay), advertising, sales promotion (working with retailers), publicity, tour support, and street teams. Recently, record labels have created new positions dedicated to Internet and social media marketing and co-branding—creating tie-ins with non-musical products.

Basic Promotion Strategies: Push vs. Pull

There are two basic promotion strategies: push promotion and pull promotion. A push strategy involves "pushing the product through the distribution channel to its final destination in the hands of consumers." Marketing activities are directed at motivating channel members (wholesalers, distributors, and retailers) to carry the product and promote it to the next level in the channel. In other words, wholesalers would be motivated to inspire retailers to order and sell more product. This can be achieved through offering monetary incentives, discounts, free goods, advertising allowances, contests, and display allowances. All marketing activities are directed toward these channel members and are regarded as "trade" promotion and "trade" advertising. With the push strategy, channel members are motivated to "push" the product through the channel and ultimately on to the consumer.

FIGURE 2.5

Under the pull strategy, the company directs its marketing activities toward the final consumer, creating a demand for the product that will ultimately be fulfilled as requests for product are made from the consumer to the retailer, and then from the retailer to the wholesaler. This is achieved by targeting consumers through advertising in consumer publications and creating "consumer promotions." With a pull strategy, consumer demand pulls the product through the channels. Years ago, when MTV still played music videos, they employed a pull strategy to get onto more cable systems, imploring teenagers to call their cable operators and tell them "I want my MTV." The campaign was so creative that it was immortalized in Dire Strait's song "Money for Nothing."

Different record companies have different philosophies on the emphasis of the two strategies. Some companies employ a balanced combination of consumer advertising and consumer promotions (coupons and sale items) and trade promotion (incentives), while others focus on consumers and a pull strategy, relying on the sales and promotions departments to win over retailers and radio, respectively.

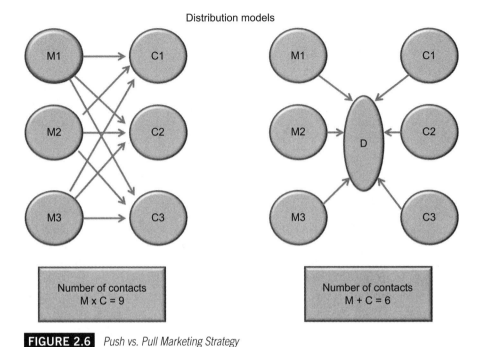

Distribution models

FIGURE 2.6   *Push vs. Pull Marketing Strategy*

Place

This aspect of marketing involves the process of distributing and delivering the products to the consumer. Distribution strategies entail making products available to consumers when they want them, at their convenience. The various methods of delivery are referred to as channels of distribution. The process of distribution in the record business is currently in a state of evolution, as digital distribution continues to encroach upon the market share for physical recorded music products. Digital delivery has had a dramatic effect on the physical marketplace, causing erosion in market share, closing of retail music chains, and causing big box stores to reduce the amount of floor space dedicated to selling recorded music.

Now the online retailers are suffering a decline of their own as consumers move from downloading to streaming. *Billboard* reported that 2013 was the first time in 10 years that digital music sales decreased. The decline in track sales of 5.7% and in album sales of .1% is attributed to the increase in popularity of music streaming services such as Pandora and Spotify (Christman, 2014).

Distribution Systems

Most distribution systems are made up of channel intermediaries such as wholesalers and retailers. These channel members are responsible for aggregating large quantities and assortments of merchandise and dispersing smaller quantities to the next level in the channel (such as from manufacturer to wholesaler to retailer). Manufacturers engage the services of distributors or wholesalers because of their superior efficiency in making goods widely accessible to target markets (Kotler, 1980).

The effectiveness of intermediaries can be demonstrated in figure 2.6. In the first example, no intermediary exists and each manufacturer must engage with each retailer on an ongoing basis. Thus, the number of contacts equals the number of manufacturers multiplied by the number of customers or retailers (M × C). In the second example, a distributor is included with each manufacturer and each customer contacting only the distributor. The total number of contacts is the number of manufacturers plus the number of customers (M + C).

Types of Distribution Systems

Three basic types of physical distribution systems are corporate, contractual, and administered. All three are employed in the record business.

Corporate systems involve having one company own all distribution members at the other levels in the channel. A record label would use a distribution system owned by the parent company, as well as a company-owned retailer in a fully integrated corporate system. In reality, the major record labels do not own record retailers, but they do own their own distribution. Such is the relationship between Mercury Records and Universal Music and Video Distribution. Both companies are owned by Universal Music Group. This type of ownership is often referred to as vertical integration or a vertical marketing system (VMS). When a manufacturer owns its distributors, it is called forward vertical integration. When a retailer such as Wal-Mart develops its own distribution or manufacturing firms, it is called backward vertical integration.

Contractual distribution systems are formed by independent members who contract with each other, setting up an agreement for one company to distribute goods made by the other company.

Distribution chain

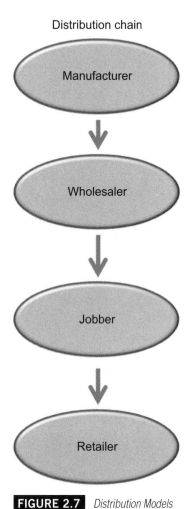

FIGURE 2.7 *Distribution Models*

Independent record labels commonly set up such agreements with independent record distributors. Before the major record labels developed in-house distribution systems in the 1960s and 1970s, nearly all recorded music was handled through this type of arrangement.

In administered distribution systems, arrangements are made for a dominant channel member to distribute products developed by an independent manufacturer. This type of arrangement is common for independent record labels that have agreements to be handled by the distribution branch of one of the major labels. However, branch distribution is on the decline as major labels shift resources away from physical product and toward digital distribution. Digital distribution systems will be addressed in the chapter on distribution and retail.

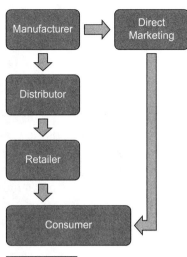

FIGURE 2.8 *Distribution Chain*

Retail

Retailing consists of all the activities related to the sale of the product to the final consumer for personal use. Retailers are the final link in the channel of distribution. There are numerous types of retailers, each serving a special niche in the retail environment. Independent stores focus on specialty products and customer service, while mass merchandisers such as Wal-Mart concentrate on low pricing. Online retailing of physical and digital download products through Amazon.com, CD Baby, and others has been growing in the past few years with iTunes now claiming the number one spot in music retailing.

Price and Positioning

Several years ago, the retail industry developed marketing tools to measure the effectiveness of product location in their stores. Universal product codes (UPC), or bar codes, and computerized scanners have helped retailers to determine the optimal arrangement of products within the store. They can move products around to various locations and note its effects on sales. As a result, retailers are charging manufacturers a rental fee for prime space in the store, including counter space, end caps, and special displays. Retail stores sometimes charge slotting fees—a flat fee charged to the manufacturer for placement of products on the shelves for a limited period of time. If the product fails to sell, the retail store has reduced its risk. A similar concept has developed in the record business, but it is also tied to promotion and favorable location for the product. It is referred to as price and position and will be discussed in Chapter 15.

SHOWROOMING AND SHRINKING RETAIL INVENTORIES

A source of tension between the digital and brick-and-mortar retail worlds is the phenomenon of Showrooming. Showrooming occurs when the consumer uses the brick-and-mortar store to gather information and gain "hands on" experience with a product before buying, then purchasing the product online, presumably at a cheaper price. This has not been a major issue for entertainment software (CDs, movies, or video games) because it can be easily sampled online, but for entertainment hardware (computers, game consoles, stereos, televisions, etc.), it is a concern for retailers that don't have an online presence and can't compete on price (including, in some cases, sales-tax-free purchasing).

Another trend impacting retailing is the ease of online access and the digitalization of old recordings by larger record labels. Chris Anderson, former editor of *Wired Magazine*, introduced the idea of "The Long Tail" in October 2004 (Anderson, 2004). Anderson proposed that everything should be made available—old recordings, obscure recordings, the song you made on GarageBand—everything. Yet brick-and-mortar retailers have reduced the space they devote to music while iTunes has become the largest seller of music. We will discuss the article in greater detail in Chapter 15, so for now, suffice to say that the cost of maintaining such huge inventories is not the only thing behind the shrinking record section at the big box retailers.

ARE THE FOUR P'S DEAD?

Seth Godin and others have argued that the Four Ps are no longer relevant in a modern world of digital marketing. We have moved from a mass media, advertising based model (interruption marketing) to a fragmented, interactive model (permission marketing) (Merzel, 2011). Like Goldhaber's (2006) attention economy, Godin points out that we are inundated with advertising and cannot attend to even a small percentage of the messages to which we are exposed. And we don't. But we are more likely to pay attention to messages we are interested in, particularly if we have given the sender permission to contact us. This does not mean the Four Ps are no longer relevant, it simply emphasizes the importance of 1) being relevant (the right product in the right place at the right price) and 2) maintaining an up to date database of contacts (targeted promotion). Both of these issues will be discussed further in later chapters.

CONCLUSION

The application of marketing theories and ideas is constantly changing. As new tools become available—the Internet or social media, for example— marketers adapt, embracing the new tools and sometimes abandoning old ones. But the underlying theories evolve more slowly than the tools that are used to implement them. This chapter has introduced the reader to some of the theories and practices from marketing that are most applicable to the music business. We have also looked at some ways that technology and innovation have changed the way those practices are applied in the industry today.

GLOSSARY

Call to action—A statement usually found near the conclusion of a commercial message that summons the consumer to act, such as "call today" or "watch tonight at 11."

Channels of distribution—The various methods of distributing and delivering products ultimately to the consumer.

Competition-based pricing—Attempts to set prices based on those charged by the company's competitors.

Consumer-based pricing—Using the buyer's perceptions of value to determine the retail price, a reversal from the cost-plus approach.

Cost-based pricing—Determining retail price based on the cost of product development and manufacturing, marketing and distribution, and company overhead and then adding the desired profit.

Diffusion of innovations—The process by which the use of an innovation (or product) is spread within a market group, over time and over various categories of adopters.

EAN—The European Article Numbering system. Foreign interest in UPC led to the adoption of the EAN code format, similar to UPC but allows extra digits for a country identification, in December 1976.

Hedonic—Of, relating to, or marked by pleasure.

Involvement—The amount of time and effort a buyer invests in the search, evaluation and decision processes of consumer behavior.

Loss leader pricing—The featuring of items priced below cost or at relatively low prices to attract customers to the retail store.

Marketing—The performance of business activities that direct the flow of goods and services from the producer to the consumer (American Marketing Association, 1960). Marketing involves satisfying customer needs or desires.

Marketing mix—A blend of product, distribution, promotion and pricing strategies designed to produce mutually satisfying exchanges with the target market. Often referred to as the Four Ps.

Price fixing—The practice of two or more sellers agreeing on the price to charge for similar products.

Product class—Products that are homogeneous or generally considered as substitutes for each other.

Product form—Products of the same form make up a group within a product class.

Product life cycle—The course that a product's sales and profits take over what is referred to as the lifetime of the product.

Product positioning—The customer's perception of a product in comparison with the competition.

Pull strategy—The company directs its marketing activities toward the final consumer, creating a demand for the product that will ultimately be fulfilled as requests for product are made from the consumer.

Push strategy—Pushing the product through the distribution channel to its final destination through incentives aimed at retail and distribution.

Showrooming—When a shopper visits a store to gather information on or physically experience a product, but then purchases the product online.

Universal Product Code (UPC)—An American and Canadian coordinated system of product identification by which a ten-digit number is assigned to products. The UPC is designed so that at the checkout counter an electronic scanner will read the symbol on the product and automatically transmit the information to a computer that controls the sales register.

REFERENCES

American Marketing Association. *Marketing Definitions: A Glossary of Marketing Terms, Committee on Definitions,* Chicago: American Marketing Association. 1960.

American Marketing Association. *Dictionary of Marketing Terms,* www.marketing power.com. 2004. Accessed November 16, 2014.

Anderson, C. "The Long Tail." *Wired,* October 2004. Accessed December 2014.

Christman, E. "Digital Music Sales Decrease for First Time in 2013." www.bill board.biz, January 3, 2014. Accessed November 15, 2014.

Clover, J. "Amanda Palmer's Accidental Experiment with Real Communism." *The New Yorker.* N.p., October 2, 2012. Accessed November 15, 2014.

Fiegerman, S. Kickstarter responds to critics of Zach Braff's campaign. May 10, 2013. http://mashable.com/2013/05/10/kickstarter-zach-braff-critics/.

Finch, J. E. *The Essentials of Marketing Principles,* Piscataway, NJ: REA. 1996.

Goldhaber, M. "The value of openness in an attention economy." *First Monday,* November 6, 2006.

Hefflinger, M. Report: U.S. Music Consumption Up in 2007, Spending Down, *DMW Daily.* Accessed 2008. http://www.dmwmedia.com/news/2008/02/26/ report:-u.s.-music-consumption-2007,-spending- down.

International Federation of the Phonographic Industry, The. Accessed July 23, 2014. http://www.ifpi.org/facts-and-stats.php.

Kickstarter.com. N.p. Accessed Sept. 22, 2013. Path: www.kickstarter.com/blog.

Kotler, P. and Armstrong, G. *Marketing: An Introduction, Upper Saddle River*, NJ: Prentice Hall. 1996.

Kotler, P. and Armstrong, G. *Principles of Marketing*, Englewood Cliffs, NJ: Prentice Hall. 2014.

Merzel, D. "The 4 P's Marketing are Dead!" David Merzel's Blog. Ed. David Merzel, The International. 2011.

Moe, W. and Fader, P. Tangible Hedonic Portfolio Products: A Joint Segmentation Model of Music CD Sales. 1999. http://www.atypon-link.com/AMA/doi/abs/10.1509/jmkr.38.3.376.18866.

Padgett, B. L. and Clyde P. Rolston, C. P. "Crowd Funding: A Case Study at the Intersection of Social Media and Business Ethics." *Journal of the International Academy for Case Studies*. 2014.

Palmer, A. (blogpost) May 13, 2012. http://amandapalmer.net/blog/.

Peoples, G. Amanda Palmer Q&A: why pay-what-you-want is the way forward, and more. January 28, 2013. https://www.billboard.com/biz/articles/news/indies/1533797/amanda-palmer-qa-why-pay-what-you-want-is-the-way-forward-and-more.

Rogers, E. M. *Diffusion of Innovations* (4th edition), New York: Free Press. 1995.

CHAPTER 3

Market Segmentation and Consumer Behavior

Before a marketing plan can be designed, it is necessary to fully understand the market—who your customers are. One should not make any marketing decisions until a thorough examination of the market is conducted. This chapter will explain how markets are identified, segmented, and how marketers learn to understand groups of consumers and their shopping behavior. We will conclude by looking at the process and some influences on the consumer's decision-making process.

MARKETS AND MARKET SEGMENTATION

A market is defined as a set of actual and potential buyers of a product or service (Kotler and Armstrong, 2014). The market includes anyone who wants or needs your product and has the ability to buy. Markets are identified by measureable characteristics of their members. Not all consumers of a product class (music) are potential buyers of your product offering (alternative rock). The basic goal of market segmentation, the subdividing of a market, is to determine the target market for your specific product.

Marketers segment markets for several reasons:

1. It enables marketers to identify groups of consumers with similar needs and interests and get to know the characteristics and buying behavior of the group members.

2. It provides marketers with information to help design custom marketing mixes to speak to the particular market segment.

3. It is consistent with the marketing concept of satisfying customer wants and needs.

On the most basic level, music markets can be segmented into three sections: 1) current fans, 2) potential fans, and 3) those people who are not now, nor ever likely to be, fans. Perhaps this third group includes people who do not particularly care for the genre that your artist represents. It may include people who do not consume music, people who are unwilling or unable to pay for music, or those without access to become consumers. For example, if I don't have a computer or Internet access I am not part of iTune's target market. Businesses focus on the first two groups.

MARKET SEGMENTATION

Because most markets are so complex and composed of people with different needs, wants and preferences, markets are typically subdivided so that promotional efforts can be customized—tailored to fit the particular submarket. For most products, the total potential market is too diverse or heterogeneous to be treated as a single market. To solve this problem, markets are divided into smaller, more homogeneous sub-markets. Market segmentation is defined as the "process of dividing a the market according to similarities that exist among the various subgroups within the market. The similarities may be common characteristics or common needs and desires" (Dictionary of Marketing Terms, 2000). The members of the resulting segments are similar with respect to characteristics that are most vital to the marketing efforts. This segmentation may be made based on demographics, behavior, geography, psychographics, or some combination of two or more of these characteristics.

The process of segmenting markets is done in stages. In the first step, segmentation variables are selected and the market is separated along those partitions. The most appropriate variables for segmentation will vary from product to product. The appropriateness of each segmentation factor is determined by its relevance to the situation. After this is determined and the market is segmented, each segment is then profiled to determine its distinctive demographic and behavioral characteristics. Then the segment is analyzed to determine its potential for sales. If the segment meets the criteria for successful segmentation (below), the company's target markets are chosen from among the segments determined at this stage. If the segments do not meet the criteria the process can be repeated looking at different segmentation variables.

Take, for example, the market for radio listeners. Wikipedia lists over 60 music formats for radio. The 2010 Broadcast and Cable Yearbook listed over 90 ("2010 Broadcast Yearbook"). No single radio station could possibly serve all of the diverse musical tastes in the country. Instead, station owners, like Citadel Broadcast Corp. and iHeartMedia, Inc., segment the local market based on musical preference and offer multiple stations, each with a format corresponding to the musical tastes of a subset of the local market.

In order to be successful, segmentation must meet these criteria:

Substantiality—the segments must be large enough to justify the costs of marketing to that particular segment. Costs are measured by how much is spent to reach each member of the market and the *conversion rate*—the percent of those you reach who follow through on the purchase.

Measurable—marketers must be able to conduct an analysis of the segment and develop an understanding of their characteristics. Historically, marketing decisions were made based on knowledge gained from analyzing the segment using surveys and other traditional research methods. Today, the Internet allows for data mining and behavioral targeting, creating market segments based on what Internet users purchase online and what types of sites they visit. For example, based on searches of weather reports and restaurant listings online, a search engine company can determine where someone lives. And based on searches they have conducted on the Web and what keywords they have used in those searches, the company can determine what products that person might be interested in receiving information about (Jesdanun, 2007).

Accessible—the segment must be reachable through existing channels of communication and distribution. The Internet has opened up accessibility to all marketers to reach members of their target market—as long as they are Internet users. It has also lowered the costs to reach target members, lowering the barriers to entry and allowing small, undercapitalized companies to compete with major players.

Responsiveness—the segment must have the potential to respond to the marketing efforts in a positive way, by purchasing the product. The use of Apple Pay, PayPal, Braintree, and gift cards sold through a multitude of outlets has opened up new payment methods, allowing businesses to expand their customer base to Internet shoppers who don't have access to credit cards (pre-teens) or may be reluctant to give out their credit card numbers online (the elderly).

Unique—the segments must be unique enough to justify separate offerings, whether the uniqueness calls for variety in product features or simply variety in marketing efforts. Media fragmentation has allowed for more tightly defined market segments. The explosion of specialty magazines, TV channels and websites, along with the ease of searching the Internet, has allowed marketers to more tightly target their messages to different audiences.

Market Segments

There is more than one way to segment markets, so marketers look for segmentation strategies that will maximize potential income. This is done by successfully targeting each market segment with a uniquely tailored plan—one that addresses the particular needs of the segment. Markets are most commonly segmented based upon a combination of geographic, demographic, and personality or psychographic variables, and actual purchase behavior. The bases for determining geographic and demographic characteristics are quite standardized in the field of marketing because they are easily measurable. Psychographics, lifestyle, personality, behavioristic, and purchase characteristics are not as standardized, and the categorization of these variables differs from textbook to textbook. Psychographics includes personality, beliefs, social class, and sometimes lifestyle. Behavioristic includes both attitudes toward the product and actual purchase behavior.

Traditionally, marketing has relied on demographic, geographic or psychographic variables, either alone or in combination (using some demographics combined with some psychographics). Driven in large part by the ease of collecting online behavior (shopping) data, market segmentation has evolved to include more purchase behavior. Behavioral segmentation is a more effective way to segment markets because it is more closely aligned with the propensity to consume the product of interest.

Geographic Segmentation

Geographic segmentation involves dividing the market into different geographical units such as cities, states, or regions. Markets may also be segmented based upon population density (e.g., urban, suburban, or rural) or even weather. Location may be used as a proxy for differences in income, culture, social values, and types of media outlets or other consumer factors (Evans and Berman, 1992). Companies like Claritas (owned by the Nielsen Co.) combine demographic, behavioral, and geographic data into behaviorally distinct segments within zip codes. These geo-demographic profiles, called PRIZM, include media consumption. While detailed descriptions are expensive, anyone can access brief profile descriptions on their website: www.prizm.com. Media research companies such as Nielsen Audio use geographic units called the *area of dominant influence (ADI)* or *designated market area (DMA)*. DMA is defined by Nielsen as an exclusive geographic area of counties in which the home market television stations hold a dominance of total hours viewed. The American Marketing Association describes both ADI and DMA as the geographic area surrounding a city in which the broadcasting stations based in that city account for a greater share of the listening or viewing households than do broadcasting stations based in other nearby cities (American Marketing Association, 2014). Following is an index chart for contemporary hit radio (CHR) listening by geographic region.

This chart shows the American national listenership to CHR radio by geographic region. With the national index, or average, being 100, it is easy to see which regions of the country tend to prefer CHR and which ones do not listen to the format as much as the average.

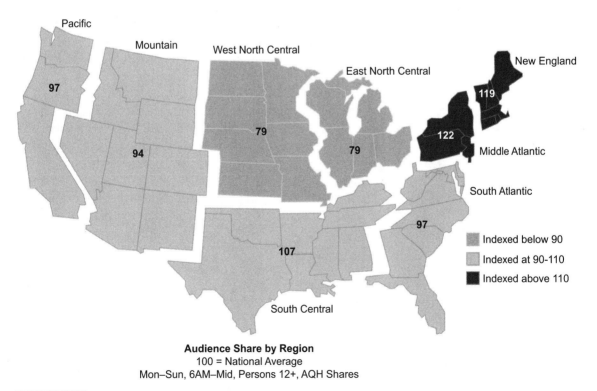

Audience Share by Region
100 = National Average
Mon–Sun, 6AM–Mid, Persons 12+, AQH Shares

FIGURE 3.1 *Audience share by region—CHR*

Source: Nielsen Audio

When the marketer of recorded music and live performances uses geographic segmentation in this manner, it can help justify the use of resources in support of the marketing plan. Marketing strategies are then easily tailored to particular geographic segments. For example, the label may use geographic segmentation to determine in which cities tour support money would be most effective.

Demographic Segmentation

Demographics are basic measurable characteristics of individual consumers and groups such as age, gender, ethnic background, education, income, occupation, and marital status. Demographics are the most popular method for segmenting markets because the information is easier and cheaper to measure than more complex segmentation variables, such as personality or consumer behavior. Fortunately, groups of people with similar demographics tend to have similar needs and interests that are distinct from other demographic segments.

Age is probably the demographic most associated with changing needs and interests. Consumers can be divided into age categories such as children, teens, young adults, adults, and older adults. The segment of "tweens" (preadolescents between the ages of 8 and 13) has been added to the mix because of their enormous spending power. In the U.S., tweens are expected to number 23 million by 2020 (Jayson, 2009). This age group accounts for $43 billion in disposable income, over half have a Facebook account and 78% have a cell phone. The typical tween spends 8 to 12 hours per day consuming media of one form or another and more than half of the girls age 10 to 12 want to be famous (Younger, 2014).

Gender is also a popular variable for segmentation, as the preferences and needs of males are perceived as differing from those of females. Differential needs based upon gender are obvious for product categories such as clothing, cosmetics, hairdressing, and even magazines. But even in the area of music preferences, differences in taste exist for males and females. Nielsen Audio reports that males are more likely to listen to alternative, rock or news/talk radio, while women prefer top 40, country, adult contemporary, and contemporary Christian radio (Arbitron Radio Today, 2010).

Income segmentation is popular among certain product categories, such as automobiles, clothing, cosmetics, and travel, but is not as useful in the recording industry. *Educational* level is sometimes used to segment markets. Well-educated consumers are likely to spend more time researching

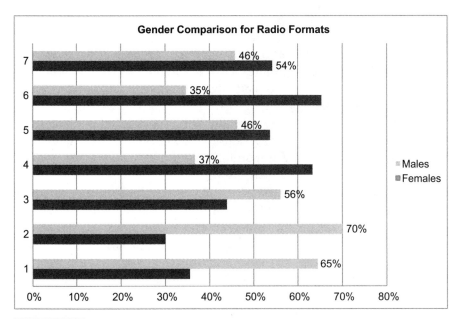

FIGURE 3.2   *Gender Comparison for Radio Formats*

Source: Nielsen Audio

purchases and are more willing to experiment with new brands and products (Kotler, 1980).

Multivariable Segmentation

The process of combining two or more demographic or other variables to further segment the market has proven effective in accurately targeting consumers. By considering age, gender and income together, marketers can better tailor the marketing messages to reach each group. For example, older males may prefer news/talk radio, whereas younger males prefer alternative rock radio.

Psychographic Segmentation

Psychographic segmentation involves dividing consumers into groups based upon lifestyles, personality, opinions, motives, or interests. Psychographic segmentation is designed to provide additional information about what goes on in consumers' minds. While demographics may paint a picture of what consumers are like, psychographics adds vivid detail, enabling marketers to shape very specific marketing messages to appeal to the target market.

By understanding the motive for making purchases, marketers can emphasize the product attributes that attract buyers. For example, if consumers are driven by price, pricing factors such as coupons can be emphasized. If another segment is driven by convenience, this issue can be addressed through widespread product distribution. Lifestyle segmentation divides consumers into groups according to the way they spend their time and the relative importance of things in their life. Imagine that your primary residence, your home, is on a piece of lakefront property. How would your life be different? You would likely have a boat tied to a dock and all of this would be a short walk out your back door. You could be on the water in the middle of the lake in minutes, as opposed to having to hook the boat trailer up to the car and drive across town, wait for your turn at the boat launch and then park the SUV before you could be in the boat on the lake. And if you lived on the waterfront, your house is much more likely to have a nautical theme to it. If you played organized team sports as a child your family probably built a lifestyle around traveling to practices, games, and tournaments that influenced much of your weekend activities and drove purchase and consumption behavior—ice chests, lawn chairs, suntan lotions, ballpark food, and hotels all become part of the lifestyle, at least during the season. Even the family's choice of cars was probably influenced by your participation in team sports!

Psychographics are more difficult to measure and require constant updating to stay abreast of changes in the marketplace. One system that uses psychographics to identify market segments is the VALS™ segmentation framework, developed by the Stanford Research Institute. Originally, the segmentation was based on values (V) and lifestyles (LS), thus the acronym VALS, which is still used for branding purposes. But the version of VALS in current use segments consumer markets on the basis of selected psychological traits.

The VALS™ system places consumers into three self-orientation categories and four levels of resources and innovations. The resulting eight segments provide insight into consumers based on motivations, beliefs, lifestyles, and resources. For example, *thinkers* are conservative and motivated by ideals; *experiencers* are young and enthusiastic and motivated by self-expression, excitement, and innovation. More information on the VALS segments is available at http://www.strategicbusinessinsights.com/vals/.

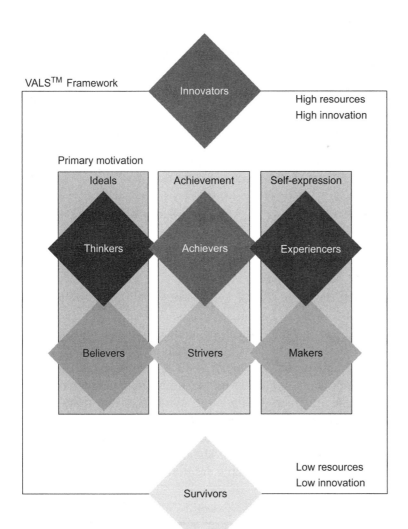

FIGURE 3.3 *Values and Lifestyle Segments*

Source: SRI Consulting Business Intelligence

THE QUEST FOR COOL

Marketers use a variety of research techniques to spot lifestyle trends as they develop, so that their products will be in the marketplace ahead of the demand and the competition. Known in vernacular terms as "cool hunting" or the "quest for cool," these marketers are hired by well-known brands to track and understand youth culture. Then the principles of diffusion theory are applied to target opinion leaders in the hopes of penetrating the market (Grossman, August 31, 2003). Cool hunting involves observing alpha consumers (cool people) talking, eating, dressing, or shopping and then predicting what the rest of the market will be doing a year from now. One such company, Look-Look (2004), uses a variety of research techniques including surveys, field observations, ethnographies, mall intercepts and focus groups. They document their findings with photographs and video recordings and report their conclusions to a variety of major companies who are targeting the youth market (pbs.org 2015).

Since the early days of rock and roll, the music industry has attempted to reach trendsetters in the hope that they would positively influence sales of the acts they were marketing. Nowadays, not just record labels, but a wide variety of companies are targeting the urban hip-hop market (marketresesarch.com).

Personality Segmentation

Researchers at the University of Texas have found that personal music preferences can be linked to personality traits. P.J. Rentfrow and S.D. Gosling (2003) found that people's music preferences typically classify them into one of four basic dimensions: 1) reflective and complex, 2) intense and rebellious, 3) upbeat and conventional, or 4) energetic and rhythmic. Preference for each of the following music dimensions is differentially related to one of these basic personality traits. While this information may seem obvious and not very useful, knowing the personality of your target market allows the marketer to make informed decisions about where and how to promote specific songs and artists. Should the promotion be more cerebral (jazz or blues) or appeal to the fans desire to fit in with the status quo (pop or country).

Genres	Personality
Classical, jazz, blues, folk	Reflective and complex
Alternative, rock, heavy metal	Intense and rebellious
Country, pop, religious, soundtracks	Upbeat and conventional
Rap/hip-hop, soul/funk, electronica/dance	Energetic and rhythmic

Behaviorial Segmentation

Behavioral segmentation is based on actual customer behavior toward products. Behavioral segmentation has the advantage of using variables that are closely related to the product and its attributes. Some of the more common behavioral segmentation variables are: product usage, benefits sought, user rate, brand loyalty, user status, readiness to buy, and purchase occasion. Let's look at a couple of behavioral segmentation variables particularly relevant to the music industry

Product Usage Segmentation

Product usage involves dividing the market based on those who use a product, potential users of the product, and nonusers—those who have no usage or need for the product. How a product is used (and for what purposes) is also of importance. For example, Kodak found that disposable cameras were being placed on tables for wedding guests to help themselves to. As a result, Kodak modified the product by offering five-pack sets of cameras in festive packaging (Nickels and Wood, 1997). The recording industry has adapted to usage situations by repackaging music that is customized for usage occasion, such as wedding music compilations, party mixes, romantic mood music (complete with recipes for romantic dinners), and so forth. (See section on hedonic responses to music in chapter 2.)

Segmentation Based on Benefits Sought

It is said that people do not buy products, they buy benefits. They buy aspirin to alleviate pain or reduce the chances of a heart attack; toothpaste to whiten their teeth or prevent cavities; and music to elevate mood or create atmosphere. Benefit segmentation divides the market according to benefits sought by consumers. In the music market, some consumers shop for classical music because it drowns out distracting noise while they work or study. Others buy particular artists because they believe it will create the right atmosphere for a romantic evening.

Brand Loyalty

Brand loyalty is defined as the degree to which a consumer will repeatedly purchase a company's product or service. As a segmentation variable, brand loyalty can be used to divide the market into those who are and those who are not brand loyal. Artists' hardcore fans are brand loyal—they will buy most anything with the artist's name attached to it, often with little consideration of any other factors. The ad agency Saatchi & Saatchi calls extreme brand loyalty "lovemarks." Not surprisingly, the Beatles tops their list of the 50 top music and radio brands. A demonstration of the power of brand loyalty can be seen in the Voyager Company's decision to make the Beatles "A Hard Day's Night" the first full-length commercial movie to appear on CD. When the Voyager Company chose to release "A Hard Day's Night" in May 1993, the number of U.S. households with computers had increased from 8.2% to just under 23%, but many of those computers still did not have CD-ROM drives (http://www.statista.com/). The Voyager Company chose "A Hard Day's Night" in large part because they knew loyal Beatles fans would buy the discs even if they could not play it on their computers yet.

User Status

User status defines the consumer's relationship with the product and brand. It involves level of loyalty and propensity to become a repeat buyer. Stephan and Tannenholz (2002) identified six categories of consumers based on user status: sole users, semi-sole users, discount users, aware non-triers, trial/rejecters, and repertoire users.

USER STATUS (BRAND A)

Sole users are the most brand loyal and require the least amount of advertising and promotion.

Semi-sole users typically use brand A, but have an alternate selection if it is not available or if the alternate is promoted with a discount.

Discount users are the semi-sole users of competing brand B. They don't buy brand A at full price, but perceive it well enough to buy it at a discount.

Aware non-triers are category users, but haven't bought into brand A's message.

Trial/rejecters bought brand A's advertising message, but didn't like the product.

Repertoire users perceive two or more brands to have superior attributes and will buy at full price. These are the primary brand switchers; therefore, the primary target for brand advertising.

The Millennial Generation

Neil Howe and William Strauss examined the emergence of a Millennial generation, which they define as the generation of young adults, the first of whom have "come of age" around the time of the millennium (Howe and Strauss, 2000). This segment includes people born after 1980, according to Pew Research, although Howe and Strauss defined the segment as those born between 1982 and 1995. As of 2012, millennials numbered about 80 million in the U.S. and, as they come of age, they are becoming a huge economic force (Schawbel, 2012). This segment is characterized by a sharp break from Generation X, and hold values that are at odds with the Baby Boomers. Pew Research (2014) described Millennials as "relatively unattached to organized politics and religion, linked by social media, burdened by debt, distrustful of people, in no rush to marry and optimistic about the future." Bock (2002) also found them to be optimists and team players who follow the rules and accept authority.

R. Craig Lefebvre (2006), a Professor at George Washington University, characterizes the millennial generation as more involved in peer-to-peer communication than previous generations. He states ". . . reliance and trust in nontraditional sources—meaning everyday people, their friends, their networks, the network they've created around them—has a much greater influence on their behaviors than traditional advertising." He goes on to state ". . . the challenge for marketers is how to create peer to group exchanges that feature their brands, products, services and behaviors. The question is no longer 'what motivates someone to change' but rather 'what motivates someone to share something they find intrinsically useful and valuable with their most trusted friends and colleagues?'

TARGET MARKETS

Once market segments have been identified, the next step is choosing the segment or segments that will allow the organization to most effectively achieve its marketing goals—the target markets. Simply stated, a *target market* is a group of persons for whom a firm creates and maintains a product mix that specifically fits the needs and preferences of that group (Dictionary of Marketing Terms, 2000). With the target markets identified, the company begins the task of positioning the product offering, relative to the competition, in the minds of the consumers using the tools available to the market, primarily advertising.

CONSUMER BEHAVIOR AND PURCHASING DECISIONS

According to the American Marketing Association, consumer behavior is defined as "the dynamic interaction of affect and cognition, behavior, and environmental events by which human beings conduct the exchange aspects of their lives" (2014). More simply put, *consumer behavior* is the buying habits and patterns of consumers in the acquisition and usage of goods and services. Entire courses on consumer behavior exist because it is one of the most interesting areas of marketing. Below we will discuss only a few key concepts in consumer behavior as they relate to music purchasing and consumption.

Needs and Motives

When a product is purchased, it is usually to fill some sort of need or desire. There is some discrepancy between the consumer's actual state and their desired state—*I want that!* In response to a need, consumers are motivated to make purchases. **Motives** are internal factors that activate and direct behavior toward some goal. In the case of shopping, the consumer is motivated to satisfy the want or need, as outlined in chapter 2. Understanding motives is a critical step in creating effective marketing programs. Psychologist Abraham Maslow developed a systematic approach of looking at needs and motives.

Abraham Maslow (1943) states that there is a hierarchy of human needs. More advanced needs are not evident until basic needs are substantially met. Maslow arranges these needs into five categories of physiological, safety, love, esteem, and self-actualization. Marketers believe that it is important to understand where in the hierarchy the consumer is before designing an effective marketing program.

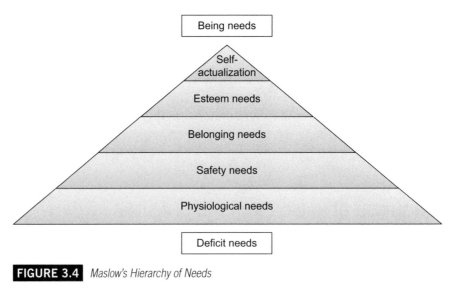

FIGURE 3.4 *Maslow's Hierarchy of Needs*

Source: Abraham Maslow

Physiological needs are the basic survival needs that include air, food, water, and sleep. Until these needs are met, there is little or no interest in fulfilling higher needs. Survivors in the VALs system fit into this category. Safety needs deal with establishing a sense of security and may motivate consumers to seek political and religious solutions. Until these needs are substantially met, the need for love and belongingness are absent. Love, approval, and belongingness are the next level, along with feeling needed and a sense of camaraderie. Products oriented toward social events sometimes appeal to this need and this includes music, particularly live performances. Esteem needs drive people to seek validation and status. Self-actualization involves seeking knowledge, self-fulfillment, and spiritual attainment.

An individual may temporarily sacrifice a lower-level need to achieve a higher level need, like skipping a meal and donating the money to a worthy cause (like buying tickets for your favorite band's show) or fasting for spiritual reasons, but in the long-term the lower level needs must be fulfilled to achieve the higher-level needs.

Nobody needs music in the Maslovian sense. We can live without it. We might not like it and certainly life is better with music, but we won't die without it. The challenge to music marketers is to make consumers *feel* like they need music and specifically the music they are selling at the time.

Converting Browsers to Buyers

One of the greatest challenges facing marketers is converting potential customers who are browsing the merchandise into actual purchasers. Retail stores refer to a *conversion rate*—the percentage of shoppers who actually make a purchase. The actual rate varies and depends upon the level of involvement and risk, with small-ticket items having a higher conversion rate than expensive items such as appliances or jewelry. Conversion rates can be improved by reducing the purchase risk through sampling (e.g., listening stations and free downloads), liberal return policies, as well as offering a level of service that meets or exceeds expectations. None of this matters, however, if the product is not in stock or can't be found. When shoppers don't buy, it's often because they were unable to find what they wanted or the purchasing process was too difficult. Making the product easy to find is equally important in both the physical and digital world.

Once the shopper has made their choice then the checkout process must be quick and simple—don't give them a chance to change their mind! People have been known to spend hours shopping for just the right item only to abandon their shopping cart, physical or digital, because the check out process took too long. This is why many online stores such as Amazon.com offer accounts to customers "one click buying." With their address and credit card information already stored in the system the consumer can purchase with a single click of a mouse button. Marketers must also make it easy to buy through banner ads, videos, apps, concerts, and at brick-and-mortar stores or risk losing sales.

HIGH AND LOW-INVOLVEMENT DECISION MAKING

The construct of involvement usually refers to the amount of time and effort a consumer invests in the search, evaluation and decision process of consumer behavior. The level of involvement in the purchase depends on the economic and social risk of making the wrong decision. Consumers who search for information about products and brands in order to evaluate them thoroughly are engaging in *high involvement decision making*. They want to know as much as possible about all choices before making their decision.

A closely related concept is product involvement (Bloch and Richins, 1983). Product involvement may take the form of situational involvement, enduring involvement, or response involvement. Consumers that are product involved may be less involved with the purchase because they already know much of the information needed to make a decision. For example, if you are really into classical music, then you probably subscribe to several magazines on the subject, visit the important blogs and websites on a regular basis, and listen to the latest releases as soon as possible. When it comes time to buy, you already are up to date on the market and, therefore, don't spend much time gathering information or weighing choices. Being product involved actually decreases the level of purchase involvement because the buyer already has most of the information required to make the decision.

In low *involvement decision-making*, the consumer perceives very little risk, low self-identification with the product and little or no personal relevance. Low involvement decision-making is often habitual, associated with inexpensive consumer package goods, (like chewing gum or soft drinks which are purchased routinely) and characterized by brand familiarity and loyalty.

Several factors can influence a consumer's level of involvement in a purchase process:

1. Previous experience—If consumers have previous experience with the product or brand, they are more likely to have a low level of purchase involvement since they are already familiar with the advantages and disadvantages of selecting a particular product.

2. Interest—The more personal interest consumers have in the product, product class, or benefits associated with the product, the more likely they are to engage in high involvement behavior.

3. Perceived risk of negative consequences—If the stakes are high and the consequences of making the wrong decision are dire, consumers engage in high involvement behavior. Price is a variable associated with risk. High-priced purchases bear more risk. As price increases, so does the level of involvement.

4. Situation—Circumstance may play a role in temporarily raising the level of involvement for a particular product. A food or beverage purchase casually made under most circumstances may call for more scrutiny if guests are coming to dinner. A CD bought for personal use is a rather low involvement purchase; when given as a gift it has higher involvement because of the risk of a bad decision.

5. Social visibility: Involvement increases as the social visibility of a product increases. Designer clothing falls into this situation. Laundry appliances carry little social risk, although they may be considered a financial risk and, therefore, invoke a high-involvement decision process.

DECISION-MAKING PROCESS

The stages involved in the consumer's decision-making process are described with the basic AIDA model: Awareness, Information search, Decision and Action (see chapter 2). This is part of the hierarchy of effects theory, which involves a series of steps by which consumers receive and use information in reaching decisions about what actions they will take (e.g., whether or not to buy a product).

A more elaborate model involves the following six steps:

1. Problem recognition—The consumer realizes that a purchase must be made. Perhaps a new need or want has emerged (I want that new song), or perhaps the consumer's supply of a commonly used product has run low. For example, you have just learned that your favorite artist will be performing locally and you don't have tickets.

2. Information search—The consumer will seek out information about the various options available for satisfying the want, need, or desire. In the concert example, you are now entering the phase of doing research: finding out how much you can afford to spend on tickets and what seats are available in your price range.

3. Alternative evaluation—The consumer will look for alternate options for the product under consideration. Should you buy tickets

at all? Is there an even better concert coming to town that you should save your money for?

4. Purchase or acquisition—The consumer makes the purchase.

5. Use—The consumer uses the product (attends the show), thereby gaining personal experience.

6. Post-purchase evaluation—Experience with the product leads to evaluation to determine if the product meets expectations and should be purchased again next time the need arises. If you enjoy the show you may well buy tickets the next time the artist is in your area.

Cognitive vs. Emotional Decisions

Much research has been done in the field of advertising on the relative roles of affective (emotional) factors vs. cognitive factors in motivating consumers. Cognition-based attitudes are thoughts and beliefs about something—an "attitude object." Affect-based attitudes are emotional reactions to the attitude object. Attitudes can have both of these components, and can be based more or less on either cognitions or affect. While there is probably an element of both in each purchase decision, the relative contribution of each, as well as the process involved, may differ depending upon the type of product and the level of involvement.

The model of the hierarchy of effects has been subsequently modified to include three constructs: *cognition* (awareness or learning), *affect* (feeling, interest, or desire), and *conation* (action or behavior)—CAB. The order in which these steps are taken is subject to the type of product and the level of involvement. According to Richard Vaughn, product category and level of involvement may determine the order of effects, as well as the strength of each effect (Vaughn, 1986). His model uses involvement (high/low) and think/feel (cognitive or affective components) as the two dimensions for classifying product categories and ordering these three steps.

To begin with, high-involvement situations suggest that the *cognitive* stage and the *affective* stage usually appear first, and these two stages are followed by the *conation* (behavioral) stage. In low-involvement situations, advertising may create awareness first, but attitudes or brand preferences are formed after product trial or experience. Thus, conation or action occurs before opinions are formed about the product. In other words, the thinking and feeling occur before the action in high-involvement situations, whereas opinions are formed only after trial of the product in low-involvement situations. Low involvement products are more subject to impulse purchases.

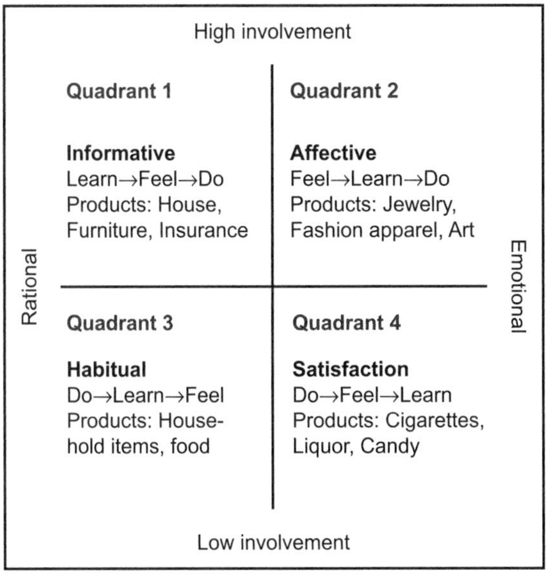

```
                        High involvement

        Quadrant 1          │    Quadrant 2

        Informative         │    Affective
        Learn→Feel→Do       │    Feel→Learn→Do
 R      Products: House,    │    Products: Jewelry,     E
 a      Furniture, Insurance│    Fashion apparel, Art   m
 t                          │                           o
 i      ────────────────────┼────────────────────       t
 o                          │                           i
 n      Quadrant 3          │    Quadrant 4             o
 a                          │                           n
 l      Habitual            │    Satisfaction           a
        Do→Learn→Feel       │    Do→Feel→Learn          l
        Products: House-    │    Products: Cigarettes,
        hold items, food    │    Liquor, Candy
                            │
                        Low involvement
```

FIGURE 3.5 *Hierarchy of Effects*

Source: Foote, Cone and Belding

The resulting model is a four-quadrant grid with quadrant one (informative) containing high involvement cognitive products, quadrant two (affective) for high involvement emotional purchases, quadrant three (habitual) for low involvement rational purchases, and quadrant four (satisfaction) for low-involvement emotional purchases. Music purchases generally fall into the satisfaction quadrant, although as prices increase, so does the level of involvement. Some music purchases can be considered impulse buys, if correctly positioned at retail. Other purchases may require creating the emotional (affective) response in the customer. Since music by its very nature evokes emotional responses, exposure to the product is known to increase purchase rate.

CONCLUSION

This chapter has provided a brief introduction to some of the fundamental theories and concepts in consumer behavior and market segmentation. The music business is just like any other business when it comes to the basics of market segmentation and consumer behavior, but a deep understanding of these basic theories and constructs, and being able to apply them to your specific artist or album, will help separate you from everyone else in the business. The best music marketers will know their consumers as well as they know their artists.

GLOSSARY

ADI—Area of Dominant Influence (see DMA).

Affect—A fairly general term for feelings, emotions, or moods.

Behavioral segmentation—Based on actual customer behavior toward products.

Cognition—All the mental activities associated with thinking, knowing, and remembering.

Conation—represents intention, behavior, and action.

Consumer behavior—The dynamic interaction of affect and cognition, behavior, and environmental events by which human beings conduct the exchange aspects of their lives.

Cookies—A small file sent by a server to a web browser and then sent back by the browser each time it accesses that server. The main purpose of cookies is to identify users and store information about them, including preferences specific to the website.

DMA—Designated Marketing Area. The geographic area surrounding a city in which the broadcasting stations based in that city account for a greater share of the listening or viewing households than do broadcasting stations based in other nearby cities.

Geographic segmentation—Dividing the market into different geographical units such as town, cities, states, regions and countries. Markets also may be segmented depending upon population density, such as urban, suburban, or rural.

Involvement—The amount of time and effort a consumer invests in the search, evaluation and decision process of consumer behavior.

Market—A set of all actual and potential buyers of a product. The market includes anyone who has an interest in the product and has the ability and willingness to buy.

Market segmentation—The process of dividing a large market into smaller segments of consumers that are similar in characteristics, behavior, wants or needs. The resulting segments are homogenous with respect to characteristics that are most vital to the marketing efforts.

Millennials—People born between the years 1982 and 1995. Also called the "Net Generation" because they grew up with the Internet.

Personal demographics—Basic measurable characteristics of individual consumers and groups such as age, gender, ethnic background, education, income, occupation, and marital status.

Psychographic segmentation—Dividing consumers into groups based upon lifestyles, personality, opinions, motives, or interests.

Target market—A set of buyers who share common needs or characteristics that the company decides to serve.

User status—The consumer's relationship with the product and brand. It involves level of loyalty.

Web analytics—The use of data collected from a web site and its visitors to assess and improve the effectiveness of the website.

Web data mining—the process of analyzing navigation patterns and other behavior of user's on the web in order to improve their navigation and interaction with the website.

REFERENCES

American Marketing Association Dictionary. 2014. https://www.ama.org/.

Arbitron Radio Today. "How America Listens to Radio." 2010. http://futureofradio online.com.

Bloch, P. H. and Richins, M. L. "A Theoretical Model for the Study of Product Importance Perceptions," *Journal of Marketing,* Vol. 47, No. 3 (Summer 1983), pp. 69–81.

Bock, W. Web. http://www.mondaymemo.net/010702feature.htm. 2002.

Evans, J. R. and Berman, B. *Marketing* (5th edition), New York: Macmillan Publishing. 1992.

Grossman, L. "The Quest for Cool," *Time Magazine.* August 31, 2003.

Howe, N. and Strauss, W. *Millennials Rising: The Next Great Generation,* New York: Vintage Books. 2000.

Imber, J. and Toffler, B. A. "Market Segmentation." *Dictionary of Marketing Terms.* 3rd ed. Hauppauge, NY: Barron's Educational Series, Inc. 2000.

"Interview: Dee Dee Gordon and Sharon Lee." *Frontline.* PBS.org, n.d. Accessed June 22, 2015.

Jayson, S. "It's Cooler than Ever to be a Tween, but is Childhood Lost?" *USA Today,* February 2, 2009.

Jesdanun, A. "Ad Targeting Grows as Sites Amass Data on Web Surfing Habits." *The Tennessean,* December 1, 2007.

Kent, S.S. and Tannenholz, L. B. (2002). *"Advertising Age" Marketing segmentation and the marketing mix: Determinants of advertising strategy.* McGraw-Hill Online Learning Center. Accessed http://highered.mcgraw-hill. com/0072415444/student_view0/chpater5/els.html Accessed June 20.

Kotler, P. *Principles of Marketing,* Englewood Cliffs: Prentice-Hall. 1980.

Kotler, P. and Armstrong, G. *Principles of Marketing,* (15th edition), Upper Saddle River, NJ: Pearson Education. 2014.

Lefebvre, R. C. "Communication Patterns of the Millennium Generation." Web. http://socialmarketing.blogs.com/. 2006.

Look-Look Company Website. Accessed http://www.look-look.com/. 2004.

Maslow, A. "A Theory of Human Motivation," *Psychological Review,* 50 (1943), pp. 370–396.

"Millennials in Adulthood: Detached from Institutions, Networked with Friends." *Pew Social Trends.* Pew Research Center, March 7, 2014. Accessed June 27, 2015. http://www.pewsocialtrends.org/2014/03/07/millennials-in-adulthood/.

Nickels, W. G. and Wood, M. *Marketing,* New York: Worth Publishers. 1997.

Rentfrow, P. J. and Gosling, S. D. "The Do-Re-Mi's of Everyday Life: Examining the Structure and Personality Correlates of Music Preferences," *Journal of Personality and Social Psychology*, 84 (2003), pp. 1236–1256.

Schawbel, D. "Millennials vs. Baby Boomers: Who Would You Rather Hire?" *Time Magazine*. March 29, 2012. Accessed November 15, 2014.

"2010 Broadcast Yearbook." *American Radio History*. Ed. David Gleason. N.p., n.d. Accessed 22 June 2015.

Vaughn, R. "How Advertising Works," *Journal of Advertising Research*. 1986. www.statista.com/statistics/184685/percentage-of-households-with-computer-in-the-united-states-since-1984/

Younger, S. "Tweens by the Numbers: A Rundown of Recent Stats." www.chicagonow.com Accessed June 17, 2014.

CHAPTER 4

Marketing Research

Research is formalized curiosity. It is poking and prying with a purpose.

—Zora Neale Hurston

INTRODUCTION

Marketing research is the function that links the consumer, customer, and public to the marketer through information—information used to identify and define marketing opportunities and problems; generate, refine, and evaluate marketing actions; monitor marketing performance; and improve understanding of marketing as a process. **Marketing research** specifies the information required to address these issues, designs the method for collecting information, manages and implements the data collection process, analyzes the results, and communicates the findings and their implications. (American Marketing Association, 2013). Marketing research and **market research** are not the same thing: market research is a subset of marketing research that looks specifically at the size, location, and makeup of a product market.

RESEARCH IN THE MUSIC INDUSTRY

Research is not a major division of any record label. Some of the largest labels have a small staff dedicated to consumer and industry research while others

CONTENTS

handle research chores ad hoc or hire the task out to an independent research firm. The more sophisticated research departments produce work that can be used to, for example, persuade major corporations of the benefit of teaming up with a particular artist in pursuit of a common consumer. The labels continue to rely mostly on syndicated data collected after a song is released. This is not to say that the labels don't collect information on an artist before they sign them. They do. But it is often limited to checking with others in the industry, social media numbers, sales numbers (if they are available), attending a few live shows, and then going with their "gut feelings."

Radio stations perform most of the research in the music industry. They purchase syndicated research as well as measure the appeal of new music through regular online, telephone, or auditorium testing.

TYPES OF RESEARCH

Custom verses syndicated: One way to categorize research is by the purpose for which it is collected. **Custom research** is done to answer a specific question such as "How much are consumers willing to pay for an advertising-free streaming music service?" Since the research is initiated and paid for by a single company or organization, they can customize the research specifically for their needs. Research that is done on an ongoing basis and sold to subscribers in the form of periodic reports is called **syndicated research**. Examples of syndicated research in the entertainment industry include Arbitron (radio listenership), Nielsen (television viewership), Broadcast Data System's (BDS) airplay monitoring, and SoundScan's retail sales monitoring. Because syndicated research has many users and is often used for longitudinal purposes (comparisons over time), the information is standardized and may be more general. The results are published periodically, ranging from daily to quarterly, as a report; however, subscribers may elect to pay for more detailed data or access to "raw" data from which they can create their own reports. For example, if you want to know what the best selling song is for this week you can simply open up *Billboard* magazine where the results from the SoundScan research will be reported in a simple sales chart form, but if you want to know in which areas of the country the song sold best, you would need to subscribe to SoundScan to get the more detailed information.

Purpose of the research: Research can also be categorized by the kinds of questions the researcher is trying to answer. This is commonly referred to as research design. The three types of research design—exploratory, descriptive, and causal—are discussed below.

OVERARCHING RESEARCH ISSUES: VALIDITY AND RELIABILITY

Steps must be taken to assure that the research results are valid and the process is reliable. There are many levels of validity, the scope of which is beyond the purpose of this text, so for our purposes let us define **validity** as the "truth." That is we want our measures to reflect true differences among the subjects being measured and not some random error or bias. To do this, it is very important to be sure the researcher asks the right questions, starting with the problem definition and ending with the interview, focus group or survey questions. Subtle differences in the wording of a question can lead to different results: "You don't really like Bluegrass music, do you?" will get a different response than the less biased, "Do you like Bluegrass music?"

Reliability is the researcher's ability to get similar results from repeated applications of the measures or from independent but comparable measures of the same trait or construct. If you measure how tall you are using a tape measure under similar conditions (no shoes, stand up straight each time), you expect the difference in measurements to be attributed to changes in the object being measured—you grew! But if your tape measure was printed on stretchy material, you might get different results each time because you have a bad measuring instrument.

There is a clock on the shelf in my office. It reads 10:35. The battery died years ago and I just never replaced it. Is the time on the clock valid? Reliable? Think about your answer before you continue reading. The clock is very reliable. No matter when I look at it I always get the same information—"similar results from repeated applications of the measures." 10:35. Unfortunately, it is only valid twice a day! For research results to be useful they need to be both valid and reliable.

Unlike height, much of what we want to measure in the music business is not physical or real—you can't touch or see it. They are latent or abstract **constructs** such as "engagement," "identification," or "emotional response" (Stewart, 2013). These constructs are much harder to accurately measure than sales or spins because they are complex and abstract, making it more difficult to construct valid and reliable measurement instruments.

THE RESEARCH PROCESS

There are several stages to the research process once a problem is recognized, all of which are equally important and demanding: problem definition, determining the research design, choosing and designing the data collection, actual data collection, data analysis, and communicating the results. Although presented as discrete stages, it is important to understand that this is not a linear process. At any stage of the research process you may have to go back to a previous stage and start over or modify the research due to new information or complications.

Problem definition—The first step is to determine what questions need to be asked. The questions may seem obvious, but the researcher must make sure they are getting to the root of the problem. Let's say that an artist or label manager has come to you and wants to know why the artist's last album didn't sell as well as the one before. The obvious answer is because not as many people bought copies of the album, but the real question is why? The end goal of the problem definition stage of the research process is to determine the objective of the research. What is it that you want to learn by doing the research?

The second step in the research process is determining the research design. You could look at the three research designs, exploratory, descriptive and causal, as a continuum. The less you know about the problem, the greater your need for basic information, the more likely you are to use an exploratory research design. Alternatively, the more you know already, the greater the probability that a descriptive or causal research design will be appropriate.

In some situations you may not be sure exactly what the problem is or what questions you should be asking. In that case, an **exploratory research** design may be helpful in clarifying and defining the research objectives. Exploratory research may be done to help better understand the situation, screen alternatives or discover new ideas and results in qualitative data. "The focus of qualitative research is not on numbers but on stories, visual portrayals, meaningful characterizations, interpretations, and other expressive descriptions" (Zikmund and Babin 2007, p. 84).

The primary methods for exploratory research are focus groups and depth interviews, discussed below.

Descriptive research is done to describe the existing characteristics of a defined target market or population (Hair, Bush, and Ortinau, 2003). The primary tool of descriptive research is the survey because of its flexibility, speed, and costs. As mentioned before, the key to a good survey is asking the right questions. Because the survey may be administered without the researcher present (via email or online, for example) clear, concise questions are essential. Finally, **causal designs** are rigidly specified experiments designed to determine cause and effect and are rarely used in the marketing side of the entertainment business.

Data Collection Methods: Once you have an understanding of what you want to know (problem definition), you should research what has already been learned on the topic and what information is already available. The answer to your question may already exist thanks to prior research, which will negate the need for further, more difficult, and expensive research. Whenever you use somebody else's research to answer a question you are doing **secondary research**. Even if the answer isn't already out there, examining existing research will help you avoid repeating the same mistakes and may give you valuable insight into the problem you are investigating. It may also help you determine which research design will give you the best answers to your questions.

If you don't find the answer to your question in the existing research then you will have to conduct additional research. The primary methods used to collect data in the music business are focus groups and surveys. A popular choice for exploratory research, the **focus group** is a face-to-face interview of six to ten people that allows the researcher to delve deeply into participants' responses by asking follow-up questions. Focus groups may appear to be unstructured and free-flowing, but the well-planned interview is designed to discover new information while maintaining the flexibility to pursue interesting answers and topics as they arise. Focus groups are good for testing new music and finding out why consumers like or dislike certain songs, but because they involve a small number of people they are not good for generalizing results or providing statistical data about the market. In other words, you should not make marketing decisions based on a single focus group (although it is often done!). Automakers also use focus groups to get feedback on design changes and new model features, but they conduct dozens of focus groups before drawing any conclusions from them. **Depth interviews** are similar in structure to focus groups but are done on a one-on-one basis.

Research panels: A research panel is like an on-going focus group except in most cases the panel does not meet face to face. Data is collected from the same participants on an ongoing basis, either via mailed surveys, online, or perhaps passively by an app on a phone. Panels may be asked to provide more detailed data than can be asked in a one time survey, or they may be asked to provide data over a long period of time for a longitudinal study. Nielsen Mobile collects data on music and entertainment exposure using a mobile phone app that collects information from the same group of participants on an ongoing basis for as long as they choose to remain a member of the panel. Panel members are recruited to fit the profile desired by the researcher and if a member of the panel drops out for some reason, they can be quickly and easily replaced.

Survey research may be used to further test information gathered from focus groups or as a stand-alone research tool. Survey research is deceptively difficult. How hard can it be to ask somebody a few questions? The challenge comes in not only asking the *right* questions, but asking them the right way to the right people.

It is tempting to take short cuts and ask questions like, "On a seven point scale, please indicate how much you like the music and lyrics of this song?" But how do you answer the question if you love the lyrics but hate the melody? Other problems might not be so obvious. A researcher might focus on the artist's music as the cause of declining sales when the actual cause may come from their appearance or their behavior. This problem could have been solved by good exploratory research and asking the right questions in a focus group. It is also important to make sure you are asking your questions to the right people, the artist's potential audience (target market).

Online Survey Tools

Several companies offer free survey tools with limited capabilities to entice users to buy the full packages. These free versions may be all you need for a short survey with a small sample size. Some of these tools are linked to or packaged with email services like Mail Chimp or MyEmma to make questionnaire distribution easier. Free survey tools include:

Survey Monkey	www.surveymonkey.com	(MyEmma)
Survey Gizmo	www.surveygizmo.com	(Mail Chimp)

CHOOSING A METHOD

Experiments require that all the variables, the factors that might influence the outcome of the experiment, be controlled. This is a monumental task under the best of conditions. Sometimes variables we didn't even know existed are later discovered to influence the outcomes. Nobody in the music business wants to do experiments with their artists because to control and manipulate the variables (so that you can see the consequences) means some part of the target market isn't going to get the full marketing exposure. If we withhold advertising here (but not everywhere else) or we use in-store promotions in one region but not the others, we may hurt sales in those areas, and nobody wants to do that. It is difficult enough to get record company executives to do any kind of research without the threat of decreased sales. Labels and managers are sometimes willing to do other kinds of research, primarily focus groups and surveys. So when should each be used?

When focus groups are appropriate: Focus groups are considered exploratory research but can be conducted either before or in conjunction with a survey, or any time the company wants to probe an issue more in depth. Focus groups are often conducted in advance of a survey in order to gain a better understanding of the issues or of the target market for the survey, allowing the researcher to better understand the context, vernacular, and how survey questions might be worded better. Care must be taken to make sure that the participants in the focus group are representative of the same target market that will participate in the survey. You would not want to conduct a focus group of eight- to sixteen-year-old girls on the subject of their favorite artists and then give the follow-up survey only to adults 25 and older. In short, the focus group should be used to inform the survey research.

Other times a focus group will be conducted without any follow-up survey or after a survey in order to probe deeply into a few research questions or to get consumer feedback to aid in decision making. Management may have been surprised by the results of the survey research and want to find out why consumers answered the way they did. Although it is unlikely they can go back to the exact same people that responded to the survey, they will be able to go back to the same target market, the same sampling frame, and draw a representative focus group that should be able to shed light on the answers given in the survey.

Let us caution you once again that a single focus group should not be used to make decisions. The ideal procedure would be to conduct multiple focus groups until no new information was gained, but this isn't the ideal world and decisions are often made based on the feedback from a single focus group.

When surveys are appropriate: Surveys are best suited to situations where the research questions can be answered in a straightforward manner, when more information about aggregate consumer groups is needed, and when that information will need to be generalized to a larger population. Surveys are good for identifying characteristics of target markets, describing consumer purchasing behaviors, and measuring consumer attitudes. Surveys provide a relatively inexpensive, efficient, and accurate means of evaluating information about a market by using a small sample and extrapolating the result to the total population or market.

That said, writing a good survey question can be very difficult. All the ambiguity must be removed because you probably won't get a chance to explain what you are trying to ask. Because of this, entire textbooks have been written just on survey design. Whenever possible the researcher should use existing scales to measure abstract concepts such as attitude and personality, because these scales will have been validated by previous research. Complete, validated and reliable scales are published in journals and books like *The Handbook of Marketing Scales*. If a valid and reliable scale is used and an appropriate sampling procedure is followed, your survey results should convey, with a high degree of confidence, opinions and characteristics that the entire market or population exhibit.

Simple surveys are conducted on a routine basis in the form of warranty registrations, bounce back, or customer response cards. The record labels once placed survey cards inside CD packaging and incentivized buyers to respond by entering them into a contest for a t-shirt or something autographed by the artist, but as the industry has contracted and shifted toward digital sales this data collection method has disappeared. It should be a mortal sin if the label doesn't at least get the buyer's email address, but additional questions can be asked that can be analyzed on many levels, including by artist and geographic area. Because these cards were small only a very few questions could be asked, but an online version can go in greater depth. But be careful about overgeneralizing the responses. This is not a random sample and research indicates that there are real differences between the people who send back the cards and those who don't. Still, the responses are beneficial in gaining an understanding of the market.

Data collection: Hair et al. (2003) defined data as "facts relating to any issue or subject." Data is the answers we get by conducting experiments, focus groups, and surveys. Once the research design (experiment, descriptive, or qualitative) has been determined and the method (focus group or survey) has been chosen, the next step is to figure out the sampling frame and sample size. You would get the best information if you asked everyone in the target population your questions, but a census is not practical or realistic. Instead, we seek to take a representative sample of the population that will allow us to draw accurate conclusions. Keep in mind that we are not talking about the entire population of the U.S. or the world. We are likely only to be interested in potential buyers of our artists' music or tickets and that market is a subset of the bigger population. So our sampling frame would be all potential buyers of a music genre or some other target market. From that sampling frame we would draw a sample, a subset, of people to actually take our survey or be in our focus groups. The exact size of sample can be calculated depending on the desired precision and confidence, but that is beyond the scope of this text. Suffice it to say that a research company testing songs for a radio station can calculate the sample size needed to give the programmer statistically precise information with great confidence, provided they get the right people to participate in the research and that is, arguably, more important than the size of the sample. You don't want to ask a group of 60-somethings their opinion on DubStep—they aren't likely to be the right audience.

Surveys can be administered online if the questions are straightforward and the population you want to sample is small or widespread. Surveys can also be administered in person. If you have the permission of the promoter, surveyors may "intercept" concertgoers entering or leaving the venue. Not exactly a simple random sample, but if you are interested in capturing the opinions of the artist's fans, this is a quick way to reach a lot of them at one time. Surveys may also be included as part of one on one, depth interviews. Depth interviews, like focus groups, allow the researcher to dig more deeply into a topic, but they are done with individuals rather than small groups.

Data analysis: Once the data is collected it must be put into useful form. The first stage of data analysis is to edit and code the data. **Editing** is the process of checking the data forms for errors, omissions, and consistency. The data is then coded. **Coding** refers to the systematic process of interpreting, categorizing, recoding, and transferring the data to the data processing program. Much of the process of editing and coding has been simplified through the use of computers. Online surveys minimize the possibility of coding errors because the human element can be minimized.

When the answers to the questions can be represented by numbers (i.e., for question five the respondent chose answer two and for question six they chose answer seven) powerful statistical programs like SAS and SPSS can be used to look for relationships among the answers or commonalities within the respondents.

Not all data can be reduced to a series of numbers to be put into a computer program for analysis. Focus groups and depth interviews may result in long transcripts that will need to be carefully read and analyzed to identify key results and trends. More and more, even this kind of data can be analyzed using specialized software.

Analysis of the data is not the final step. The results of the analysis must be interpreted and given meaning and that information must be communicated to the decision makers in the form of conclusions and recommendations. This is normally done in a formal research report, although in the music business the report or presentation may be less formal than in other industries. The purpose of the report is to communicate the results, findings, and recommendations to the marketing client. The written report should begin with a one or two page executive summary that covers how the research was conducted and the highlights of the findings. The purpose of the executive summary is to give a quick but thorough overview of the research report for a busy executive who does not have time to read the full report. The full report should address each question asked in the research and how the findings are different than expected or otherwise interesting. In the conclusions of the report the researcher should explain what was found and their interpretation of the meaning of the responses. Finally, recommendations should be made based on the results of the research.

Online vs. in person: One of the advantages of doing surveys on computers is that it reduces error during data collection and coding. There are other advantages and disadvantages to computer-based and online research. Smartsurvey.co.uk gives the following ten advantages to online surveys over paper and pencil (P&P) surveys:

Faster: The time it takes to complete an online survey is one-third of the time it takes to complete a paper and pencil survey. Turn around time is shortened making the data and analysis more timely.

Cheaper: Online surveys can cut your costs in half. Fewer administrators to train, no printing costs, no travel or telephone costs (the researcher can take a global sample from the comfort of their office) and the responses are collected automatically and are immediately accessible.

More accurate: As we stated before, there is less error because "participants enter their responses directly into the system."

Quick to analyze: Because the responses go directly into a database and less time is needed to fix errors, analysis can begin sooner.

Easy to use for participants: Ninety percent of people who have access to the Internet prefer to answer surveys online instead of using the telephone. With an online survey, participants can pick a time that suits them best, and the time needed to complete the survey is much shorter. Questions that are not relevant to a particular participant can be skipped automatically (this assumes, of course, that the respondent has easy access to a computer).

Easy to use for researchers: Since the answers are already in the database, the researcher can quickly pull the data into a statistical analysis software for more detailed analysis. Charts and graphs can be easily generated for visual presentation.

Easy to style: If desired, the survey can be branded to match the company doing the research. Graphics and fonts can be easily changed and audio and video can be easily included in the survey.

More honest: Market researchers have found that participants in online surveys usually provide longer and more detailed answers. Because participants feel safe in the anonymous environment of the Internet, they are more likely to open up and give a more truthful response.

More selective: A more thorough screening can be done and only relevant questions can be asked of each respondent thanks to preprogramming of the survey software.

More flexible: One issue with long surveys is participant fatigue. By the end of the survey respondents just want to be done and may not think about their responses as much, or even just tick off answers just to be done. Online surveys allow the researcher to randomize the order of questions thus avoiding the possibility that the questions at the end of the survey are never seriously considered (www.smartsurvey.com).

DISADVANTAGES OF ONLINE SURVEYS:

Sample quality: Have you heard the term "the digital divide"? It refers to the difference between those who have access to the Internet and those that do not. This is something that must be considered in any online survey. If your target market is suburban, middle-class Caucasian teenagers, then you are probably safe doing an online survey. However, if your target market is inner-city youth, you may want to reconsider using online surveys because they may not have the same level of access to computers and the Internet.

Clarity and follow-up: Online surveys don't allow for direct interaction between the researcher and the respondent, so there is no second chance to explain what you are asking or to encourage respondents to stick it out to the end. This makes pretesting the survey for clarity and length even more important.

Perception and response rate: Email solicitation of respondents will probably be treated as junk mail and ignored unless there is a third-party endorsement. Getting a known entity, like a well-respected university, a popular blogger, or a celebrity, to endorse the survey will help increase the response rate—the percentage of people invited to participate who actually complete the survey.

Technical issues: While problems can arise in any type of survey, the more technology involved the bigger the problem may be. A respondent may lose power or Internet access during the survey and be unable to finish. The researcher will not know this and may think they just quit halfway through unless they are willing to spend the time and money to create a survey that will allow respondents to return to the survey at the same point where they were interrupted. If the proper precautions are not taken, the same person may take the survey repeatedly, thus skewing the responses. Finally, the survey itself must be tested and retested to make sure that every possible combination of responses is glitch free.

Sources: http://www.utexas.edu/learn/surveys/disadvantages.html and http://shlee. myweb.uga.edu/onlinesurvey/valueofonlinesurveys.pdf. Web. August 8, 2013.

ONLINE SURVEY SERVICES

Survey Monkey—surveymonkey.com
Zoomerang—www.zoomerang.com
Instant Survey—www.instantsurvey.com
E-customer Survey—http://ecustomersurvey.com
My Survey Lab—www.mysurveylab.com
Vovici—www.vovici.com/
Survey Gizmo—www.surveygizmo.com
Wufoo—http://wufoo.com/
Qualtrics—www.qualtrics.com

SYNDICATED RESEARCH

The buzzword in research circles and business today is "big data." **Big data** has been defined as "a collection of data from traditional and digital sources inside and outside [the] company that represents a source of ongoing discovery and analysis" (Arthur, 2013). New companies like Next Big Sound and BuzzAngle specialize in making the mountains of data, especially Internet data, manageable for the music business industry. They will be discussed in depth in the Technology and the Music Business chapter. Both these companies are collecting detailed information ranging from CD sales to artist mentions on Facebook and Twitter, to contribute to charts, and for analysis by their subscribers. What sets them apart from previous information providers in the entertainment industry is the breadth of data and the user's ability to drill deep into the details of the data.

Nielsen SoundScan is arguably the most important company providing research data and information to record labels. They are self-described as "an information system that tracks sales of music and music video products throughout the United States and Canada. Sales data from point-of-sale cash registers is collected weekly from over 14,000 retail, mass-merchant and nontraditional (online stores, venues, etc.) outlets." Weekly data from sales are compiled by SoundScan and made available every Wednesday. "Nielsen SoundScan is the sales source for the *Billboard* music charts" (SoundScan, 2005).

Since the introduction of SoundScan and BDS, the use of syndicated research has become a valuable tool for making marketing decisions in the record business. Chapter 16 illustrates how SoundScan data can be used as a basis for more in-depth research to detect sales trends and the impact of marketing strategies. Data from BDS can be merged with SoundScan to determine a more precise impact of radio airplay on record sales than was possible before. The use of SoundScan, and primary research such as the business reply cards, syndicated research from other sources, and occasional focus groups, are combined for predicting marketplace performance of new releases, tour analyses, target market definition, and to persuade radio stations to increase airplay.

Nielsen also owns Broadcast Data Systems (BDS). BDS provides airplay tracking for the entertainment industry using a digital pattern recognition technology. Nielsen BDS captures in excess of 100 million song detections annually on more than 1,600 radio stations, satellite radio, and cable music channels in over 140 markets in the U.S. (BDS, 2009). More information on BDS is available in the chapter on radio.

MediaBase is another company that tracks radio airplay (see chapter 8). A division of iHeartMedia, Inc., MediaBase 24/7 monitors and provides research to nearly 1,700 affiliate radio stations in the U.S. and Canada on a barter subscription basis. The data collected from radio stations is used not only by record labels and radio stations, but to compile airplay charts reported in U.S. *Today* and countdown shows such as *American Top 40 with Ryan Seacrest* and *CMT Country Countdown USA with Lon Helton*. In addition to their subscription-based radio service, MediaBase offers a consumer-oriented service called MediaBase Music. Music fans are invited to rate popular songs on their site www.ratethemusic.com (http://www.mediabasemusic.com/). Radio programmers use the information gathered to aid in deciding what songs to play on their stations and how often.

Arbitron, Inc. provides information on radio listening audiences, and much of that information is valuable to the record business. Arbitron not only determines how many listeners each radio station has, they also break down the audience demographically. This information is useful not only to record labels trying to work a new single to the radio audience but to advertisers trying to reach the same target audiences.

Others

Trade Association Research

All the major trade organizations provide research for their members. The Music Business Association (formerly the National Association of Recording Merchandisers or NARM) publishes its monthly *Research Briefs*, where they give the results of studies they have commissioned. Music Business Association provides research findings at its annual convention on a variety of current and ever-changing industry topics. The Recording Industry Association of America (RIAA) provides annual data on shipments, and contracts with an outside research firm to conduct its annual consumer profile. The International Federation of Phonographic Industries (IFPI) collects data from member countries and publishes an annual report called *The Recording Industry in Numbers*. The IFPI also releases periodic reports on digital music and global piracy. The associations tend to conduct issue-oriented research of benefit to all members of the record industry.

Billboard publishes MarketWatch in its weekly magazine. Market Watch provides a weekly synopsis of record sales both for the previous week and year-to-date sales, and compares this with sales figures from one year earlier. Generally, the reports are provided to dues paying members of the trade organizations or sold to anyone willing to pay the substantial fee for them.

CUSTOM RESEARCH FIRMS

Custom research firms collect and provide data and analyses to answer a specific question for a client. These research firms may specialize in a specific area, such as Internet consumers, and several cover the recording industry and technology. These same companies are also contracted by the various industry associations to conduct specialized research that is then made available to association members.

Forrester Research is one of the major market research firms focused on the Internet and technology; the company conducts research for the recording industry on all aspects of music and the Internet. Forrester also offers custom research and consulting services to its clients. Jupitermedia Corporation is a top provider of original information, images, research, and events for information technology, business, and creative professionals. The associations often hire Jupitermedia to conduct and report on online music consumers. Edison Media Research is a leader in political, radio, and music industry research with clients that include major labels and broadcast groups. Music Forecasting does custom research projects on artist imaging and positioning.

The NPD Group, recently acquired by Ipsos, provides marketing research services through a combination of point-of-sales data and information derived from a consumer panel. NPD's research covers music, movies, software, technologies, video games, and many other product groups. ComScore offers consulting and research services to clients in the entertainment and technology industries and conducts audience measurements on web site usage through its Media Metrix division. Taylor Nelson Sofres, a UK firm, provides both syndicated and custom research of media usage and consumer behavior. Based in France, IPSOS is a global group of researchers providing survey-based research on consumer behavior. BigChampagne (owned by Clear Channel) tracks online P2P usage and reports, among other things, the most popular songs on P2P networks.

TRACKING CONSUMER BEHAVIOR ON THE WEB

Before the emergence of the Internet, marketing researchers used a variety of techniques to learn more about consumer behavior. Many of these studies were not comprehensive; shopping behavior may have been measured on one group of consumers while advertising exposure was measured on another. It is difficult to conduct a comprehensive measurement program without being intrusive and the act of collecting the data may actually have an effect on the subjects' behavior. One company attempted to measure media consumption and consumer purchases in the same household. Participants had to subject themselves to extensive monitoring and extraordinary procedures to collect the data. In some ways, that made them unlike the general marketplace to which the results would be generalized. So the measuring had a tendency to get in the way of the natural consumer behavior. With the Internet, data collection is more transparent—Web users are not really aware that their movements through the Web are being recorded and analyzed. Of course, this is not a complete picture of their consumption behavior, either.

An important aspect of gathering information on your consumer base involves monitoring the traffic to your site. Web traffic refers to the number of visitors to your web site and the number of pages visited. Oftentimes, it is measured to analyze the importance of its individual pages and elements. By including a bit of programming code on each page of the website, the webmaster can learn a lot about the visitors to the site. This helps the webmaster and other marketing professionals understand what product information and which products are considered valuable to its visitors and which are not.

Marketers can also determine whether or not the email they sent you was opened by embedding code in it that sends a request to their server (thus signaling that the email was in fact opened). This method can be used to test different email campaigns or messages for effectiveness.

Quantcast

Quantcast, a targeted online advertising company, provides information on Web traffic combined with demographic data to present a clear depiction of a website's visitors. The service combines sample-based information and analytics of web behavior to present a profile of a website's traffic, including information on: age, gender, ethnicity, income, education, repeat vs. casual traffic, monthly traffic volume, and a comparison of other sites that your visitors frequent, like, and search for. Analytics and "big data" are discussed further in chapter 16.

In the lifestyle analysis, Quantcast analyzes what other sites your fans visit when they are not on your site. This helps develop a profile of what else your visitors are interested in when they are not on your site. For example, visitors to Kelly Clarkson's website also visit bravadousa.com and AmericanIdol.com.

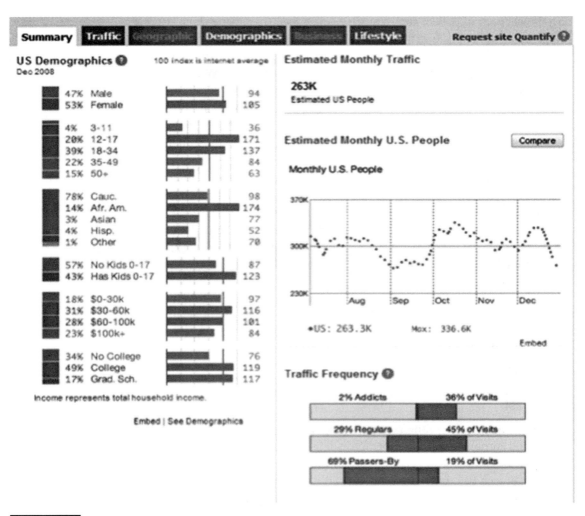

FIGURE 4.1 *Quantcast Information (courtesy of Quantcast)*

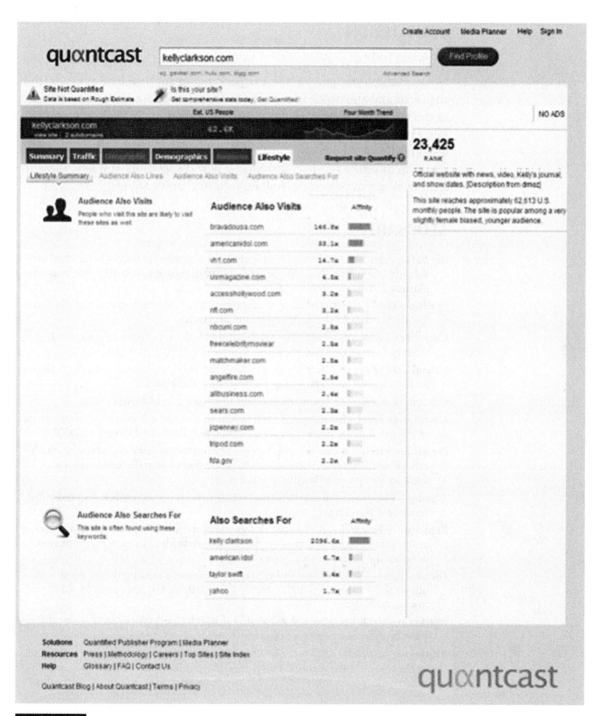

CONCLUSION

Quality research is deceptively difficult and requires careful planning. Different research designs are used depending on the research question to be answered and the depth of information desired. Most of the data collected in the music business is for syndicated reports such as Nielsen reports and *Billboard* charts. The Internet has allowed researchers to more easily reach respondents, reduce costs and time needed to complete the research process. The Internet has also introduced new competitors and new measures into the research process, as will be seen in chapter 16.

GLOSSARY

Big data—A collection of data from traditional and digital sources inside and outside [the] company that represents a source of ongoing discovery and analysis (Arthur 2013).

Coding—The systematic process of interpreting, categorizing, recoding, and transferring the data to the data processing program.

Construct—A theory or concept representing ideas that cannot be measured directly.

Causal design—A research design in which the major emphasis is on determining a cause-and-effect relationship (AMA online dictionary).

Custom research—Research undertaken to answer a specific question.

Descriptive research—Research undertaken to describe the existing characteristics of a defined target market or population (Hair, Bush, and Ortinau. 2003).

Depth interview—A one-on-one, face-to-face interview done out of the researcher's office that allows the research to probe deeper into the respondent's answers by asking follow-up questions.

Editing—The process of checking the data forms for errors, omissions, and consistency before coding.

Exploratory research—A research design focused on better understanding a situation, screening alternatives, or discovering new ideas thereby clarifying and defining the future research objectives.

Focus group—A face-to-face, interactive interview of a small group of people that allows the researcher to delve deeply into participants' responses by asking follow-up questions.

Market research—A subset of marketing research that looks specifically at the size, location, and makeup of a product market.

Marketing research—The function that links the consumer, customer, and public to the marketer through information—information used to identify and define marketing opportunities and problems; generate, refine, and evaluate marketing actions; monitor marketing performance; and improve understanding of marketing as a process. Marketing research specifies the information required to address these issues, designs the method for collecting information, manages and implements the data collection process, analyzes the results, and communicates the findings and their implications (American Marketing Association, 2013).

Problem definition—Stating what problem is to be solved and what questions need to be asked to obtain the information needed to determine the solution to the problem.

Qualitative research—Research used to gain an understanding of how underlying attitudes, opinions, and motivations impact consumer behaviors.

Reliability—The researcher's ability to get similar results from repeated applications of the measures or from independent but comparable measures of the same trait or construct.

Research design—A plan that guides the collection and analysis of research data.

Research panels—A group of people put together by a researcher from which data is collected on an ongoing basis.

Secondary research—Using existing research data in an attempt to answer your own research question.

Survey research—A research design that collects data using identical questionnaires administered either in person or electronically.

Syndicated research—The information collected on a regular basis that is then sold to interested clients (American Marketing Association Online Dictionary).

Validity—The extent to which differences in results in the measurements reflect true differences among the objects or characteristics being measured rather than constant or random errors.

REFERENCES

American Marketing Association. http://www.marketingpower.com/aboutama/pages/definitionofmarketing.aspx. Web. June 25, 2013.

Arthur, L. "What Is Big Data?" *Forbes,* August 15, 2013. Web. November 10 2014.

BDS. www.bdsonline.com/about.html. 2009.

Hair, J., Bush, R., and Ortinau, D. *Marketing Research Within a Changing Information Environment* (2nd edition). Boston: McGraw-Hill. 2003.

Smartsurvey.com. http://www.smartsurvey.co.uk/articles/10-advantages-of-online-surveys/#.UgPpeVPpaww. Web. August 9, 2013.

SoundScan. www.soundscan.com/about.html. 2005.

Stewart, S. M. "Artist-Fan Engagement Model: Implications for Music Consumption and the Music Industry," (Doctoral dissertation). 2013. Retrieved from ProQuest Dissertations & Theses database. (UMI No. 3612137).

Zikmund, W. G., and Babin, B. J. *Essentials of Marketing Research*. 3rd ed. Mason, OH: South-Western. 2007.

CHAPTER 5

The Value of Branding in the Music Business

Contributed by Tammy Donham

Tammy Donham received her undergraduate degree in Marketing from Western Kentucky University and her Master's degree in Business Administration from Middle Tennessee State University. Donham worked for Fruit of the Loom, Inc. in several marketing capacities before moving to Nashville in 1996 where she worked for nearly seventeen years for the Country Music Association (CMA).

She held various marketing-related positions within CMA ultimately rising to Vice-President of Marketing. While in her post as VP, she oversaw all marketing, creative services, and research efforts for the CMA Awards, CMA Music Festival and CMA Country Christmas events and television specials, including broadcast, digital, radio, out-of-home and print initiatives. She was CMA's lead liaison with ABC Television Marketing, Synergy and Affiliate teams and worked closely with these and other event partners to maximize promotional and brand-building opportunities for CMA properties across all platforms. Donham is a graduate of Leadership Music, as well as a member of the Academy of Television Arts & Sciences and the Country Music Association.

Donham began teaching Marketing of Recordings and Digital Strategies for the Music Business in the Recording Industry Department at Middle Tennessee State University in the fall of 2013.

The American Marketing Association defines a **brand** as a "Name, term, design, symbol, or any other feature that identifies one seller's good or service as distinct from those of other sellers" (AMA, 2014). Branding an artist involves clearly communicating who that artist is and what they

CONTENTS

CONTENTS

represent. An artist's brand is how he or she is perceived by consumers, corporations, and the media. Branding strengthens and coalesces the perceptions of these groups when they hear or see an artist's name.

By building a strong brand, an artist is able to leverage their name to generate additional sources of revenue from licensing deals, to endorsements, to business ventures where the artist owns the company to help offset the declining revenue. The music company also benefits from that strong brand since it opens the door to branding and positioning opportunities for the artist and their recorded music outside the traditional music retail space. More exposure in the media and **non-traditional outlets** executed through strategic partnerships creates a broader awareness of the artist which leads to a bigger audience and, in turn, the opportunity for increased sales and overall brand equity for the artist, all resulting in greater profitability for the music company.

Platinum-selling artist Jay-Z is a music celebrity on the top of the *Forbes* top celebrities list because of the income generated not just by his music sales, but from his corporate partnerships and business ventures like Roc Nation, a full service entertainment and sports management company, a chain of night clubs, and a branded cognac (*Forbes*, 2014a). In addition, a partnership with Samsung for his *Magna Carta Holy Grail* album resulted in the company purchasing 1 million copies to give to owners of its smartphones before the album's official release (Houghton, 2013).

FIGURE 5.1

Forbes Celebrity List Rank*	Artist	Earnings	Major Businesses, Partnerships, and Endorsements
			Top Earning Artist Brands in Music
1	Beyoncé Knowles	$115 M	Fragrance line (Heat, Pulse, and Rise), House of Deréon clothing line, H&M endorsement, and Pepsi sponsorship
3	Dr. Dre	$620 M	Beats by Dr. Dre headphones and Beats Music (recently acquired by Apple)
6	Jay-Z	$60 M	Entertainment and sports company Roc Nation; 40/40 Clubs; D'Usse Cognac, and Armand de Brignac Champagne
8	Rihanna	$48 M	Women's and men's fragrance lines, Vita Coco coconut water endorsement
9	Katy Perry	$40 M	CoverGirl and PopChips endorsements, and her own fragrance Killer Queen

Source: Pomerantz, 2014
*Rankings are based on fame as well as income.

Taylor Swift changed the way the music industry thought about album promotion and the value of partnerships with an unprecedented effort to promote the launch of her album, *Red*. According to Nielsen SoundScan, the album sold 1.2 million copies in its first week of release. These were also the highest first week sales figures in the U.S. in 10 years. Arguably, this was due in large part to the leveraging of her brand partnerships across multiple product categories. The extensive promotional effort included a branded display in the front of Walgreen's stores, an album-and-pizza promotion with Papa John's that included album art on box tops, as well as a special offer to purchase the CD with every pizza order, her own shoe line for Keds promoted across multiple media outlets, a deluxe exclusive and **branded display** with Target, a promotion with American Greetings, and a four-sided corrugated cardboard floor display in Wal-Mart's **Action Alley**. Swift also renewed her presence in fragrance departments with the launch of a new fragrance for Elizabeth Arden to follow the success of the artist's Wonderstruck perfume (Hampp, 2012). The buzz and excitement generated around the album launch was largely credited to the comprehensive marketing campaign. The exposure value of the partner promotions was worth millions, it helped secure the album's successful launch, and it increased Swift's brand equity.

For her album *1989* released in 2014, Swift once again leveraged her promotional partnerships and massive social audience to sell nearly 1.3 million albums in the first week alone (Mansfield, 2014). This should be noted as an unusual feat in the challenged music sales environment. Swift's promotional efforts around the album included a stream on Yahoo!, a number of "secret sessions" giving her most passionate fans a private listening party of the album before its release, teasers across all social media platforms, and support from corporate partners American Greetings and Keds. In addition, Diet Coke partnered with Subway for a month-long promotion leading up to the album release supported by in-restaurant promotion, as well as a media campaign on TV, radio, social, and online channels (Lukovitz, 2014).

Jay-Z, Taylor Swift, and many other artists have discovered the power of leveraging their names or brands into successful business ventures where the revenue potential can augment and sometimes dwarf their earnings from recorded music.

Branding is not just important to superstars. A strong brand can be leveraged by artists at every level—from local artists who are attempting to grow their audience or secure a partnership with a local retailer to mid-sized acts that are on the cusp of breaking out. Importantly, partnerships between artists and corporate brands can be critical to artists who are not seeing mainstream radio or sales success by generating awareness of and interest in the artist and their music. These partnerships can also help newer artists by providing **in-kind support**.

Many industry professionals believe branding will be a key focus for the future of music, so it is important to understand both the history of branding and how branding impacts marketing strategies at companies across all business spectrums.

HISTORY OF BRANDING

Branding is often thought of having originated with cattlemen who would burn a distinctive symbol into an animal's skin with a hot branding iron as a way of differentiating one's cattle from another's. Historically, however, all manner of goods and crafts were often imprinted by branding with the trademarks of the tradespeople who produced or crafted the product. Among other uses, these marks were a way to assure customers of the quality and the origin of a product.

While trademarks were applied to goods in other ways besides branding irons—and the practice of actually branding a trademark onto barrels, bags, sacks, and goods fell out of favor—the term "branding" remained.

The modern concept of branding products originated in the early 19th century when the advent of industrialization created the need for manufacturers who were mass-producing goods in factories to promote their goods visually to a larger number of consumers who may have been unfamiliar with their products. A few of the first manufacturers to embrace this "branding" concept were Campbell's Soup®, Coca-Cola® and Ivory™ Soap (AEF, 2014).

BRANDING BASICS

Brands like Apple, Disney, MTV, and Sony are just a few examples of strong identities whose brand marks, according to *Forbes*, are valued at millions of dollars and in the case of Apple, more than $100 million (*Forbes*, 2014b).

Individuals also have their own brands to manage, which is often referred to as **personal branding**. An individual artist's or group's brand involves how they are perceived by the public, fans, corporate America, and members of the press. The artist's brand has to do with all aspects of their public image from how the artist's website looks, to the album art, to style and brand of clothing that is worn. Logos, fonts, the look and feel of marketing materials and press kit, live performances, concert merchandise, hair style, videos, wardrobe, what the artist says and does in public, and the brands with which the artist is affiliated all play a role in the development of the artist's brand (McQuicklin, 2011).

What are the images and words that come to mind when you think of your favorite music artist? What about Robin Thicke? Lady Gaga? The Rolling Stones? Katy Perry? Justin Bieber? Each of these artists has a strong brand that is vastly different. A strong brand increases a product, company, and an artist's marketability and value. In fact, a brand can be one of the most valuable components on a corporation's balance sheet even though it is an intangible asset. For example, Flowers Foods paid bankrupt Hostess Brands $350 million for its bread brands (including Wonderbread) and attributed 55%, or $193 million, of the purchase price to identifiable intangible assets, mainly trademarks (Lavin, 2014).

BRAND ARCHITECTURE

Brand architecture is the structure and interrelationship of brands across an entire organization. It outlines the roles and relationships between all of the brands in the portfolio showing how the corporate brand and **sub-brands** relate to each other, support each other, and reinforce the main mission of the corporate brand. The way brands in a company are connected helps a company introduce new products by leveraging the brand equity and consumer loyalty of that product line and helps prevent the company from introducing new products that do not support the main brand's core mission. This also helps to provide clarity for the consumer enabling them to easily differentiate one product from another in the line of offerings.

In their book *Brand Leadership* authors David Aaker and Erich Joachmisthaler outline the main brand relationship strategies (2009). The two most popular of these are a **branded house** or a **house of brands**.

A Branded House—In this approach the master brand is the dominant brand with sub-brands being descriptive in nature. Virgin is often used as one of the best examples of Branded House. Under the Virgin umbrella is Virgin Airlines, Virgin Music, Virgin Radio, Virgin Hotels, and Virgin Mobile Phones among others. All offerings under this approach include the master brand so the branded house offers clarity and leverage with the consumer. All products carry the overarching brand name and the brand values and attributes that come with it.

This approach enables companies to maximize resources by putting the majority of marketing effort into supporting and building the master brand. Although the synergy of this approach offers improved efficiencies that same close connection between brands also increases risk; if the master brand is damaged the negative connotations are extended throughout the entire portfolio.

House of Brands—Under a house of brands approach, the primary brand is used minimally on package and in communication with consumers. Instead, the sub-brands are the focus and are treated as distinct, independent entities with the parent company receiving little or no mention. The brands are independent and unconnected. Proctor & Gamble is probably the most famous example of a house of brands structure. Under the P&G "House" there are dozens of brand offerings including Charmin, Cascade, Pampers, Prilosec, Duracell, Gillette, and Tide, just to name a few. These brands are recognized by their brand names alone and not by "P&G's Pampers" or "P&G's Gillette razors." This approach is often used to allow a company to position multiple products differently even in the same category.

There are also hybrid approaches to brand architecture where brands are combined in some way to be complementary. Aaker and Joachmisthaler note that using an **endorsed brands** strategy offers credibility to a brand while a sub-brand strategy allows a parent brand to compete in new categories that would have previously been inconsistent with the brand's positioning.

Endorsed Brands

Endorsed brands take independent brand names and attempt to strengthen them by adding an organizational brand (e.g., Fairfield Inn by Marriott). An endorsement may be used to alter the image of the endorsed brand but its main role is to add credibility to the endorsed sub-brand in the eyes of the consumers. The endorsement signals to the consumer that the company will deliver on its brand promise particularly when the endorser company has established credibility in the same product category. With an endorsement both parties (the endorser and the endorsee) will each have their own name, logo, personality, and brand image. They will also each have their own marketing plans and messaging to consumers. Although each brand is distinct, the endorsing brand must have the higher-level brand promise to signal to the consumer that the endorsed brand is worthy of their consideration, but the endorser brand generally plays a minor driver role.

Sub-Brands

Sub-brands are brands connected to a master brand that add associations to make a brand seem more appealing to consumers or can be used to stretch a master brand, allowing it to compete in a different category. The master brand is the primary reference point but the sub-brands allow for additional associations. A sub-brand can also signal that a new product is newsworthy as it did when Crest introduced its Pro-Health™ line of products.

Automotive companies often use a sub-brand approach. Ford Escort, Ford Mustang, Ford Fusion, and Ford Explorer are just a few offerings under the Ford umbrella. By using the Ford name, the underlying brand promise is conveyed but each sub-brand is allowed to develop its own identify.

Determining the right company structure and brand positions in the family of brands is an important decision that is based on many factors. Often, one of the main points to consider is how much the brand in question needs to be in the driver (or lead) role (Aaker and Joachimsthaler, 2009).

BRANDING IN THE MUSIC INDUSTRY

In the music industry different structures and blended strategies are often utilized. Most label groups have different sub-labels or imprints that each target a different segment of the market. For example, the Universal Music Group (UMG) brand family includes diverse offerings like Blue Note Records that focuses on Jazz music and was formed to target the adult music consumer while the company's Def Jam Recordings focuses primarily on a younger audience with Hip Hop and Urban music.

Building strong brands has become a central focus of marketers across the music landscape. Radio stations, radio personalities, concert venues, musical instrument manufacturers, and artists each have their own unique personality and attributes for which they are or want to be known. Through the use of imagery, words, slogans, logos, and style, these entities can create a powerful brand that consumers recognize and trust.

Why is branding such a cornerstone of marketing artists? Because a strong brand increases an artist's value to the media, to fans, and to brand partners.

SUCCESSFULLY CREATING AN ARTIST'S BRAND

Strong artist brands do not happen overnight. It takes years of cultivation on behalf of the artist and the artist's team to develop a strong brand. To be successful at branding one must have a firm understanding of how to create the artist's brand, how to promote and build the brand, and how to maintain the brand.

DEFINING THE ARTIST AND DEVELOPING A BRAND IDENTITY

The first step to developing an artist brand is to truly understand who the artist is. What values do they hold dear? Are they focused on family or a free spirit who loves adventure? What is their personality and style? Is the artist polished and fashion-forward, or rugged and disaffected? Are the artist's live shows full of energy and high production or more laid back and intimate? Does the artist drink alcohol or publicly denounce any kind of vices?

What makes the artist stand out from others in the genre? What is their story? What makes the artist interesting to fans? What makes them interesting to the media? These and other such questions provide insight into the characteristics and personality of the artist. Unique characteristics can often be embraced as those differentiators help a consumer more easily recognize that brand when compared to others. Defining the artist and being clear about who they are authentically is the first step to building an image and developing a strong brand (Dahud, 2012).

FIGURE 5.2

DEFINING THE ARTIST'S AUDIENCE

In addition to needing to know who the artist is, it is important to know who that artist serves. In other words, who is buying the music? Who is attending the concerts? Who is engaging with the artist on social networks?

Research plays a critical role in this step. One of the best methods of developing a deep understanding of the artist's primary audience is through the use of surveys. Online surveys may be sent to the artist's email list, mobile list, ticket purchaser database, or social network asking basic demographic, geographic, lifestyle, and behavior questions and intercept surveys may be conducted by a team of researchers as concert goers enter the building (See chapter 4 for more information.)

Through this research activity artist marketing teams can gain understanding of what their fans want from them and what characteristics and traits they love about the artist and the artist's music, which will help the artist better meet their needs. They can also find out more about the audience that will help them better target both existing and potential fans. With this research a **target market** and **customer profile** can be developed. A customer profile is a persona or description of the ideal customer that includes demographic, geographic, and psychographic characteristics, as well as consumption patterns. This profile can be used to identify the best places to place paid advertising promoting new music and identify which categories and brands they like most. This information, in conjunction with the artist's profile, helps the artist team identify which companies and brands would be a good **brand fit**; in other words, a company that shares the same personality and target audience.

PROMOTING AND BUILDING THE BRAND

Building a strong brand, goes beyond creating a logo or visual mark and placing paid advertising. It is about consistently shaping and promoting all of the associations of the brand both mental and physical.

Strong Brands are Authentic and Consistent

To build a strong brand, artists must be clear and consistent about who they are and what they represent. The artist's brand identity should reflect the values, attitude, and personality of the artist in all communication with the public and industry. This step is critically important because of the strong bond an artist has with their fan base. That relationship is built on trust and an understanding of knowing whom the artist is.

The importance of consistency extends to all creative elements surrounding the artist. Logos, copywriting, tone of voice, and imagery all should remain consistent across all platforms and media (McQuilkin, 2011).

Strong Brands Are Active

Artists should actively communicate who they are. They should engage and interact with their audience on a regular basis and embrace social networks to communicate on a one-to-one basis particularly while building their fan base. They should be mindful of the "voice" used in all communications. It should also be consistent with the artist's brand regardless of whether or not it is the artist himself speaking or a representative.

Labels and artists should actively look for opportunities to showcase the artist's brand from including the logo and marks across all available media, to walk on roles in popular television programs, to getting the artist involved in a local charitable effort, all of which help build awareness and brand equity.

Strong Brands Develop a Deep, Emotional Connection with the Consumers

The best artists know how to meaningfully connect with their fans by sharing a story, a special moment, or just communicating an everyday frustration. These interactions are opportunities to promote the artist's authenticity and create a lasting bond with loyal supporters. The fan base of the popular band One Direction has often been described by the media and insiders alike as one of the most passionate groups in the music business. The group often sends updates on projects to its social audience, recognizes its strongest supporters publicly, and engages with fans in creative ways like asking fans to submit video of themselves for a chance of being featured in a movie about them.

MAINTAINING THE BRAND

Strong Brands Monitor What Is Being Said About Them

TweetDeck, Hootsuite, Social Mention, Google Alerts, and Google Trends are just a few of the services and tools that make it easy to monitor social conversations about a brand. Set up a command post to monitor the artist's name, album names, song titles, related hashtags and other strategic keywords to engage with consumers who are talking about the artist and to protect the brand by quickly diffusing any negative comments that may damage the brand.

Protecting the Brand

Pop Music Superstar Britney Spears certainly had her share of instances with the potential to damage her brand. From a head-shaving incident, to a 55-hour marriage to a high school friend, to a stint in rehabilitation, to other brand-damaging incidents, the artist and her team have had to work hard to protect her public image. Even with all of the negative publicity, Spears slowly rebuilt her brand over several years leading to a popular and lucrative run in Las Vegas. Then in July 2014, a YouTube clip that reportedly featured a Britney Spears song that had not yet been auto-tuned went viral. With criticism mounting and her brand again at risk, the song's producer quickly issued a statement in Spears' defense noting that the leaked track was actually not an official take but was a vocal warm-up session. The record company also took swift action by having the leaked video removed from YouTube to prevent further sharing.

While social media has made connecting with fans easier than ever before, today's connected world means artists and the artist's team must be vigilant to ensure that all messaging is "on brand" and that potentially damaging incidences are controlled quickly.

Linking Brands to Build Brand Equity

There are many opportunities to leverage the strength of two or more brands through strategic partnerships. These relationships take different forms with some of the most popular being co-branding, sponsorships, and artist endorsements.

At times it makes sense to partner with another entity to leverage the full strength of both brands. The Nike+iPod Sport Kit is an example of **co-branding**, which is when two or more brands from different organizations create a joint product for which each company plays a major role. While having multiple partners involved increases the exposure opportunity because more resources are being dedicated to the product's marketing effort the design and implementation process can become very complex because each brand wants to ensure that the product is representative of their brand image and brand promise (Aaker and Jachimsthaler, 2009).

Sponsorships occur when an organization pays a sum of money toward the cost of an activity or event (such as a music festival, concert tour, or album launch) in return for the right to advertise and participate in the event. Sponsorships allow companies to connect with consumers by becoming part of an event. The brand hopes by being associated with an event, an audience likes and enjoys that the positive feeling about the event may get transferred to the brand. This is referred to as the association effect (Dimofte and Yalch, 2011).

Another way of linking brands is the **celebrity endorsement**. A celebrity endorsement is the act of giving one's public support or approval to someone or something. With both sponsorships and endorsements in the music industry a company is essentially renting an artist's audience in exchange for money and/or promotional support (Erogan, 1999).

THE ROLE OF CORPORATE PARTNERSHIPS IN THE MUSIC INDUSTRY

Music and brand partnerships continue to gain in popularity because they offer benefits for the brand, the label, and the management team and artist.

From the Brand's Perspective

Corporate brands value partnering with music acts and events to build brand awareness among the artist's fan base, engage with consumers, and build brand loyalty. Corporate brands understand the deep connection artists have with their fans, and they hope that by being involved in the artist's activities and events they can enhance the brand's image or improve the attitude toward the brand.

In recent years, the increased use of social networks to develop a strong direct relationship with a brand's consumer base has made music partnerships even more attractive. Brands today are embracing **content marketing** strategies at a rapid rate seeking to build goodwill and drive a deep emotional connection to their customers by creating great content (primarily video) around topics in which their customers are most interested rather than pushing out sales messages. Since music partnerships offer both the connection aspect and video it is a natural choice.

The energy drink company Red Bull for instance has built a strong brand in recent years around music. The company's efforts include the underwriting of a music festival (the Red Bull Music Academy), a record label that works with up-and-coming artists, and a music publishing arm. For 20 years skateboard shoe manufacturer Van's has also created a bond with music fans through sponsorship of the Warped Tour (Diaz and Pathak, 2013).

The most successful music partnerships are those where there is a strong brand fit where artist and brand values and personalities align. The partnerships where the artist speaks to the brand's customer base and where the brand appeals to the artist's fan base provide a win-win scenario that each party can fully leverage for their benefit.

From the Record Label's Perspective

In a world where consumers have an almost insatiable appetite for music and want to pay as little as possible for it artists and labels alike recognize the importance of building and leveraging a strong brand. Record labels have been seeing decreasing CD and single download sales as consumer behavior continues to migrate rapidly toward streaming. And the small payments offered on streams have not yet generated meaningful revenue for most artists or their labels.

For record labels, the importance of developing and building the artist's brand has become increasingly critical. Because of decreased revenue, record labels are struggling to find a financial model that works. Marketing budgets for most albums are conservative and often consist of the advertising and promotional elements that have a strong correlation to sales, which typically includes radio, spot television, and online ads. In addition, traditional retail space in mass merchandising stores like Wal-Mart and Target continue to shrink and most record stores have either merged or gone out of business entirely. Labels today are actively seeking alternative sources of revenue and non-traditional promotional opportunities in order to generate awareness and ultimately sales.

Increasingly, when an artist is signed to a label brand partnerships are top of mind. Early on the partnerships team at the label will ask the artist a series of questions to gauge his or her affinity for certain products, interests, and values in order to begin to formulate a list of potential corporate partners for the album launch. The team wants to find a good brand fit for the product because that can help ensure an artist's commitment to the partnership (or buy-in), which is a critical component to its success.

Labels look at corporate partnerships favorably because of the potential to generate additional revenue from a sponsorship fee. Notably, most agreements now offer the opportunity for the label to share in most of the artist's revenue streams, not just sales of the recorded product. Equally, and perhaps more importantly, the music industry sees value in the wide distribution and significant promotion that corporations can bring to an artist and recorded music. In today's market labels and artists often develop partnerships with corporations where the media and marketing support the corporate brand brings to the table are the only "currency" involved in the deal.

What is important is that in this scenario the artists and label may not be paid money at all. The media and promotional assets provided by the partner brands bring value to the artist which they might not otherwise have access to given the label's limited marketing budget. Many corporate brands spend millions of dollars each year on in-store displays, television ads, radio campaigns, direct mail, outdoor advertising, and digital efforts dwarfing the amount generally spent annually by the label to promote a recording. By incorporating promotion of the album into the corporate brand's messaging, the corporations offer the opportunity to expose the artist and the music to a wider audience, which translates into greater awareness; increased revenue from ticket sales, merchandise sales, record sales and streams; and the building of the artist's brand.

For instance, consider that a billboard in Times Square costs tens, and sometimes hundreds, of thousands of dollars per month; a thirty second spot on NBC's The Voice costs more than $200,000 and a full page, four color ad in People magazine has an **open rate** of more than $330,000 and you can start to see the tremendous value that brands can bring to the partnership table. These high-priced promotional vehicles are more easily attained by an artist with a limited budget when acquired through corporate partnerships.

It is common for each partner (artist/label camp and brand) to approach a potential partnership by throwing each of their assets on the virtual table and seeing which ones are interesting and valuable to the other party. Any cash payment involved as well as which assets are offered up depend on what the artist is asked to do (private concerts, commercial shoots, meet and greets, etc.). Most often, trade-only or in-kind deals are more popular with up and coming or mid-tier artists; some sort of cash and trade combination is standard for a well-known, mainstream artist.

From the Artist and Manager's Perspective

As important as partnerships are for the labels in promoting the recorded music sales to the masses, managers and artists are also seeing the value brands provide in generating awareness of the artist, expanding the fan base, driving sales, and building brand equity. As noted, partnerships can be used to promote the artist in non-traditional **product categories** and places such as in the cosmetics counter of a department store, or on the shelves of a grocery or sporting goods retail store. This not only helps stimulate sales but also provides valuable impressions that help build brand equity. Partnerships may also provide an alternative source of revenue to help offset low royalty payments.

FINDING THE RIGHT BRAND PARTNER

Artists build strong brands in part by remaining consistent about who they are and partnering with brands that are a good brand fit. A bad fit can damage an artist's brand just as much as a good fit can help build it.

As noted previously, a firm understanding of an artist's brand identity, as well as the artist's audience profile are the first steps to being able to hone in on the right category and brand for a potential partnership. Primary research of the fan base can provide valuable insight. In addition to primary research of the audience labels and management groups use other tools and services to help them develop the artist's profile and identify the best brand fit for possible partnerships.

An alternative data collection source is using a third party provider of consumer behavior. There are several companies in the United States that specialize in surveying consumers and selling the results to interested parties. These parties include corporate brands, media outlets, agencies, and yes, record labels. Three of the most widely utilized databases in the advertising and marketing industry are GfK Mediamark Research and Intelligence (MRI), Nielsen Scarborough, and The NPD Group.

GfK MRI gathers information on consumer behavior through annual in-person, at home interviews of 26,000 consumers about their "media choices, demographics, lifestyle and attitudes, and usage of more than 6,000 products in 550 categories" (GfK MRI, 2014). The company licenses this database to companies and organizations that can pull data specific to their need (most major advertising and promotional agencies subscribe to MRI). For instance, a record label looking to secure a brand partnership for an EDM artist might run a crosstab report of all adults ages 18–45 who purchase or listen to EDM music with the brands and products they purchase, what television shows they watch, and what activities in which they like to participate during their spare time. The reports can be used to identify the most likely prospects for a partnership or to strengthen a partnership proposal.

Nielsen Scarborough is another market research service that offers consumer demographics and insight on media consumption, attitudes, and purchase behavior. Nielsen Scarborough captures consumer data on more than 2,000 product categories and brands collected from a sample size of more than 210,000 adults in 77 Designated Market Areas (DMAs) across the United States (Scarborough, 2014). The main difference between MRI and Scarborough is that Scarborough specializes in local market insight. For example, while MRI can tell you how many people who go to hockey games stream rock music heavily, Scarborough can tell you how many people who go to Seattle Seahawk games do so. This localized information is particularly helpful to radio stations that often use the data to sell ads to local businesses.

The NPD Group (formerly National Purchase Diary) is a consumer market research company that uses sales data from retailers and distributors, as well as consumer-reported purchase behavior through its online survey panel of more than two million consumers to provide its clients with "a comprehensive view of consumer behavior and attitudes across all distribution channels and demographics." Of particular interest to the entertainment industry is that the company also tracks the awareness and product usage of more than 100 celebrities allowing artist managers and labels to provide detailed product usage and behavior information for fans of these specific artists (NPD, 2014).

THE IMPORTANCE OF BRAND FIT

Imagine a fan who is a passionate supporter of a popular artist. The fan was there from the beginning of the artist's career—attending multiple concerts, buying all of the artist's albums and following the artist on major social networks. The fan has a good idea of the artist's brand: young, rebellious, taking pleasure in being the bad boy. He drinks, he smokes, and he lives life in the fast lane. His look is rugged; he has tattoos and piercings. His look is "bad" in a good way. The artist knows who he is and embraces his rough and tough image.

Now imagine that this artist teams up with Yoplait yogurt to sponsor his upcoming tour. He is in Yoplait's commercial promoting the health benefits of yogurt and free yogurt samples are given to fans at the artist's shows. His fans' reaction is one of confusion because the pairing seems odd. It is not a strong brand fit. The fans start to lose faith in the artist because they do not know who he is. They begin to identify with him less and less.

CREATING AN ARTIST BRAND PROFILE

An **artist brand profile** (also referred to as an artist brand presentation or artist brand deck) is a visual presentation of an artist that is often provided to corporate brands and agencies to offer some background on the artist's career and personality. The information conveyed should help these stakeholders understand the attitude, style of music, preferred activities, and other characteristics of the artist even if the brand representatives have never seen or heard of the artist. The artist brand profile helps brands and agencies determine if a partnership is a good brand fit.

The presentation is usually brief (8–10 slides) and includes the following elements:

- Links to the artist's recent and most popular videos.
- Highlights of the artist's career and successes they have obtained—radio, sales, albums released, etc. (see Figure 5.3).
- Social stats and fan reach and engagement.
- Press coverage or other industry quotes that speak to the success or appeal of the artist.
- Upcoming events including appearances in movies or television programs, album releases, collaborations with other artists, tours, etc. This is important information as it gives the brand an opportunity to compare their promotional calendar to see at which events they may be able to leverage their involvement.

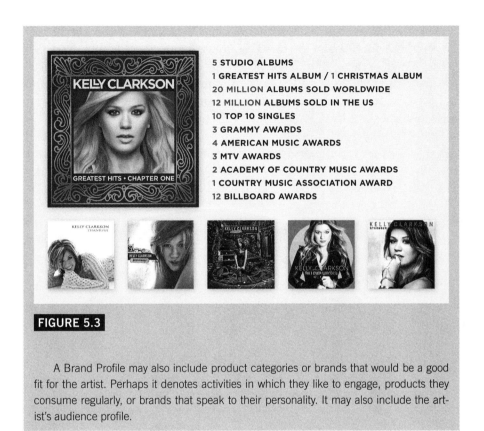

5 STUDIO ALBUMS
1 GREATEST HITS ALBUM / 1 CHRISTMAS ALBUM
20 MILLION ALBUMS SOLD WORLDWIDE
12 MILLION ALBUMS SOLD IN THE US
10 TOP 10 SINGLES
3 GRAMMY AWARDS
4 AMERICAN MUSIC AWARDS
3 MTV AWARDS
2 ACADEMY OF COUNTRY MUSIC AWARDS
1 COUNTRY MUSIC ASSOCIATION AWARD
12 BILLBOARD AWARDS

FIGURE 5.3

A Brand Profile may also include product categories or brands that would be a good fit for the artist. Perhaps it denotes activities in which they like to engage, products they consume regularly, or brands that speak to their personality. It may also include the artist's audience profile.

CAUSE MARKETING WITH A BRAND

While most artists have individually taken on a cause because it is something they believe in or by which they have been personally affected, many have found success in partnering with a corporate brand to jointly promote a cause. Corporate brands have money and media assets (magazine ad, television spots, and large social networks) that can be used to support an effort about which an artist is passionate. This **cause marketing** is often a more comfortable way for artists to create an affiliation with a brand, particularly those that fear their fans will see a corporate endorsement as the artist "selling out." The brand often gets the benefit of creating goodwill amongst its customers and the artist's fans for its support of a project. It is a win-win-win scenario for the artist, corporate brand, and the cause.

For instance, Granola products manufacturer Nature Valley formed the Preserve the Parks program to bring attention to and improve our open lands. Their project, The Quietest Show on Earth, was a special concert featuring the folk singers Andrew Bird and Tift Merritt performing in the Mojave Desert Land Trust. The concert was taped and distributed online to raise awareness for the Parks (Ferreiro, 2013).

Big Machine Label Group partnered with General Mills and Feeding America for the Outnumber Hunger campaign, which included special concert events and the label's roster of artists being featured on selected boxes of General Mills cereal (General Mills, 2013).

FIGURE 5.4

THE FUTURE OF MUSIC PARTNERSHIPS

Artist-Owned Brand Opportunities

Increasingly, instead of an artist renting or licensing their name to an organization, the artist chooses to take more of an ownership role by owning and controlling the brand. While the risk with this approach is greater for the artist the potential rewards are substantial.

Jimmy Buffet's expansion of the Margaritaville brand is one example or this. Buffett has built an empire from the leveraging of the name, which includes massive merchandising efforts and a national restaurant chain.

Avril Lavigne has spent years building a brand that she was able to leverage into several businesses, including her own line of clothing and a line of fragrances (Sauer, 2013).

Instead of opting for another endorsement deal multi-platinum artist Kenny Chesney create his own Caribbean Premium Rum Company, Blue Chair Bay, inspired by his music and island lifestyle.

FIGURE 5.5

Chesney was able to use his connection with his loyal fan base to promote the brand. In fact, one of the key marketing tactics for the launch of Blue Chair Bay was a co-sponsorship of Chesney's No Shoes Nation tour in 2013 where fans could sample the three flavors of rum before the show on the deck of a specially-created 20-foot sailboat, reaching an estimated 1.25 million fans in 42 markets nationwide (O'Malley Greenburg, 2013). The sampling program continued with the 2014 Nowhere But Here tour hitting 25 markets (Blue Chair Bay, 2014). Today, the spirit is cited as one of the fastest growing new brands in the liquor business (Bieler, 2014).

Product Placement in Videos

As record labels look to videos as not only a marketing vehicle for selling recorded music but also as a viable revenue stream, product placements continue to grow at a rapid pace. Whether it is Avril Lavigne answering her Sony Xperia phone that was sitting in a glass of water during her video for "Rock and Roll," Brittney Spears' integration of a dating site in "Hold it Against Me" or Lady Gaga's numerous product integrations into her videos, artists are offering brands the opportunity to be featured in music content. Whereas previous product placements in music videos have been blatantly obvious, the best product integrations today tend to apply a more subtle approach of showing the products being used naturally by the artists.

True Partnerships

Indeed, an ongoing trend in music partnerships is getting the brand organically involved in the artist's career and initiatives so that there is authenticity in the partnership.

Pepsi, a longtime supporter of music initiatives, signed megastar Beyoncé as a "brand ambassador" in a deal that was reportedly worth $50 million. The agreement was heralded as the future of music partnerships because of the star appearing on Pepsi products and commercials, having a guaranteed performance in the Pepsi-sponsored Super Bowl, and being provided a multimillion-dollar fund for Beyoncé to use to generate content from live events, for videos, or humanitarian efforts. The move fosters an even deeper spirit of collaboration amongst the parties by allowing the artist to drive the decision regarding how the funds will be used while Pepsi benefits from the brand equity and sales boost generated from the partnership (Coffee, 2012).

CONCLUSION

The importance of branding for an artist cannot be overstated. Today's environment is challenging, retail space is scarce, and revenue from album sales is down resulting in significantly less profit for the label and artists alike.

A strong artist brand:

- Assists an artist in building a passionate following.
- Generates interest among the media resulting in increased publicity opportunities.
- Allows the artists themselves to capitalize on the brand name they've developed and generate additional revenue.
- Attracts the attention of corporate partners looking to expose their products to a passionate fan base. The increased marketing impressions lead to more brand equity for the artists.

Increased consumer awareness from broader exposure helps build brand equity that can be leveraged in many ways for greater profitability by both the artist and the label.

GLOSSARY

Action alley—a term used most commonly within the Wal-Mart community of employees and suppliers that refers to the main aisles down the center of the store that separate the departments.

Artist brand profile (aka artist presentation or artist deck)—a visual presentation usually created in software such as PowerPoint or Keynote that is created to provide a snapshot of an artist's background, lifestyle, sales success, career highlights and brand preferences. This presentation is most often used to educate members of the media, advertising agencies and brand marketers on the artist's brand.

Brand—a name, term, design, symbol, or any other mark that uniquely identifies a seller.

Brand architecture—the structure and organization of brands across an entire organization that outlines how the company's various brands interact with and support one another.

Branded display—a special section in a retail environment, usually at the end of an aisle or in the middle of a wide aisle that separates departments, that features select products and is branded with that product's logo and photos.

Branded house—a brand relationship strategy where the master, or main brand is the dominant one with other brands being descriptive in nature, e.g., Virgin Music, Virgin Hotels, Virgin Mobile, Virgin Airlines, etc.

Brand fit—refers to how well two partnership brands align. A good brand fit refers to a partnership where the artist's personality and fan base fit the personality of the corporate brand and vice versa.

Cause marketing—a form of marketing where an artist teams up with a charitable organization to raise awareness or resources for a social issue or problem.

Co-branding—when two or more brands from different organizations combine to create a special product.

Content marketing—a marketing tactic by which companies create, gather and distribute valuable and relevant content to its existing and desired target audience. The objective of content marketing is to establish goodwill, which will eventually lead to sales or other customer action.

Customer profile—a "persona" or detailed description of the ideal customer. Whereas a target market might be 18–21-year-old females who make a salary of less than $25,000 and live in suburban markets, a customer profile might be a 19-year-old female college student who likes top 40 pop music, shopping at Forever21, and dancing.

Endorsed brand—a brand relationship strategy where independent brand names are strengthened by attaching an organizational brand, e.g. Courtyard by Marriott allows the mid-priced hotel chain to leverage the quality association of the word "Marriott."

Endorsement—the act of an artist giving his or her public support or approval to an individual, brand, or company (e.g., Jessica Simpson and Proactiv).

House of brands—a brand relationship strategy where the master, or main company brand is used minimally on packages and in advertising. The brands in the company's portfolio each have their own distinct image and positioning, e.g., Facebook owns Instagram and Whatsapp but each brand is independent and is seen by most consumers as separate entities.

In-kind support—goods or services provided to an artist instead of money. Common examples include music equipment, lighting for stages, and travel services.

Master brand—the dominant brand identity under which all of the company's other products and services fall.

Non-traditional outlets—online and offline stores where music is not generally sold, e.g., grocery stores, sporting goods stores, and clothing retail stores.

Open rate—the highest rate charged by an advertising medium such as a magazine or newspaper for placement in the publication. Contract customers, or those placing multiple ads throughout a specified timeframe, receive quantity discounts but non-contract advertisers (such as those who only run one ad) are charged the open rate.

Personal branding—the branding and self-positioning of the individual person. A personal brand involves all of the attributes and characteristics associated with an artist including their style, demeanor, and reputation.

Product category—a group of similar, like, or related items, e.g., Ford, Toyota, and GM all make products that fall under the "Automotive" category.

Sponsorships—the act of an organization paying a sum of money toward the cost of an activity or event such as a tour or music festival in exchange for assets surrounding the event like inclusion in advertising, on-site signage like stage banners, and the ability to engage with consumers onsite via booths and product sampling.

Sub-brand—brands connected to a master brand that add associations to make a brand seem more appealing to consumers or to stretch a master brand allowing it to compete in a different category, i.e., a record label that markets predominantly country music might add a separate sub-brand or sub-label to focus on the pop genre.

Target market—a broad group of potential customers usually defined by ranges (age, income, etc.) and is the group of people most likely to purchase a product of service.

REFERENCES

Aaker, D., and Joachmisthaler, E. *Brand Leadership*. New York: The Free Press. 2009.

"A brief overview of the history of branding." *AEF Advertising Educational Foundation*. AEF, 2009. Accessed November 14, 2014. http://www.aef.com/index.html.

American Marketing Association. www.ama.org. 2014.

Bieler, K., "High tide for Blue Chair Bay." Beverage Media Group Blog, April 1, 2014. Accessed November 14, 2014. www.beveragemedia.com/index.php/2014/04/high-tide-for-blue-chair-bay.

Blue Chair Bay. "Blue Chair Bay Rum Launches 25-City Spring into Summer Sampling Tour." April 24, 2014. http://www.prnewswire.com/news-releases/blue-chair-bay-rum-launches-25-city-spring-into-summer-sampling-tour-256529591.html.

"Celebrity Endorsement: A Literature Review." *Journal of Marketing Management*, Vol. 15, No. 4 (1999), pp. 291–314.

Coffee, P. "Pepsi and Beyoncé: The New Sponsorship." *PR Newser*, December 11, 2012. http://www.mediabistro.com/prnewser/pepsi-and-beyonce-the-new-sponsorship-model_b52383.

Dahud, H. "The 3 Cs of Effective Artist Branding." *Hypebot*, August 6, 2012. Accessed August 14, 2014. www.hypebot.com/hypebot/2012/08/the-3-cs-of-effective-artist-branding.html.

Diaz, A.-C., and Pathak, S. "10 Brands that Make Music Part of their Marketing DNA." *Advertising Age*, September 30, 2013. www.adage.com/article/special-report-music-and-marketing/licensing-10-brands-innovating-music/244336.

Dimofte, C. V. and Yalch, R. F. "The Mere Association Effect and Brand Evaluations," *Journal of Consumer Psychology*, Vol. 21 (2011), pp. 24–37. http://www-rohan.sdsu.edu/~dimofte/jcp11.pdf.

Erdogan, B Z. "Celebrity Endorsement: A Literature Review." *Journal of Marketing Management* 15.4 (1999): 291–314.

Ferreiro, L. "Andrew Bird Teams up with Nature Valley for the Quietest Show on Earth to Preserve National Parks." *Music for Good*, November 8, 2013. http://musicforgood.tv/2013/11/andrew-bird-teams-up-with-nature-valley-for-the-quietest-show-on-earth-to-preserve-national-parks.

Forbes. Web. www.forbes.com/celebrities/list. 2014a.

Forbes. Web. www.forbes.com/powerful-brands. 2014b.

General Mills. "General Mills, Big Machine Label Group, and Feeding America Join Forces to Outnumber Hunger." March 30, 2013. http://www.businesswire.com/news/home/20120330005615/en/ADDING-MULTIMEDIA-General-Mills-Big-Machine-Label#.VYnL3E3wuM8.

GfK MRI. www.gfkmri.com. 2014.

Hampp, A. "From this Week's Billboard: How Taylor Swift's *Red* is Getting a Boost from Branding Mega-deals." *Billboard*, October 22, 2012. Accessed November 14, 2014. www.billboard.com/biz/articles/news/branding/1083319/from-this-weeks-billboard-how-taylor-swifts-red-is-getting-a.

Houghton, B. "Samsung Buys 1 Million Copies of Jay-Z New Album." *Hypebot*, N.p., June 17, 2013. Accessed November 14, 2014. www.hypebot.com/hypebot/2013/06/samsung-buys-1-million-copies-of-jay-z-new-album-magna-carta-holy-grail.html.

Lavin, K. "What's in a Name? Just ask Hostess Brands—The Value of Branding in Bankruptcy." *ABF Journal.* www.abfjournal.com/articles/whats-in-a-name-just-ask-hostess-brands-the-value-of-branding-in-bankruptcy. 2014.

Lukovitz, K. "Subway, Diet Coke Team on Taylor Swift Promo." *Media Post.* September 27, 2014. http://www.mediapost.com/publications/article/234991/subway-diet-coke-team-on-taylor-swift-promo.html.

Mansfield, B. "Taylor Swift's '1989' Sells 1.287 Million in First Week." *USA Today*, November 4, 2014. www.usatoday.com/story/life/music/2014/11/04/taylor-swift-first-week-million-sales/18480613.

McQuicklin, J. "Understanding Image and Branding as a Musical Artist." *Artist Development Blog*, N.p., January 27, 2011. Accessed September 9, 2014. http://artistdevelopmentblog.com/advice/understanding-image-branding-musical-artist/.

NPD. www.npd.com. 2014.

O'Malley Greenburg, Z. "Rum Diaries: Kenny Chesney Diversifies with Blue Chair Bay." *Forbes*, July 1, 2014. http://www.forbes.com/sites/zackomalleygreenburg/2013/07/01/rum-diaries-kenny-chesney-diversifies-with-blue-chair-bay.

Pomerantz, D. "The World's Most Powerful Celebrities." *Forbes*, June 30, 2014. Accessed November 16, 2014. Forbes.com.

Sauer, A. "Avril Lavigne May Have Just Pulled off the Greatest (and Worst) Product Placement of All Time." *Brand Channel.* August 20, 2013. http://www.brandchannel.com/home/post/Avril-Lavigne-Product-Placement-Rock-N-Roll-082013.aspx.

Scarborough. www.scarborough.com. 2014.

CHAPTER 6

The Marketing Plan

Meticulous planning will enable everything a man does to appear spontaneous.

—Mark Caine

INTRODUCTION

A **marketing plan** is a single or album specific plan that describes activities selected to achieve specific marketing objectives for that product, within a set period of time.

Record labels create an initial, written marketing plan like you might see at any other company. The document allows the label to develop a timeline and coordinate the plan with the artist, their management and other external partners. For a record label, the marketing plan is the blueprint for each release, but after the release is launched the plan may be adjusted in response to the market. Every record released has its own unique marketing plan based on expectations of sales performance and tailored to the target market.

There are two basic target markets for the marketing plan: The consumer and the trade. The trade consists of those people within the industry for which business-to-business marketing is done. This includes radio program directors, journalists, editors, distributors, retail buyers, and others involved in the **push strategy**.

CONTENTS

WHO GETS THE PLAN?

The plan is distributed to everyone involved in marketing the record, including people both inside and outside the label. The label will provide an edited version of the marketing plan to trade partners (distributors and radio) to indicate the seriousness of their effort to sell this particular album.

Internally, it is important that everyone at the record label understands what is being done in other departments, and how synergy is created when all of the elements come together. The chart in Table 6.1 gives some examples of who may be involved in the execution of the plan.

There are other departments that benefit from, and are included in the plan, including grassroots marketing (street teams), new media (Internet), and, if the artist has a "360 deal," merchandising and tour support. Coordination is necessary for synergy to occur. The publicity department may develop materials necessary to send to retail accounts and radio (such as press kits). Advertising needs to be coordinated with retail accounts for sales in the marketplace and with the promotions department for trade advertising aimed at radio. The sales department works with the promotion department to ensure that product is available in all markets where

Table 6.1	Departments Involved in the Marketing Plan		
Function	**Internal**	**External**	**Goals**
Publicity	Label publicist	PR firms, press and media outlets	Getting reviews, features, interviews, photos and appearances
Radio	Promotion department	Indie promoters, radio programmers	Work with radio to get airplay
Sales	Sales staff	Distribution, account buyers	Works with distributors and retailers on retail promotions
Advertising	Developed by creative services, coordinated by marketing department	Outside firms sometimes hired to develop and implement programs	Much of advertising is coordinated with retail accounts
Social media/ Grassroots	Social media and Internet specialists, street team coordinators	Web design firms, Internet promotions	Online promotion and sales
Video	Creative, promotions, coordinated by marketing department	Production, possibly distribution and promotion	Production is external to the label, while promotions may be either
Touring	Publicity, retail	Production, possibly distribution and promotion	Label ensures there is product at retail, press coverage

airplay is prevalent. Advertising and publicity go hand-in-hand, sometimes combined in a "media plan," because in many instances, the same media outlets are targeted for both. Outside of the company, summaries of marketing plans may go to account buyers in retail, program directors in radio, the artist's booking agency, and they are always presented formally to the artist and their manager. Often, the manager will help develop the plan if corporate sponsors are involved or if there are special events that could be mined for extra marketing impact.

FIGURE 6.1

WHAT'S IN THE PLAN?

The plan provides basic information about the release, and a description of what is occurring in each department or area. A timetable is often included to ensure that timing and coordination are optimal. Financial specifics are often eliminated from the copies circulated at the marketing meetings, but are an integral part of the plan. Sales goals and shipment quantities, however, are often included in the plan. The following is a list of some basic information included in a marketing plan:

- Project title and artist name
- Street date
- Price, dating, and discounts
- Song selections
- Overview or mini-bio
- Configurations and initial shipment quantities
- Selection number and bar codes
- Marketing goals and objectives
- Target market
- Promotion (radio)
  - Names and release dates of singles
  - Promotion schedule
  - Promotional activities
- Publicity
  - Materials
  - Target media outlets
  - Specific goals (interviews, appearances, reviews, etc.)
- Advertising
  - Materials
  - Target media outlets and coordination with retail accounts
  - Ad schedule
- Sales
  - Sales forecast
  - Account-specific promotions, P&P
  - POP materials
- Video
  - Production information
  - Distribution and promotion

- New media/Internet promotions
- Street teams/grassroots marketing
- Lifestyle/promotional tie-in with other companies/products
- Tour support (usually tied in with retail and publicity)
- Tour dates—help with spread of product at retail and tie-ins with radio
- International marketing
- Artist contact information (manager, publicist, and so on)
- Comprehensive timetable
- Budget (not included in copies that are distributed to employees or contract agents)

Sections of the Plan

The first section of the plan contains specific technical information about the release, including the configurations available, pricing information, initial shipment numbers, **SKU** numbers, list of tracks, production information, release date, contact information and a description of the project and goals. If the plan is for the artist's first release, versions of the plan given to external partners may include a brief bio. The next section is generally radio promotion and contains information on the singles that have been selected to release to radio. Since radio promotion of the first single usually precedes the release by 8–12 weeks, this information needs to be presented early in the plan. The goal is to get substantial radio airplay leading up to the release date for the album, thereby creating a demand. The (radio) promotion section will also feature any special promotions geared toward radio, such as promotional touring, promotional contests, or showcases. Specific radio markets may be targeted in this section, although sometimes targeting is done nationally but concentrated on specific formats. Then, in no particular order, the sections on advertising and publicity (sometimes grouped together in a media plan) and retail promotions follow.

The publicity section will have goals, materials being created, and targeted media outlets, complete with a timetable. Remember that some publications, particularly print magazines, have a long lead-time and must be worked several months in advance of the placement you are seeking. Publicity materials, including photos and a bio, are usually created well in advance of the release date. A press release is sent out closer to the release date to generate additional media interest.

The retail section of the plan includes particular price and position programs for each chain of stores, and may include particular information about costs for these programs, as well as specific features such as exclusive releases or content, endcap placement, or other positioning, and newspaper and coop advertising.

An advertising section may have information on specific media being targeted for advertising and information on advertising materials being generated (such as TV commercials, radio spots, etc.). Different labels have different philosophies on advertising. Some focus almost exclusively on advertising to consumers, while others budget for a healthy amount of trade advertising. Sometimes, radio and TV ads will be developed so that they can be produced on short notice to give the record a boost in the marketplace or capitalize on some unforeseen opportunity.

The section on video may include production information and promotional plans, including placement on national, regional and local television shows and any Internet viral marketing campaigns, although that may be found in the new media section. The new media section may be next and include information on specific online promotions, targeted Web partners (such as video channels, online media publications and online retailers). Online consumer promotions, contests, email campaigns, and online street teams are a part of this section.

A tour section will include a list of tour dates and venues and any local media and retail partners that will be included in the plans. A lifestyle section contains information about sponsorships and co-branding with other products. A timeline of all events and deadlines can sometimes be found at the end of the marketing plan, allowing for a quick glance to reassure that all events are coordinated and taking place in a timely fashion. This timeline may not be shared with external partners. A detailed marketing plan for a major artist is included in the appendix of this chapter.

TIMING

Timing is crucial to the success of an album, including the time of year the album should be released. The album release date is called the **street date**. Albums are strategically released on Fridays to take advantage of a full week of SoundScan sales numbers. Release dates are subject to change for many reasons, not the least of which is what other artist might release a new album on that particular Friday. Given consumers' limited budgets, it would be unwise for a label to release a new pop artist on the same day as the long awaited release from an established major act. The time of year that a new album is released also varies depending on the stage of the artist's career. Established artists frequently release albums in the fourth quarter to take advantage of holiday sales. Retailers have frequently expressed their displeasure at the glut of releases at the end of the year when their stores are already full of shoppers and the lack of releases in other times of the year when a strong release may lure shoppers into stores. New artists are advised to release during those times of the year when competition is less, but then so is the number of shoppers. An album may be strategically timed to coincide with a concert tour or a major media event for the artist. Valentine's Day is also a popular time of year for music purchases.

THE IMPORTANCE OF STREET DATE

It is extremely rare for any label to release an album in the U.S. on any day other than Friday. There are conflicting stories for why street date was on Tuesdays, but since that day was chosen the charts and sales reporting were built around it on July 7, 2015 new releases world-wide were moved to Fridays. By releasing a new song or album on Friday, it gets a full week of sales before the first chart report. This is important because of the publicity opportunities afforded by doing well on the charts. If you drive around Music Row in Nashville, you'll see plenty of banners celebrating #1 songs and albums, but you won't see any celebrating being #2 or #3 or #10. How many sales does it take to be #1 on the charts? One more than the guy that's at #2! So every sale and every spin matters and you don't want to have that first week's numbers split over two reporting periods. (Kelly, 2010)

In the following illustration, some retailers must have violated the street date by selling early. Missy Elliot sold 1,484 units the week before the album was officially released. Then on the week of the release, the album sold 143,644 units—enough to place it at #13 on the chart. But if the previous week's sales not been made before street date and instead sold during the first week, the additional 1,484 units would have been enough to rank the album at #12 for the week, above Nelly, except for the fact that he also had some early sales.

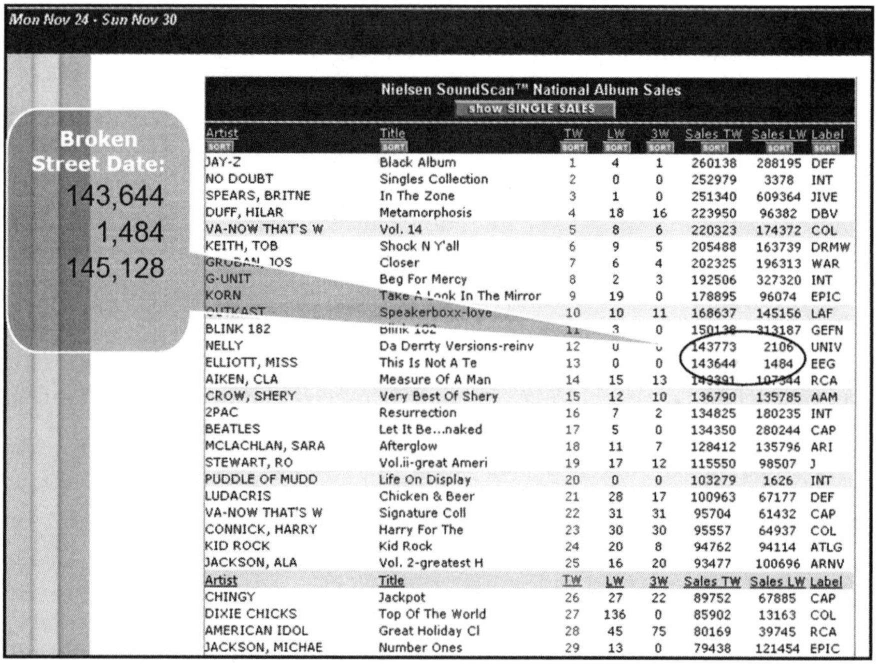

Nielsen SoundScan™ National Album Sales

show SINGLE SALES

Artist	Title	TW	LW	3W	Sales TW	Sales LW	Label
JAY-Z	Black Album	1	4	1	260138	288195	DEF
NO DOUBT	Singles Collection	2	0	0	252979	3378	INT
SPEARS, BRITNE	In The Zone	3	1	0	251340	609364	JIVE
DUFF, HILAR	Metamorphosis	4	18	16	223950	96382	DBV
VA-NOW THAT'S W	Vol. 14	5	8	6	220323	174372	COL
KEITH, TOB	Shock N Y'all	6	9	5	205488	163739	DRMW
GROBAN, JOS	Closer	7	6	4	202325	196313	WAR
G-UNIT	Beg For Mercy	8	2	3	192506	327320	INT
KORN	Take A Look In The Mirror	9	19	0	178895	96074	EPIC
OUTKAST	Speakerboxx-love	10	10	11	168637	145156	LAF
BLINK 182	Blink 182	11	3	0	150138	313187	GEFN
NELLY	Da Derrty Versions-reinv	12	0	0	143773	2106	UNIV
ELLIOTT, MISS	This Is Not A Te	13	0	0	143644	1484	EEG
AIKEN, CLA	Measure Of A Man	14	15	13	143391	107344	RCA
CROW, SHERY	Very Best Of Shery	15	12	10	136790	135785	AAM
2PAC	Resurrection	16	7	2	134825	180235	INT
BEATLES	Let It Be...naked	17	5	0	134350	280244	CAP
MCLACHLAN, SARA	Afterglow	18	11	7	128412	135796	ARI
STEWART, RO	Vol.ii-great Ameri	19	17	12	115550	98507	J
PUDDLE OF MUDD	Life On Display	20	0	0	103279	1626	INT
LUDACRIS	Chicken & Beer	21	28	17	100963	67177	DEF
VA-NOW THAT'S W	Signature Coll	22	31	31	95704	61432	CAP
CONNICK, HARRY	Harry For The	23	30	30	95557	64937	COL
KID ROCK	Kid Rock	24	20	8	94762	94114	ATLG
JACKSON, ALA	Vol. 2-greatest H	25	16	20	93477	100696	ARNV
Artist	Title	TW	LW	3W	Sales TW	Sales LW	Label
CHINGY	Jackpot	26	27	22	89752	67885	CAP
DIXIE CHICKS	Top Of The World	27	136	0	85902	13163	COL
AMERICAN IDOL	Great Holiday Cl	28	45	75	80169	39745	RCA
JACKSON, MICHAE	Number Ones	29	13	0	79438	121454	EPIC

FIGURE 6.2 *Effect of Breaking Street Date on Chart Position*

Timing of the individual elements is also crucial. The first single is usually released weeks before the album, to create an initial demand for the album. Then a second single is released to further support the album once it has been in the marketplace for a while. Some new artists on major labels may release a single to radio in the fall to introduce the artist into the marketplace, but wait until early the following year to launch the album and release the second single. The actual timeline varies for different genres. Rap/hip-hop albums have a shorter life cycle and the marketing activities may be more compressed. Country music has a longer life cycle and the release of singles may come at longer intervals. A study of the charts in 2000 found that it took a popular rap single only 10.6 weeks to rise to a "top 5" chart position, whereas it took 14.8 weeks for R&B, 13 weeks for top 40, 26 weeks for adult contemporary (AC), and 18.6 weeks for rock. It took a country single 19.5 weeks to enter the top 5. In 2000, top pop singles spent the most time on the Billboard chart, at 29 weeks, with rap singles spending the least amount of time (Hutchison,

2010). A study of the country charts in 2010 found that it was taking the typical country single 25 to 30 weeks to peak on the charts. The same study also found that the number of #1 singles on the charts had dropped from 49 in 1989 to less than 25 in the years 1999 to 2004 (Heine, 2010). It is difficult to break a new artist when you cannot place more than one or two singles on radio in a year.

Rock and metal acts that depend more on touring may plan events around the Ozzfest, Bonnaroo, or other summer tours and festivals. Smaller indie labels may depend more upon touring to sell records, since getting radio play is often too expensive.

Some publicity materials are sent to print vehicles in advance of the release date so that album reviews and features will appear during the week or month of the release. Trade advertising and publicity to radio occurs prior to the release of each single; and to retail, advertising is timed with the release of the album. All the elements are timed and coordinated to create a comprehensive snowball effect, generating a buzz in the marketplace and convincing gatekeepers and consumers that this project is the "hottest thing" in popular music.

MARKETING STRATEGY

The marketing strategy is the fundamental marketing plan by which the company intends to achieve its marketing objectives. It consists of a coordinated set of decisions based on target markets, the marketing mix, and the marketing budget. The marketing strategy is the "road map" or guide to growth and success.

The marketing plan for a company should start with an assessment of the company's internal strengths and weaknesses, followed by an examination of the market environment. This situation analysis is sometimes referred to as a *SWOT analysis* (strengths, weaknesses, opportunities, and threats). It is a review of the company's present state, and an evaluation of the external and internal factors that can affect future success.

As a part of the SWOT analysis, the **internal marketing audit** is designed to examine the company's areas of strength (the company's competitive advantages) and weaknesses (the company's vulnerable areas and underperforming units). Then, an assessment is conducted on the external opportunities (areas for growth) and threats (roadblocks and challenges) that exist for the company. The resulting investigation yields the SWOT analysis—strengths, weaknesses, opportunities, and threats. A SWOT assessment allows the marketing department to determine:

- Where the company is right now: What is our position in the marketplace? Our market share? Which products are currently successful? What are our areas of strengths and what are our weaknesses?
- Where the company needs to go: Where are the opportunities? Where are we likely to hold a competitive advantage in the marketplace in the future? What is our game plan for the future? What threats should we be aware of?
- How to best get there: What action needs to be taken to achieve our goals?

A SWOT analysis for a record label may yield some of the following findings.

Table 6.2	The SWOT Analysis	
Strengths		**Weaknesses**
■ Strong roster		■ Underperforming artists on roster
■ Strong catalog		■ Expenses are too high
■ Skilled label personnel		■ Unhappy artists leaving label (or suing)
■ Effective distribution		■ Artists legal problems
■ Abundant finances		■ Loss of label personnel
■ Reputation		
Opportunities		**Threats**
■ New technologies		■ Piracy and P2P file sharing
■ New markets (international or domestic)		■ Censorship
■ Piracy controls		■ Competition for entertainment dollars
■ Trends swing your direction		■ Economic downturn
■ International trade agreements		■ Loss of retail opportunities
■ New ways to monetize music consumption		■ New technology

The SWOT helps the marketing department determine marketing strategy, including the identification of suitable markets and appropriate products to meet the demand. The company can then engage in sales forecasting, budgeting, and projection of future profits.

In the book *Marketing: Relationships, Quality, Value*, Nickels and Wood use Motown Records in their example of a SWOT analysis (Nickels and Wood, 1997). Motown's main strength is its strong catalog of 1960s and 1970s music, including the Supremes, Four Tops, The Jackson Five, Michael Jackson, and Boyz II Men. A weakness is their lack of start-up urban and rap artists compared to other labels such as Interscope. Opportunities exist in the nostalgia fads that periodically emerge, creating a demand for the old recordings and may include a resurgence of interest based upon the release of a movie or movie soundtrack. In 2006, a movie based upon the Supremes was released called *Dreamgirls*. The soundtrack slipped into the number one position on the *Billboard* chart in January 2007. A threat Nickels and Wood mention is "a shift in popular interest away from the label's music" as old-school R&B gives way to newer artists.

The SWOT analysis helps the company discover its core competencies, which enables the company to surpass the competition. However, it is the product **marketing plan** that helps to coordinate the strategic efforts to promote a particular product. It is a carefully planned strategy "with specialists in the areas of artist development, sales, distribution, advertising, promotion, and publicity joining forces in a coordinated effort to break an artist and generate sales" (Lathrop and Pettigrew, 1999). The term *marketing plan* is used by record labels to refer exclusively to the plan designed for each album release and is generally not associated with an overall company strategy.

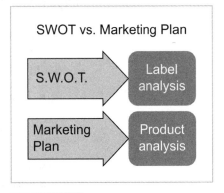

FIGURE 6.3 *SWOT vs. Marketing Plan*

CONCLUSION

Lewis Carroll said "If you don't know where you are going, any road will get you there." If you want to get to the top of the music business it is best to have some goals and a plan to get there. In this chapter we have reviewed basics of marketing planning used by record labels to promote artists and their recordings. Every label will bring their unique philosophies, strengths, and weaknesses to the process but they will all have the same goal—to have a hit record. The plan gives the marketing team a starting point and an adaptable framework to achieve that goal. In the chapters that follow we will discuss the specifics of each of the major marketing tools in greater detail.

APPENDIX: MAKING SENSE—*MAJOR ARTIST MARKETING PLAN*

Below is a detailed marketing plan for a major artist. The names and dates have been changed, but everything else is real—the timing, the numbers. This plan, created the week after the album street date, was an update to the original marketing plan to keep the sales and marketing teams in the field updated on current events.

Title: *Silver Pennies*
Artist: *Making Sense*
Release Date: February 7

ROADMAP

10/10: Single 1 "Lucky" premiered online, single serviced to Pop & Hot AC radio, free radio station download
10/11: "Lucky" made available on iTunes

- Heavy online buy and social media outreach 10/14–11/7: U.S. Promo Tour
- Covered 36 radio stations, VH1, PBS, Cumulous, Vevo, IHeartRadio, NY showcase, LA radio/licensing showcase
- B-Roll crew covering entire trip
- Online content constantly rolling out 10/18: NY Showcase at Angel Orensanz
- College reps, Epic, press, radio, video and licensing attended (and real fans!)

10/24: Pop & Hot AC impact date for "Lucky" 10/28: LA showcase at Sayers Club, 8pm

- Licensing & press attended, also KBIG lounge 11/7: "Lucky" video delivered
- iTunes look 11/15, followed by Vevo 11/17, VH1.com, AOL, Myspace, and Yahoo
- Social media outreach
- Serviced EVERYWHERE after premiere 11/15–11/23: Australia promo (incl. X Factor)

11/21: "Lucky" rotation began at VH1, MTV, MTVU, & MTV Hits
11/28: "Lucky" was added to TeenNick 11/30: WalMart sound check taping

11/30–12/17: U.S. radio shows

12/7: Beginning 12/7, fans were able to "check in" to the Merch Table at the holiday shows and receive a special discount for the preorder using Foursquare.

12/8: NYC Democratic Party charity gig, press covered

12/11: "Lucky" in ESPN Ravens special

12/20: iTunes album pre-order

- Today appearance 12/22
- Artist hospital visit while in NYC
- Video for "Go Go" rolled out week of 12/20/1/2012: Grey's Anatomy promo sync

1/16–18: Canada promo 1/19–20: NY promo

- Satellite Radio Tour, Sirius, WFAN, I Heart Radio, VH1, Baeble. com and Music Choice

1/22–28: UK promo

1/31: iTunes exclusive listening stream (7 days)

2/4–5: Direct TV & tailgate Superbowl performances 2/7: Silver Pennies U.S. album release (#1 at iTunes!) 2/7–17: LA and New York TV appearances

2/13: Video shoot for "Money Talks" 2/2012: VH1's Posted Artist of the Month online

2/21: "Lucky" will be the featured Song of the Day on Amazon MP3. The track will be $0.99 2/21–2/28.

2/16–28: U.S. theater tour

3/12 (week of): video premiere for "Money Talks" (tent) 3/13: adult radio impact date for "Money Talks"

4/3: National Anthem at Final Four

SALES/MARKETING

- *Silver Pennies* sold 14,675 TW with 169,766 sold to date and is currently #135 on iTunes
- "Lucky" sold 19,675 TW with 457,756 sold to date and is currently #160 on iTunes
- *Silver Pennies* debuted selling 89,634; secured spot #4 on the *Billboard Top 200* and was #1 at iTunes!
- "Lucky" sold 49,690 and reached #11 on the iTunes single chart on its first week

- Direct TV campaign started in January; Making Sense performed at Direct TV's pre-game Superbowl party 2/4
- 2/3: *Silver Pennies* album mini included in the next Sony Valentine's catalogue; distributed to 3.5 million households
- 2/5: NFL pre-game tailgate performance
- Verizon VCast ran banners through their service 10/26 through 11/21

iTunes

- Targeting video premiere, session or another look for their curated room
- Deluxe edition available with 5 cover versions as bonus tracks
- Single up 10/11
- Video premiere 11/15
- Album pre-order up 12/20 (with 2-week placement through Xmas buying season), includes LP on deluxe, "Go Go" as IG track
- Artist curated room up week of release
- Album listening party 1/30–2/6
- 360 online ad campaign runs 1/31–2/13

Target

- Possible Easter Target circular 4/8
- Exclusive special edition with 3 live tracks and acoustic Springsteen track
- Band to play 3/20 Spring mktg meeting in exchange for 2 spots on the Grammys

Spotify

- Banners ran throughout November
- "Lucky" is also featured in On Your Radar playlist (14k subs) and the Ultimate Singer Songwriter playlist (7k subs) and will be added to the Top of the Charts playlist as well as it climbs the charts (20k subs)
- Gold package exchange for exclusive live EP (includes pop-up light box on front page)

Wal-Mart

- Band filmed WalMart soundcheck 11/30 in Boston, with live audio included as download card with album purchase

- "Lucky" premiered early on walmart.com, with rest of session and b-roll to roll out 2/7
- Band signed 750 booklets for wm.com pre-order

Best Buy

- Shot HD content for Best Buy HD wall
- Circular 2/5

Amazon

- "Lucky" will be the featured Song of the Day on Amazon MP3. The track will be $0.99 2/21–2/28.

Indies

- Value-add 7" with indies for added online and in-store visibility

D2C

- Boxed t-shirt with download card, sold through the holidays
- Pre-order laminate sold at radio Christmas shows starting 12/7

RADIO

First single "Lucky" shipped 10/10, impacted 10/24

Hot AC BDS 10–14; 1745x (–470x), 5.9 mil (–1.7 mil)
Hot AC MB 10–13; 1961x (–448x), 7.8 mil (–1.8 mil)
AC BDS 24*–24*; 89x (+1x), 422K (+1K)
AC MB 21*–22*; 147x (flat), 696K (–4K)
Hot 100—No. 125; 12.4 mil (–6.0 mil)

- Next single: "Money Talks" impacts 3/13
- I Heart Radio session to tape 1/20

VIDEO

- "Money Talks" video to deliver 3/5, premiere week of 3/12
    - BTS at video shoot also available

- "Lucky" video rollout:
  - 11/14: VEVO premiered Behind The Scenes clip (4.5 million impressions)
  - 11/15: iTunes video premiered
  - 11/17: VEVO video premiered (live at 3am EST)
  - 11/17: VH1.com video premiered (live at 3pm EST)
  - 11/17: AOL Music homepage feature (video & interview)
  - 11/18: MySpace homepage video featured
  - 11/19: VH1 Top 20 Countdown broadcast video premiered (9 am)
  - 11/21: Yahoo Music homepage video featured
  - 11/21: rotation started at VH1, MTV, MTVU, and MTV Hits
- 2/7: VEVO feature
- "Lucky" is at #18 on this weekend's VH1 Top 20 Countdown.

Music Choice

- Music video serviced, in rotation 11/30
- Band shot int'w when in NYC (1/20)

Screenvision

- "Lucky" was featured in 7K Screenvision reels in movie theaters nationwide from 1/6/12–1/26/12
- Video: VH1—#18/6x
- "Lucky" is in GUNG HO rotation on VH1 and was the #1 most played video during the week of release.
- For one week, starting Sunday 2/12, Making Sense will be the featured artist and "Lucky" will be highlighted on VH1's Top 20 Countdown. Spots (:30) will run a minimum of 10x throughout all day-parts all week long. The spot includes artist, song and album info.
- "Money Talks" video shot 2/13 in L.A.

TELEVISION

- Pre-Superbowl performance aired on Direct TV
- Pre-Superbowl tailgate party performance aired on NBC
- Letterman performance (and re-aired)
- Today Show appearance
- PBS Artists Den
- Rachael Ray: shot 2/7, aired
- Leno: aired
- Conan appearance

- Other targets: Fallon, Ellen, Extreme Home Makeover, Nightline, Colbert, CNN, E!, Access Hollywood
- Making Sense performed on VH1's Big Morning Buzz Live 10/11, and again in Feb. at 9am.
- Kimmel outdoor performance 10/27/11 (3 songs on-air)
- Today Show: 12/11 8am AND 10am performance

PRESS

- Album release went out 1/31/12
- Tour announcement 1/10/12
- Serviced press release on 10/11 announcing album and single.
- Press running in:
  - *The New York Times*
  - *New York Post*
  - *People Magazine*
  - EW
  - *Billboard*
  - *USA Today*
  - American Songwriter
  - MTV News
- Other Targets include:
  - Rolling Stone, GQ, Esquire, Men's Journal, AP, CNN, Huffington Post

ONLINE

- 55K email list subscribers [email schedule below]
  - 11/29—In House performance video, holiday pre-order, Heartbeat video
  - 12/6—premiere exclusive video content
  - 12/20—iTunes pre-order, video online this week
  - 1/3–1/31—exclusive announcements re: tour, deluxe editions, TV audiences
  - 1/19—SEM/CPC Pre-order campaign begins (runs through 2/13)
  - 1/30—Hershey Kisses—Silver Pennies promo goes live? Social sharing give-away contest got Hershey Kisses—Silver Pennies goes live
  - 2/7—album release, New website launched, Turntable.fm Album Listening Party, Release day Twitter Party/Q&A, Shoot fan tweet video replies, album listening parties go live (AOL, MSN, Myspace,

Rolling Stone, IHeartRadio, and ArtistDirect), Hershey Kisses/ Silver Pennies give-away winner announces, Pandora Interview

- 2/7—Banners go live on MOG's free service until 3/6
- 2/10—Announce Instagram Live Show Fan Video (song TBD)
- 3/12—Instagram Live Show Video goes live
- 2.7M fans on Facebook
- 145K Followers on Twitter
- Full song premiered October 10 via free song giveaway at Clear Channel radio stations
- Rolled out webisodes from October–November promo tour, rolling out December radio show webisodes now
- IHeartRadio and Vevo int'w done 10/18 and 10/19
- "Lucky" single features on AOL, Myspace, Yahoo, and Clear Channel
- To shoot The Onion/AV Club perf. in Chicago (date TBD)
- Pitching Vevo Go Show and Area Code early February around private club gig
- Beginning 12/7, fans were able to "check in" to the Merch Table at the holiday shows and received a special discount for the preorder using Foursquare.
- "Lucky" featured as a "Hot Video" on MTV.com's homepage during December
- Making Sense are VH1's Posted Artist of the Month in 2/2012
- Still targeting: Baeble session, AOL session, Pandora int'w, Onion int'w

LICENSING

- Targeting Olympic spot placement for "Money Talks"
- Grey's Anatomy promo spots ran 2/2–2/9 feat. "Lucky"
- Salmon Fishing in Yemen trailer sync (confirmed to begin 12/16, movie released 3/2012 starring Emily Blunt and Ewan MacGregor)
- "Lucky" in Direct TV spots for celebrity beach party performance tagging iTunes
- "Lucky" was featured on VH1's partnership reel for 1/2012
- "Lucky" in ESPN Broncos special which aired 12/11, narrated by Isaac Slade
- "Lucky" was featured on The X Factor on 11/2
- Target ad to run 2x during Grammys
- Direct TV promo ads ran pre-Superbowl 2/4 and 2/5

BRANDING

- October NY showcase sponsored by W Hotel and Crown Royal
- Summer Crown Royal campaign consisted of f/p ad in Rolling Stone and exclusive video content, without live downloads on specially marketed bottles of Crown Royal (w/9 other bands)
- Hershey Kisses—Silver Pennies promotion started 1/31—"Lucky" download included in "Lucky" Kisses collection

MERCH

- No merch rights other than merch that is album-cover based.

ADVERTISING

- Online buy ran around single launch 10/11/11
- Album launch advertising plan:
  - Making Sense iTunes 360 campaign (1/31–2/13)
  - Making Sense online search around Superbowl and street date
  - In-theater & in-lobby video (2/3–2/10)
  - Clearchannel street-date program (100's of stations—pop/adult/ AAA) and *USA Today* strip ad (2/7)
  - NY taxi tops
  - Baltimore digital billboards
  - Target will be running two TV spots during the Grammy's 2/12
  - In-theatre video and spot 2/3–17

COLLEGE MARKETING

- 54 Alternative College Reps in various markets around the country received posters and placed them up in their markets before they left for winter break
- They also received stickers recently to prep their markets before the release
- Reps focus on lifestyle (coffee-shops, clothing stores, tattoo parlors, etc.) and various independent retail accounts in their markets, as well as their respective college campuses
- Reps are also directing traffic to the video for "Lucky" as well as iTunes to purchase the single
- Developing artist magnets for reps in key markets

LIFESTYLE MARKETING

- Targeted Christmas radio shows (with and without Making Sense on the bill) for distribution of tools, and appropriate movie theatres e.g., New Years Eve movie 12/2011)
- Targeting females 18–25

ASSETS

Bio (10/26)

- Photos
- Single art
- Lyric video for "Lucky"
- Album art 11/1/11
- Music video "Lucky" 11/7/11
- Numerous webisodes—around Oct/Nov promo tour (8 webisodes up on the band's sites) and rollout coming soon for Xmas show webisodes
- "Lucky" video still & b-roll ready 11/18/11
- Posters and stickers

PROMO TOUR AT A GLANCE

1/16/12–1/18/12 Canada promo 1/22/12–1/27/12 Europe promo 2/6/12–2/15/12 NY & LA TV's

2/16/12–2/28/12 US Tour 3/3/12–3/13/12 Europe promo 4/11/12–5/11/12 US tour

UPCOMING U.S. TOUR

2/16	San Diego, CA
2/18	Las Vegas, NV
2/20	Oakland, CA
2/21	Los Angeles, CA
2/22	Portland, OR
2/24	Boise, ID
2/25	Seattle, WA
2/26	Vancouver, BC

2/28	Spokane, WA
4/11	Providence, RI
4/12	New York, NY
4/14	Philadelphia, PA
4/17	Chicago, IL
4/22	Louisville, KH
4/24	Ashville, NC
4/25	Atlanta, GA
4/27	Shreveport, LA
4/28	Houston, TX
4/29	Austin, TX
4/30	Dallas, TX
5/1	New Orleans, LA
5/3	Orlando, FL
5/4	Miami, FL
5/8	St. Louis, MO
5/11	Baltimore, MD

TOP TEN MARKETS

- New York, NY
- Los Angeles, CA
- Chicago, IL
- Baltimore, MD
- Philadelphia, PA
- Boston, MA
- Dallas-Ft. Worth, TX
- SF-Oakland-San Jose, CA
- Washington, DC
- Minneapolis-St. Paul, MN

BAND HISTORY

SALES

- Making Sense (Album #1)
  - Album: 2.8 million sold
  - 9.1 million Digital Singles

- Making Less Sense (Album #2)
  - Album: 901K sold
  - 4.8 million Digital Singles

GLOSSARY

Marketing plan—A single or album specific plan that describes activities selected to achieve specific marketing objectives for that product, within a set period of time.

Push strategy—Pushing the product through the distribution channel to its final destination through incentives aimed at retail and distribution.

SKU—Stock-Keeping Unit—an inventory ID that represents one or more items sold as a single unit. A vinyl and a CD version of an album would each be a separate SKU and each have their own barcode. If the CD and vinyl disk were offered together as part of a box set that would be a third SKU.

The trade—Those people within the industry (middlemen) for which business-to-business marketing is done.

Trade advertising—Paid promotions targeted and industry middlemen and designed to stimulate purchasing and, in turn, promotion to consumers.

REFERENCES

Heine, P. "Stuck on Repeat." *Billboard*, June 12, 2010: 18+. November 4, 2014.

Hutchinson, T. Unpublished research. 2010.

Kelley, F. "Why Albums Are Released on Tuesdays." *The Record: Music News from NPR*. NPR, September 8, 2010. Accessed October 3, 2014.

Lathrop, T., and Pettigrew, J. *This Business of Music Marketing and Promotion*. New York: Billboard Books. 1999.

Nickels, W. G., and Wood, M. B. *Marketing: Relationship, Quality, and Value*. New York: Worth Publishers. 1997.

CHAPTER 7

The U.S. Industry Numbers

SALES TRENDS

In the early 1990s, the recording industry was in the midst of a *replacement cycle*, where consumers were in the process of replacing their old vinyl and cassette collections with the first digital format—the compact disc (CD). In 1994, there was a 20% increase in sales, to over $12 billion. This boom was fueled by discount retailers such as Best Buy and Circuit City aggressively entering the music market, opening up a multitude of retail outlets, and using discounted CD prices as a loss leader to attract customers. These practices, along with the concept of selling "used CDs," threatened the viability of traditional music stores and caused a plateau in the number of units shipped in the mid-1990s (Phillips, 2001).

By 1995, the CD replacement cycle was nearing completion, and the impact of the changing retail landscape was beginning to take its toll on the bottom line. Traditional retail stores were struggling to compete with discount stores, and sales gains in the industry overall were modest (from $12.068 billion in 1994 to $12.320 billion in 1995). Blockbuster closed hundreds of stores, causing a massive rush of returned product. There was slight improvement in 1996, mostly on the strength of sales of Alanis Morissette's *Jagged Little Pill* and the soundtrack of *Waiting to Exhale* featuring Whitney Houston. With sales at $12.5 billion the industry sought to reexamine its strategy for selling prerecorded music, while traditional retailers continued to struggle (RIAA, 1997). The industry experienced a decline in 1997 as "the industry was responding to a smaller but healthier

Table 7.1	RIAA Data on Annual Shipments		
Year	**Dollar value***	**Units shipped***	**Price per unit**
1993	$10,046.60	995.6	$10.09
1994	$12,068.00	1,122.7	$10.75
1995	$12,320.30	1,112.7	$11.07
1996	$12,533.80	1,137.2	$11.02
1997	$12,236.80	1,063.4	$11.51
1998	$13,723.50	1,124.3	$12.21
1999	$14,584.70	1,160.6	$12.57
2000	$14,323.70	1,079.2	$13.27
2001	$13,740.90	968.5	$14.19
2002	$12,614.20	859.7	$14.67
2003	$11,854.40	789.4	$15.02
2004	$12,154.70	814.1	$14.93
2005	$12,269.50	1,301.8	$9.43
2006	$11,510.20	1,583.2	$7.27
2007	$10,372.10	1,774.3	$5.85
2008	$8,480.20	1,852.5	$4.58
2009	$7,683.90	1,851.8	$4.15
2010	$6,995.00	1,739.6	$4.02
2011	$7,133.10	1,824.9	$3.91
2012	$7,015.70	1,803.3	$3.89
2013	$6,996.10	1,685.6	$4.15

* in millions

Source: RIAA

retail base" (RIAA, 1998). But it was one faced with bankruptcy filings and consolidation. One anomaly in 1997 occurred with a 54.4% increase in the sale of CD singles, attributed to Elton John's remake of "Candle in the Wind," a tribute to Princess Diana who died in August of that year. Growth returned briefly in 1998 as shipments grew by 5.7%. While shipments of singles dropped, CD units and music video units showed a healthy increase. The RIAA attributed the increase to a steady flow of releases by top artists throughout the year and an increase in the diversity of offerings. The moderate growth continued through 1999 with a 3.2% increase in units, fueled by strong growth in the full-length CD format (12.3% in value), despite a drop-off in music video sales. Credit is given to retailers and suppliers for improved efficiency in inventory management. The RIAA stated that the music industry was successfully competing against the "ever-increasing competition for the consumer's entertainment dollar" (RIAA, 2000).

The new millennium ushered in a period of decline in recorded music sales as the industry struggled against threats on numerous fronts. The market for CD singles plummeted as the Internet took over, and peer-to-peer file-sharing services were blamed for much of the downturn. The year 1999 introduced Napster, and the concept of free downloading rocked the marketplace. Sales began its steep decline with labels and retailers trying to make up for their loss by increasing the unit cost per transaction with the height of the practice hitting in 2003 at $15.03 per unit. Adding to the rapid adoption of digital files included the introduction of Apple's iPod in 2001 and its online retail megastore, iTunes, in 2003.

What decreased the unit cost so dramatically over time in the inclusion of digital single sales along with the value of SoundExchange revenue that is not identified by configuration. If 2013 data were parsed to show only full-length albums, RIAA sales data per unit value calculations would be approximately $11.80 per unit. Most notably, since its all-time high in 1999 at $14.5 billion, 2013 revenue is less than half the value at $6.993 billion. With fewer physical retail stores operating, more and legal online portals available, and streaming outlets proliferating, consumers are acquiring music in many different ways. By contrast, transactions of music are at an all-time high, with more people engaged with music than ever, up 70% since 1993, but this engagement or experience is not translating to sales at the cash register, as reflected in the sales data.

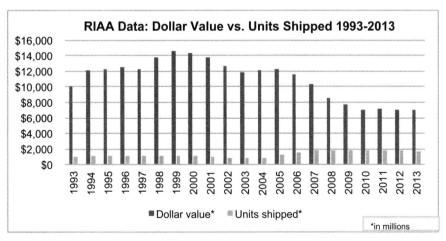

FIGURE 7.1 *RIAA total dollar value*

Source: RIAA

The period of decline began in the first half of 2000, and was attributed to the rapid drop in sales of CD singles, brought on by the Internet. (The U.S. economy also experienced a downturn at that time.) Cassettes continued to decline, while full-length CDs grew slightly in dollar value due to higher retail prices. The decline was more drastic in 2001 as total units dropped 10%, with a 6.4% drop in full-length CDs for the first time. This downturn coincided with the rapid adoption of music downloading from illegal peer-to-peer (P2P) services and ownership of CD burners, up drastically in two years.

In 2002, the market dropped another 11.2% in units and 8% in value (Christman, 2003). Every configuration saw a decrease in sales except DVDs. At this point, the RIAA began an aggressive campaign to discourage consumers from using illegal file-sharing services, but other factors also contributed to the downturn. Young consumers were spending their entertainment budget on cell phones, computers, video games, DVD collections, and other forms of entertainment. Consumers began to report (through research studies) that they perceived the cost of CDs to be too high. Upon further analysis, young consumers, who had grown accustomed to cherry-picking songs from albums through P2P services, were opposed to paying retail price for an album just to own the one or two songs they wanted. These consumers were beginning to burn their own personal mix CDs either from their own collection or from P2P services. As a result, they preferred music à la carte and were a receptive market for licensed music downloading services. But these services were slow to develop and the offerings were not sufficient to entice droves of consumers into subscription services.

Things did not improve for 2003 as the industry fell another 7.2% in units and 6% in value (Hiestand, 2004). Again full length CDs, the industry's moneymaker, fell 6.7% in value and 7% in units. The industry began to see some evidence of bottoming out, and even a slight turnaround in 2004 with total sales of $12.154 billion. This turnaround was fueled by a slight increase in CD album sales (1.9% in dollar value), digital singles (139.4 million units) thanks mostly to iTunes, and DVD videos (66% unit increase).

In 2005, the RIAA added new format categories of mobile, digital subscription, digital music video, and kiosks. "Counting all formats and all distribution channels (retail and special markets distribution), overall unit shipments of physical product decreased by 8.0 percent in 2005" (RIAA, 2006). Mobile formats (such as ringtones) shipped 170 million units, representing $421.6 million in retail value. Full-track downloads were a not significant sector at that time.

In 2007, the world and U.S. market was thrust into what has come to be known as the Great Recession, brought on by several economic factors including the subprime mortgage crisis, which was attached to a real-estate bubble nationwide. These troubling times magnified household debt and consumers withheld discretionary income purchases for fear of harder times ahead. For 2008, there was a 14% drop to 428 million albums sold in the U.S., while sales of digital downloads increased 27% to just over a billion songs. CD sales were down almost 20% from the previous year (Sisario, 2009).

Music is a luxury good and during the distressing economic downturn from 2007 to 2012 sales plummeted by 32%. Year-end data for 2013 reflect a plateau in purchases, with music sales maintaining right at $7 billion in revenue (RIAA, 2013).

SoundScan Sales Trends and Configurations

Since 1991, the most substantial change of the use of SoundScan data in compiling the *Billboard* charts occurred December 13, 2014 with the modification to the Billboard Top 200 to include not only album sales, but the additional data of TEA (track equivalent album) and SEA (streaming equivalent album) information. Whereas the RIAA measures shipments in for their annual industry numbers, SoundScan measures over-the-counter sales. In an effort to monitor recorded music sales and determine trends and patterns, the industry in general, and SoundScan in particular, have come up with a way to measure digital album sales and compare them with music sales in previous years. When SoundScan first started tracking digital download sales, the unit of measurement for downloads was the single track, or in cases where the customer purchased the entire album, the unit of measurement was an album. But this did not give an accurate reflection of how music sales volume had changed, because most customers who download buy individual songs instead of complete albums.

In an attempt to more accurately compare previous years with the current sales trend, SoundScan came up with a unit of measurement called *track equivalent albums* (TEA), which means that 10 track downloads are counted as a single album. This evaluation is based on a financial equivalent, being a $.99 download x 10 would equal a $9.99 album download or CD. Thus, the total of all the downloaded singles is divided by ten and the resulting figure is added to album downloads and physical album units to give a project picture of "total" activity reported in *Billboard*'s top 200 chart.

The value of consumption does not stop at the download. Streaming songs generate licenses and royalties from the various sites and add revenue to the bottom line of copyright holders. The *streaming equivalent album* (SEA) was introduced in 2013 to measure streaming consumption. To evaluate streaming equivalents, the current industry standard is 1,500 streams of any songs from a particular album are counted as a single album; 1,500 streams x the standard royalty generated by this airplay $.005 = $7.50, being the wholesale price of an album. All of the major on-demand audio subscription services are considered, including Spotify, Beats Music, Google Play, and Xbox Music, as well as YouTube/VEVO. Combine the streaming activity of all the singles from a particular album and divide by 1,500 and the resulting figure is added to the album downloads and physical album units to give a project picture of "total" activity, known as the top 200.

Here is an example of how this works:

Table 7.2

Rank	Artist	Title	TW Total Activity	TW Album Sales	TW Song Sales	TW Stream Activity
1	Taylor Swift	1989	154,898	110,977	439,219	0
2	Ed Sheeran	X	75,558	34,432	334,343	11,539,403
3	Nicki Manaj	Pinkprint	59,573	33,723	177,404	12,165,982
4	Sam Smith	In The Lonely Hour	56,730	29,722	199,830	10,538,792
5	Rae Sremmurd	Sremmlife	49,037	34,403	118,066	4,243,493

Source: SoundScan Top 200 January 11, 2015

Nicki Manaj: 33,723 albums
17,740 TEA = (177,404 / 10 Songs)
8,110 SEA = (12,165,982 / 1500 Streams)
= 59,573 Total Activity

Even though Rae Sremmurd *sold* more albums (at 34,403 units) this week than Nicki Manaj, Nicki sold more songs and had triple the streaming activity, which caused her album project *Pinkprint* to outperform *StremmLife* on the top 200 for this week. You may notice that Taylor Swift's *1989* does not reflect *any* streaming activity. Swift made a conscience decision to not allow her songs to be included in any streaming outlets after the street date of the album—hence no streaming activity.

The industry as a whole has found the inclusion of these consumption models to be a better barometer of the marketplace.

"Music consumption in today's marketplace is a diverse mix of access and acquisition, including on-demand streaming, track and album

*downloading, and physical product purchasing. The introduction
of this expanded scope chart brings the Billboard 200 more closely
in line with the multi-platform, multi-format experience of music
fans."—Darren Stupak, executive vice president of U.S. Sales and
Distribution, Sony Music Entertainment, quoted in "Billboard 200
Makeover: Album Chart to Incorporate Streams and Tracks"*

(Billboard Staff, 2014)

Having established TEA and SEA as new units of measurement, industry trends show the following: U.S. album sales have continued to slide
every year from a high in 2000 of 785 million units to 476.5 million units
in 2014 (including TEA and SEA. Without the addition of the TEA/SEA,
album units in 2014 were 257 million). This represents a drop of 1.9%
from the 2013 overall album sales of 486.1 million. The big news of 2014
was that on-demand streaming was up 54% with 164 million streams. But
this staggering increase in consumption did not make up for the losses in
physical and digital sales. Additionally, the vinyl LP format increased its
sales by 52%, comprising 6% of all physical sales in 2014. This spike can
partially be attributed to a breakout release by indie-rocker Jack White with
his summer hit "Lazaretto." As noted in the graphic below, SEA reflects
streaming consumption, which appears to be replacing digital single
downloading at a greater pace than album sales. From 2013 to 2014, digital tracks sales are down by 12% where albums without TEA/SEA show a
slower decrease at 10%.

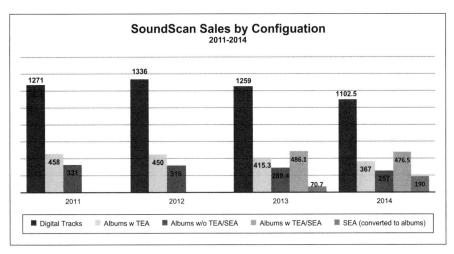

FIGURE 7.2 *U.S. Sales in millions*

Source: SoundScan™.

What has been remarkable is the fast pace in which the U.S. consumer has adopted this new technology known as streaming. In less than a decade, the industry has realized a majority of its revenue from digital music sources, based on shipment data from the RIAA. The year 2011 became the first to generate more revenue from digital sources than physical sales. And by 2012, digital music revenue exceeded $4 billion for the first time, with $1 billion of this coming from streaming alone. By 2013, $1.438 billion dollars was distributed as monies generated by digital sources. Considering that music sales are $6.99 billion industry, digital sales make up approximately 20.6% of all music sold.

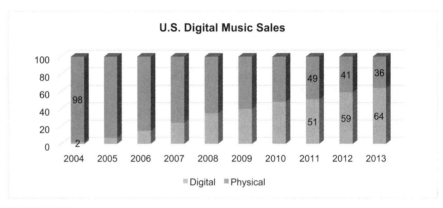

FIGURE 7.3   *U.S. Digital Music Sales*

Source: RIAA

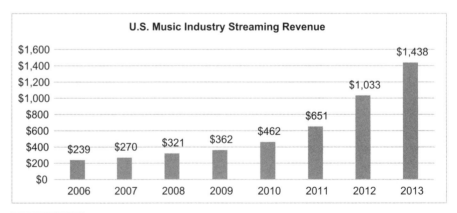

FIGURE 7.4   *U.S. Industry Streaming Revenue*

Source: RIAA

Annual Sales Trends

Music sales are seasonal, with the majority of sales occurring during the fourth quarter holiday season. For this reason, many of the superstars wait until the fourth quarter to release a new album, so they can take advantage of the holiday shopping season. And conversely, many newer artists will avoid releasing an album in this same time period because competition for retail space is more intense. Valentine's Day is the second largest record purchasing holiday (if Thanksgiving weekend is considered a part of the Christmas holiday season). Note that digital track sales caught up to album sales right after the Christmas holiday as music fans cashed in their iTunes gift cards and purchased music online. Notice the very different purchasing practices of 2014, where buyers are cherry-picking singles and the level of both album and single purchasing has dropped dramatically: from a steady 7+ million album units per week in 2008 to somewhere closer to 3 million in 2014 and an average 20 million single tracks per week.

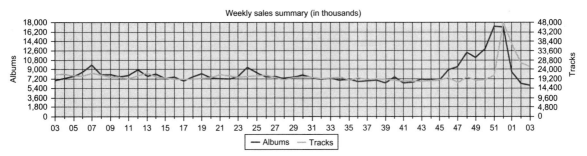

FIGURE 7.5   *Weekly Sales 2008*

Source: SoundScan™

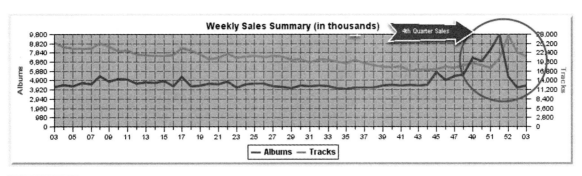

FIGURE 7.6   *Weekly Sales 2014*

Source: SoundScan™

GENRE TRENDS

The category of rock music has dominated the marketplace for many years, with a double-digit lead of about 33% of the market share for nearly a decade. With a lull in popularity in the mid-2000s, the genre began gaining momentum with the help from the popular show "American Idol," which began to feature more rock artists among the top contestants in 2007–2008, with rocker David Cook winning the 2008 contest. A driving force in the genre would be Eminem who topped the year-end best-selling charts in both 2010 and 2013 and continues to be an influential voice within the rock community.

The concept of "bro-country" made itself felt as early as 2012 with the male dominance on both the sales charts and country radio airwaves. Artists such as Eric Church, Florida Georgia Line, and Blake Shelton have all broken through as household names—especially Shelton who has been one of the judges on television's popular "The Voice" contest.

In 2011, 2012, and 2013, Drake's name has been near the top of the best-selling list and has helped to keep the R&B/Hip Hop genre buoyed. Not one artist is a singular standout, but each of these acts had a strong year in 2013: Jay Z, Kanye West, Lil' Wayne, and Lamar Kendrick, which gave the genre a spike in sales for the year.

The take-away for each genre should be how each consumer group is engaging the music. To look at the market share of each music type, check out how album sales alone versus the additional punch of TEA and SEA added to the bottom line can help or hinder the genre position. EDM, Latin, pop, and R&B/hip hop all enjoy an active consumer who engages the music not just through purchases, but also through streaming, which increases the overall share of consumption. Whereas christian/gospel, classical, country, holiday, jazz, and rock consumers are not as active online, causing their market share to diminish by as much as four percentage points in the rock category, which is significant.

The RIAA measures age and gender for their consumer profile studies. Current data is available only from 2012–2014, with no data collected from 2009–2011.

Age

Look at the total Internet population that is represented by people over the age of 13 years. When identifying strong performing groups, look to

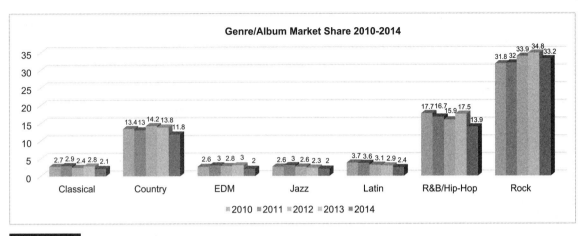

FIGURE 7.7 *Top Genres: Market Share*
Source: Nielsen Year End Reports

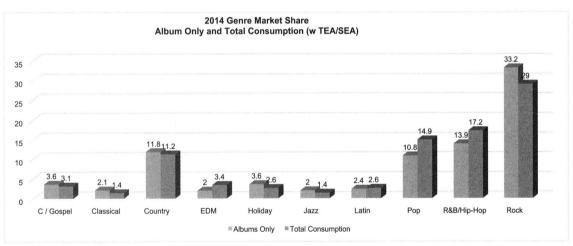

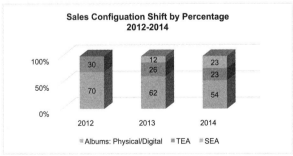

FIGURE 7.8 *Top Genre Market Share: Album Only v. Total Consumption*
Source: Nielsen Year End Reports

those numbers that outperform the average Internet population percentage. As you see, active buyers look to be between the ages of 26–50, with CD buyers aging older, which shouldn't be a surprise since this is part of their culture and habit. Digital buyers skew younger with streamers shifting to paid subscriptions more actively as the technology becomes more familiar. Not a shock, peer-to-peer downloading is dominated by the younger crowd, ages 13–25, but the practice is not forsaken by those as old as 35 years of age.

Table 7.3		Market Share by Age Group							
Year	Age	Total Internet Pop 13+	Music Buyers	CD Buyers	Digital Buyers	P2P DL	Streamers	Paid Sub	Locker
2011	13–17	11	9	9	11	21	13	7	
2012	13–17	10	10	10	13	21	14	18	
2013	13–17	10	9	8	12	17	14	5	19
2011	18–25	15	15	14	19	30	20	17	
2012	18–25	15	15	14	20	28	19	13	
2013	18–25	15	17	15	22	30	19	18	33
2011	26–35	18	20	18	24	25	23	34	
2012	26–35	18	20	18	26	26	23	29	
2013	26–35	18	18	17	21	27	21	31	23
2011	36–50	27	31	31	31	16	26	32	
2012	36–50	27	29	29	28	19	25	29	
2013	36–50	26	27	26	26	17	24	34	18
2011	51+	29	26	28	14	9	18	11	
2012	51+	30	26	30	13	7	19	11	
2013	51+	30	28	34	18	9	22	12	7

Source: The NPD Group/2013 Annual Music Study

Gender

It's difficult to correlate sales of music with the popularity of certain genres, the adoption of music consumption technology and gender and age. The population of potential music buyers is split nearly equally, with women edging out men, 51% to 49%. But analysis of music sales reveal that women out purchase men by nearly 10 percentage points year-over-year. However, as the technological gateway to the music advances, men out-engage women, specifically peer-to-peer exchanges, paid subscriptions, and locker downloading and stream ripping such as Mediafire or Rapidshare.

Table 7.4	Market Share by Age Group								
Year		**Total Internet Pop 13+**	**Music Buyers**	**CD Buyers**	**Digital Buyers**	**P2P DL**	**Streamers**	**Paid Sub**	**Locker**
2011	Men	49	46	46	49	53	46	58	
	Women	51	54	54	51	47	54	42	
2012	Men	49	45	46	45	54	46	56	
	Women	51	55	54	55	46	54	44	
2013	Men	49	46	47	47	50	49	61	65
	Women	51	54	53	53	50	51	39	35

Source: The NPD Group/2013 Annual Music Study

MARKET SHARE OF THE MAJOR

Label market share can be measured in a myriad of ways. The term "overall" business includes all format sales, but to see how well a music company is doing with new releases, one can isolate catalog sales from currents (current releases) to reveal the "state" of the competition's business. Universal has dominated the market since acquiring PolyGram Records in 1998. It is impressive that Universal has managed to grow additional market share since this acquisition and they have continued on that trajectory since the turn of the century. When reading the market share chart, note that data from EMI stops after 2011. EMI was purchased by Terra Firma, a private equity firm, and dismantled with the label imprints sold to the Universal Music Group and the EMI Publishing Group going to Sony Tree Publishing. Imagine the powerhouse labels of Universal's Interscope (Eminem, Black Eyed Peas, Philip Phillips), Def Jam (Kanye, Justin Bieber, The Roots), and Island Records (Bon Jovi, Fall Out Boy) merging with EMI imprint Capitol

Records (Katy Perry, 5 Seconds of Summer, Sam Smith), among many more. No wonder the company enjoys near 40% market share.

At one time, there were six major music conglomerates that controlled the U.S. music distribution landscape. The 2012 EMI sale was preceded by the 2004 merger of Sony and BMG, which created the second largest music company in the industry, behind Universal. Before the merger, BMG held a strong market position in the late 1990s and early 2000s due in part to their partnership with Jive Records who, at that time, was responsible for the teen hit sensations 'N Sync, Britney Spears, and Backstreet Boys. Sony enjoyed success in the late 1990s with Celine Dion, Mariah Carey, and other pop artists but saw their market share slip when the pop movement subsided. Sony now wholly owns BMG and has created efficiencies by shedding expenses and eliminating redundant departments.

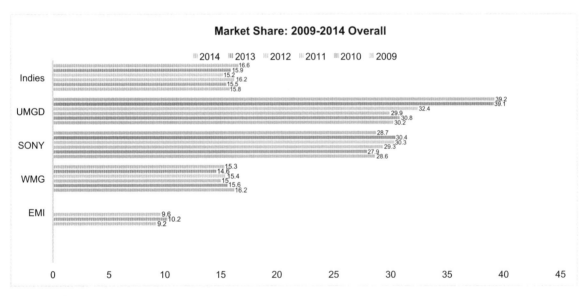

FIGURE 7.9 *Market Share of Majors: Overall Business*

Source: SoundScan

OUTLETS

It's clear that digital consumption online is impacting where consumers purchase their music. In a year's time between 2013 and 2014, the "non-traditional" retailer gained 5% in sales in a year that lost 2% overall. "Non-traditional" retailers include Internet portals, venues, direct-to-consumer, and other non-traditional outlets. The next big winner would

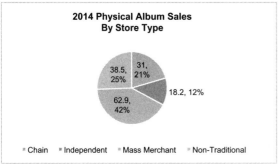

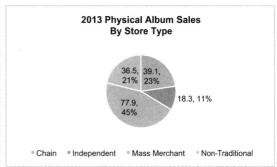

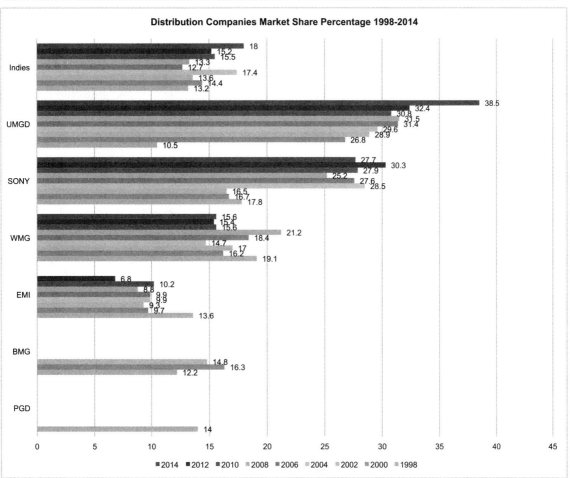

FIGURE 7.10 *Physical Album Sales by Store Type*

Source: 2014 Nielsen Music U.S. Report

be the independent store outlet that gained a 1% market share. A featured event such as Record Store Day, celebrates the indie store spirit and is traditionally held on the third Saturday of April of each year. Independent stores are working hard to feature "live" events and cater to specific clientele while offering the vinyl format to specialty shopper. Music retail chains are finding it harder to compete with the current financial model in place and mass merchants are dedicating less and less floor space to music specifically. These dynamics make for a perfect digital shift in consumption since it's harder to find physical product in the marketplace.

Comparison of All and Current Albums

In the following charts, a comparison is made for market share of all albums and market share of new releases for each label. An increase or decrease in current album market share is one indicator of the company's health, but a comparison of all albums versus current and catalog albums indicates how much a label relies on catalog sales compared to sales from its new releases.

Of the major labels, only UMG has a significantly larger market share for current releases than for all album releases, although new independent releases have also faired well during 2014. This larger piece of the pie indicates an entity's power to release and sell new hit records in the current climate. When one slice grows bigger, another slice must shrink. In 2014 Warner's current releases suffered but their healthy catalog helped to maintain a 15% market share. When reading this chart, the "current" and "catalog" numbers combine to create the "all" share.

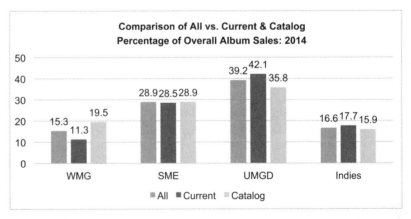

FIGURE 7.11   *Comparison of All, Current, and Catalog Album Sale by Label Market Share, 2014*

Source: SoundScan

Importantly, as market shares vary from hit to hit and year to year, remember the importance of the independent label and their relationship with the music group. Each of these music companies have "independent distributors" as part of their business models. Sony Distribution distributes records for truly independent labels through their Sony Red distribution arm. Universal's indie distributor is Caroline Distribution and Warner's is ADA. If the labels that were distributed through these major distribution arms were re-categorized with their percentage of business moved to the "Independent" category the market shares would look very different. As generated by Billboard in the summer of 2014, the value of the independents was measured by doing just this: look at the graphic below. Independents generate over 35% market share if isolated by themselves. This revenue is recognized by the major distributor, but the independent label is generating the money.

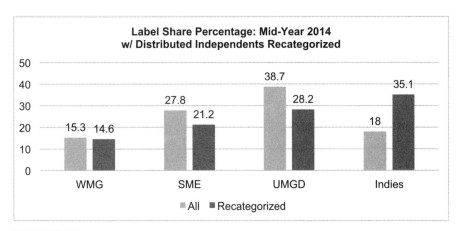

FIGURE 7.12 *2014 Label Share Percentage with Distributed Independents Recategorized*

Source: SoundScan/Billboard

CATALOG SALES

Catalog sales are defined as sales of records that have been in the marketplace for over 18 months. Current catalog titles are those over 18 months old but less than 36 months old. Deep catalog albums are those over 36 months since the release date.

When the compact disc was first introduced in the early 1980s, it fueled the sales of catalog albums as consumers replaced their old vinyl and cassette collections with CDs. This windfall allowed labels to enjoy huge profits and led to the industry expansion of the 1980s and early 1990s. However, catalog sales started to diminish in the mid-1990s as consumers finished replacing their collections. The closure of traditional retail stores also contributed to the decline in catalog sales, with customers having fewer opportunities to be exposed to the older titles.

Music companies have always looked to maximize their catalog's potential by utilizing various pricing strategies. Throughout the years, labels sought new ways to promote catalog sales through reissues, compilations, and looking at new formats. During the mid-2000s, catalog sales declined overall, but the introduction of digital downloads saw a spurt in catalog sales as consumers sought to fill in their collections with catalog albums and singles that had not been available for some time. Since then, current product dominated the sales trends, but catalog sales in both physical and digital formats have seen a resurgence recently and sales of currents and catalog have become evenly spread in 2014.

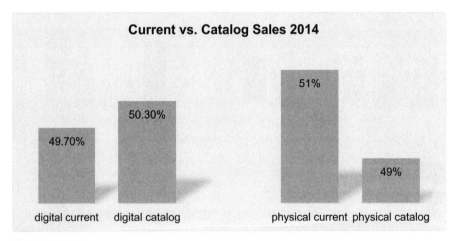

FIGURE 7.13 *Current vs. Catalog Sales for Digital and Physical Product 2014*

Source: 2014 Nielsen Music U.S. Report

As a result of this resurgence, the sales share of catalog albums in 2014 rose to account for 49% of album sales, compared to the 35.8% the category comprised in 2004.

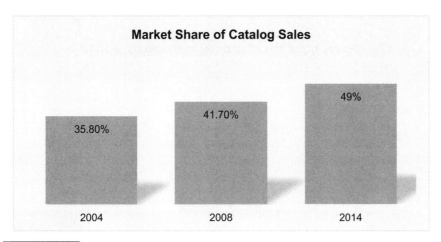

Market Share of Catalog Sales

35.80%   2004

41.70%   2008

49%   2014

FIGURE 7.14 *Market Share of Catalog Sales*

Source: 2014 Nielsen Music U.S. Report

INDUSTRY CONCENTRATION

During the 1990s and early 2000s, the industry relied on just a few massive-selling hits to drive the industry, rather than spreading the wealth around with a plethora of profitable releases. Of the albums released in 2004, only 100 titles made up nearly 50% of the 265 million units sold. Over 70% of the 265 million albums sold in 2004 came from less than 1% of the releases. But that day is no more.

Fast forward 10 years and, with the help of the Internet, crowdfunding, and the taste buds of individual consumers, the market has fractionalized and the "hit" album no longer dominates. By mid-October 2014, not one album had cracked the platinum mark and only 60 singles had sold over 1 million copies. In a scramble to the year-end finish line, only four albums made it to 1 million units sold: Pentatonix *That's Christmas To Me*, Sam Smith's *In The Lonely Hour*, Disney's soundtrack *Frozen*, and Taylor Swift's *1989*.

Why buy when you can stream? The proof of that statement continues to validate itself in the numbers. Starting in the middle of the decade, the top 200 albums began to account for a shrinking share of all albums sold. In 2009, Ed Christman of Billboard magazine wrote: "Hit album releases still account for a large but shrinking share of total sales. The 200 best-selling titles of the year have seen their share of annual sales fall from 40.1% in 2004 to 35% in 2008" (Christman, 2009). Looking at the best-selling top 200 titles in 2014, this number represents closer to 25% of total sales.

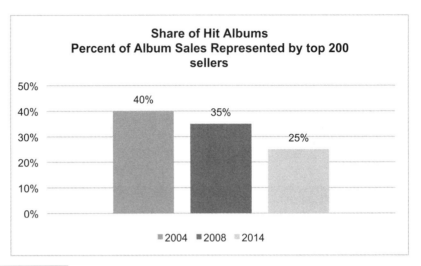

FIGURE 7.15   *Percent of Album Sales Represented by the Top 200 Titles*

Count the stark reality and how times have changed. The number one album in 2000 ('N Sync's *No Strings Attached*) sold 9.9 million copies that year; by 2004, the top selling album (Usher's *Confessions*) sold 7.9 million units. By 2007, the number one album (Josh Grobin's *Noel*) only sold 3.7 million copies that year. In 2008, the top-selling album (Lil Wayne's *Tha Carter III*) sold less than 2.9 million copies. And in 2014, the biggest

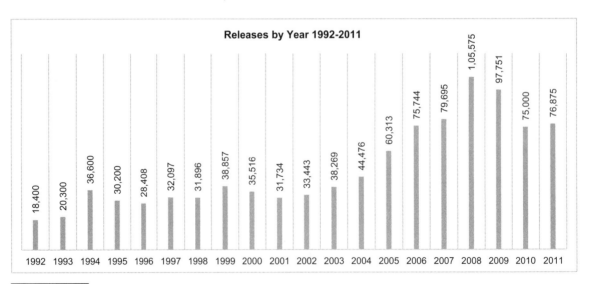

FIGURE 7.16

Source: Compiled from Various Sources, Including Nielson End of Year Reports

title that was released eight weeks prior to the end of the year sold 3.6 mil-lion units—Taylor Swift's *1989*. Meanwhile, the number of titles released each year has varied over the years, jumping from 33,433 in 2002 to an all-time high in 2008 with over 105,000 titles. Most recent data shows that fewer records are being traditionally released, with the 2011 revealing 76,875 titles in the marketplace.

CONCLUSION

This chapter presented tools and measures used to evaluate the health of the labels and the industry. We used these assessments to illuminate some trends and shifts within the U.S. recording industry. By looking at trends in music genres, the demographics of buyers, configurations, label market share, the proportion of catalog to current releases, and the pro-portion of blockbuster hits to the total number of releases, one can make inferences about the current state of the industry. Armed with an under-standing of these tools and measures the reader will be able to draw their own conclusions in the future as new data becomes available.

GLOSSARY

BMG—Bertelsmann Music Group. Used to by one of the U.S. major labels, merged with Sony and now has publishing entities in the U.S. and global presence worldwide.

Catalog—Older album releases that still have some sales potential. Recent catalog titles are those released for over 18 months but less than 36 months. Deep catalog: those titles over 36 months.

Dollar value—The monetary worth of a stated quantity of shipped product multi-plied by the manufacturers suggested retail price of a single unit. The value of shipments is given in U.S. dollars.

EMI—Electrical and Mechanical Industries. One of the major music conglomerates. Also known as EMG—EMI Music Group.

Market share—A brand's share of the total sales of all products within the product category in which the brand competes. Market share is determined by dividing a brand's sales volume by the total category sales volume.

P2P (Peer-to-peer)—Electronic file swapping systems that allow users to share files, computing capabilities, networks, bandwidth and storage.

Product configuration—Any variety of "delivery system" on which prerecorded music is stored. Various music storage/delivery mediums include the full-length CD album, CD single, cassette album or single, vinyl album or single, DVD audio, DVD, mp3 or streaming audio and video.

Replacement cycle—Consumers replacing obsolete collections of vinyl records and cassettes with a newer compact disc format.

Streaming Equivalent Album (SEA)—An industry standard used to derive an album equivalent; 1,500 streams equals one album count.

SMG—Sony Music Group. One of the major music conglomerates.

Track Equivalent Albums (TEA)—Ten track downloads are counted as a single album. All the downloaded singles are divided by ten and the resulting figure is added to album downloads and physical album units to give a total picture of "album" sales.

UMG—Universal Music Group. One of the major music conglomerates.

Units shipped—The quantity of product delivered by a recording manufacturer to retailers, record clubs, and direct and special markets, minus any returns for credit on unsold product.

WEA—The distribution arm of WMG. (Stands for Warner, Elektra, Atlantic)

WMG—Warner Music Group. One of the major music conglomerates.

REFERENCES

Billboard Staff. "Billboard 200 Makeover: Album Chart to Incorporate Streams & Track Sales." *Billboard*. N.p., November 19, 2014. Accessed January 26, 2015.

Christman, E. "Running the Numbers." *Billboard.biz*. January 17, 2009. http://www.billboard.biz/bbbiz/search/article_display.jsp?vnu_content_id=1003928717.

Christman, E. "Average Sale of Albums Dropped in '02 as Labels Released More, Sold Less," Billboard. April 26, 2003.

Christman, E. "SoundScan Numbers Show .35% of Albums Account for More Than Half of All Units Sold," Billboard. April 28, 2001.

Hiestand, J. "Music sales off in '03, but decline slows," Hollywood Reporter. January 02, 2004. http://www.hollywoodreporter.com/thr/article_display.jsp?vnu_content_id=2059949.

Nielsen and *Billboard*'s 2013 U.S. Music Report. http://www.nielsen.com/us/en/insights/reports/2014/u-s-music-industry-year-end-review-2013.html

NPD Group. Annual Music Study. 2013.

Phillips, C. "Record Label Chorus: High Risk, Low Margin," Los Angeles Times. May 31, 2001.

RIAA. "1996 Yearend Marketing Report on US Recording Shipments," RIAA press release. 1997.

RIAA. "1997 Yearend Marketing Report on US Recording Shipments," RIAA press release. 1998.

RIAA. "Recording Industry Releases 1999 Yearend Marketing Report," RIAA press release. 2000.

RIAA. Year-End Industry Shipment and Revenue Statistics. 2013.

Sisario, B. Music Sales Fell in 2008, but Climbed on the Web. *The New York Times*. January 1, 2009. http://www.nytimes.com/2009/01/01/arts/music/01indu.html?_r=3&scp=2&sq=music%20online&st=cse.

2014 Nielsen Music, U.S. Report. http://www.nielsen.com/content/dam/corporate/us/en/public%20factsheets/Soundscan/nielsen-2014-year-end-music-report-us.pdf

CHAPTER 8

Label Operations

In an era when every level of the food chain within the entertainment industry is being scrutinized as to its value, record labels are being squeezed from both sides of the equation. Record sales are diminishing and retailers are looking for more profit in the product. And as technology advances, artists see an opportunity to completely circumvent the "label deal" and sell directly to their fans. The challenge for labels is to create value for both sides of the equation: create various "products" that draw consumers into retail stores on a consistent basis AND create a loyal fan base to sell product to—which is not as easy as throwing a site onto the world wide web and expect for fans to come! It takes the creativity and business acumen of a team that is savvy to today's technology, but understands long-standing business tactics that can endure strong competition and challenging economic situations to win in today's music business.

Every record label is uniquely structured to perform at its best. Often times, the genre of music along with the "talent pool" of actual label personnel dictate the organization and inner-workings of the company. As talent is signed to a label, the "artist" will come in contact with nearly every department in the process of creating a music product for the marketplace.

A typical record label has many departments with very specific duties. Depending on the size of the label, some of these departments may be combined, or even out-sourced, meaning that the task that the department fulfills is hired out to someone not on staff of the label. But the end result is to be the same—create a viable music product for the marketplace. In the structure below, there usually is a general manager/sr. VP of marketing who coordinates the all the marketing efforts.

GETTING STARTED AS AN ARTIST

As talent is being "found" or developed, the first contact with a record label is usually the A&R Department. The Artist & Repertoire Department is always on the hunt for new talent, as well as songs for developing and existing artists to record. But before the formal A&R process occurs, the talent has to be signed to the label.

CONTRACT

A contract may take months to negotiate. Depending on the agreement and nature of the deal, the artist/manager/lawyer may sign a contract a year or more prior to street date of the first release.

ACCOUNTING

Once an agreement is reached, the accounting department will distribute "advances" to the artist, as outlined in the contract.

BUSINESS AFFAIRS

The business affairs department is where the lawyers of a label reside. Record company lawyers are to negotiate in the label's best interest. Most often, new talent will have a manager and lawyer working on their behalf, with the contract in the middle. Clearly, the label wants to protect itself and hopes to reduce risk by maximizing the contract.

Besides being the point person for artist contracts, label lawyers negotiate and execute many other types of agreements including:

- License of recordings and samples to third parties
- Negotiate for the right to use specific album art
- Point person when an artist asks for an accounting or audit of royalties
- Renegotiates artist contracts
- Contractual disputes such as delivery issues
- Conflicts with contract such as guest on another
- Vendor contracts and relations

Oftentimes, the accounting department falls within business affairs, since the two are related regarding contractual agreements and financial obligations. The accounting department is the economic force driving all the activity within the record company. It takes money to make money, and the accounting department calculates the budgets for each department as it aligns with the forecast of releases. Most record label accounting departments have sophisticated forecasting models that calculate profitability. Each release is analyzed to determine the value of its contribution to the overhead of the company. This analysis is examined in the profit & loss statement, which acts as a predictor equation as to the break even of a release and its future value over time. (P&Ls are discussed at great length in chapter 9.)

Additionally, accounting department acts as an accounts payable/receivable clearinghouse, managing the day-to-day business of the company such as payroll, leasing of the building, and keeping the lights on.

> ### A&R
> The A&R process, from securing the talent to finding the musical content can take months. Ideally, masters should be delivered four months prior to street date.

ARTISTS AND REPERTOIRE: HOW LABELS PICK AND DEVELOP ARTISTS

The truth is: there is no next model. Show me 1,000 talented musicians, each with a unique style and personality, and I'll show you 1,000 ways to make a career in music . . . there is no longer an off-the-shelf solution.

—Amanda Palmer

Being the A&R person is the job most people want at a record label. They see the glamour and the glory of being the person that discovered the next big act that changes the music scene, hanging out in the studio or backstage with big name acts and getting their picture in *Billboard* and *Rolling Stone*. What they don't see is the countless hours and late nights listening to bands that aren't that good and never get signed; hours away from the family and the high turnover rate because you are only as good as the last act you signed. Or the repertoire part of the job—helping the artists find just the right song for their next album. In a smaller music company or record label, the A&R person may also be the person in charge of creative and marketing (Passman, 2012). As the title indicates, the position has two different tasks: signing new artists and helping all the label's artists select the best songs to advance their careers.

Discovering Artists: The A in A&R

Arguably, the single most important talent of a good A&R person is to identify and sign successful artists. Their bosses are likely to overlook a project going over budget if it is a big hit, while bringing the recording in on budget will not advance your career if the record does not sell. A good A&R person is such a valuable asset that they are often considered in the valuation of a record label in a merger or acquisition. While other departments in the combined companies will probably be merged and downsized, successful A&R persons will be retained from both labels (Krasilovsky and Shemel, 2007)—as long as they continue to be successful.

One of the pluses of the A&R job is that every day is different. The primary tasks of A&R are to discover and sign new artists to the label and help all the artists find good songs. Different labels put more or less emphasis on different resources for discovering potential new artists. Although in recent years some artists have been discovered on the Internet and gone on to great success, this is only one of the many sources used by A&R to find new talent.

Producers as A&R Men

Independent and label staff producers are one way that artists land recording contracts. Producers are in a unique position to discover and promote an artist by using their talents as an instrumentalists or background vocalists on another artist's recordings, thus gaining exposure for the unsigned talent while developing a professional relationship. One of the most famous examples of this is John Hammond who, as a talent scout and producer for Columbia Records, discovered and produced Bob Dylan, Aretha Franklin, Pete Seeger, and many others (Yano, 2014).

Purchasing an already produced master reduces the label's risk. Many artists will work in conjunction with producers to create a recording, with the idea that the producer will market the master, the master will be purchased by a label, and that the artists will secure a recording contract. Again, the label knows what its getting since it has the final product, the master, in hand.

An example of this is Matchbox Twenty, who was produced by Matt Scerlatic prior to landing a record deal. As the story goes, Scerlatic felt that the band had the talent and took a "flier" on producing the act. He "shopped" the deal, landing a record contract with Lava, an imprint of Atlantic, and the rest is history. With multiplatinum selling records and top-of-chart hits, Matchbox Twenty continues to make relevant music nearly 20 years after their 1996 initial release.

Attorneys and A&R

An established entertainment attorney with connections in the industry may be willing to take on an unsigned artist in a "shopping deal." For a fee, ranging from an hourly charge plus expenses to a percentage of the artist's advance when a deal is signed, the attorney will pitch the artist to their connections in the industry. The artist is paying the attorney for their industry contacts. Before signing such a deal, the artist needs to do their research and determine the attorney's stature in the industry. Have they successfully made such deals in the past? Do they represent a number of artists and are they well respected in the industry? To the point, do they really have the contacts they claim and can they do the job you are hiring them to do?

The attorney has to believe in the artist as well because their reputation is at stake, and that is no small risk. Like any person doing a sales job (like a song plugger pitching songs to an artist or producer) their product is their credibility, so if the attorney is not passionate about you and your music you probably should keep looking. Another major deal point is the duration of the shopping arrangement. Assuming you have entered into an exclusive deal with the attorney then you need to negotiate an end date so that if he is not successful you can move on.

Publishers and A&R

Perhaps the A&R department's strongest allies are the publishers who sign and promote their singer-songwriters to the labels in hopes that a record deal will result. Publishing companies want their writers to have recording deals—it is a guaranteed outlet for getting their catalog recorded. It makes sense to "train up" a singer-songwriter that writes his or her own hit songs in the hope that their songs will run up the charts and generate both mechanical and performance rights revenues for the publishing company. So publishers have their own A&R people and sign artist-writers that they believe have the potential to be recording artists. Examples of successful artists that were developed through publishing companies include Florida Georgia Line (Big Loud Mountain Publishing) and Taylor Swift (Sony/ATV), who was signed as a songwriter at age 14.

The publisher will finance the recording of demos and pitch them to record labels. These development deals usually give the publisher six to eighteen months to secure a recording deal for the artist-writer (Brabec and Brabec, 2011).

A recent big "hitter" that falls in this category would be country rockers Florida Georgia Line, who started as writers for publisher, Big Loud Mountain Publishing and Management companies. As part of the deal, the duo were produced "in house" by Joey Moi, whose recording credits include Nickelback and Hinder, who assisted in giving the music its edge and party component that got this band noticed at a time when "bro-country" couldn't have been bigger. These songwriters-turned-artists have not slowed down since they've landed on the charts.

American Idol, *The Voice*, and Other Reality TV

American Idol, launched in 2002, has been an absolute television phenomenon placing no less than 64 finalists on the *Billboard* charts 345 times in the first 10 years (Bronson, 2012). Although they are often criticized for their iron clad, all-encompassing lifetime contracts, would-be contestants wait in line for hours for an opportunity to audition. By the time a contestant reaches the finalist stage they have hours of television exposure, thousands of fans and probably a single playing on radio. It is an A&R department's dream. Winners receive a recording contract and other finalists may be picked up, at the label's option, for a development deal or a full contract. Successful alums of "American Idol" include Kellie Clarkson, Carrie Underwood, Daughtry, Kimberly Locke, Adam Lambert, and Clay Aiken, among many others. Even some non-finalists, like William Hung, have managed post-show success (Bronson, 2012).

No other reality competition show has come close to matching the A&R success of "American Idol." The most successful contestant thus far from "America's Got Talent" is classical singer Jackie Evancho, who finished second in 2010, the show's fifth season. She was 10 years old at the time. "The Voice," which first aired in April 2011, is yet to produce the same level of post-show success for any of its contestants. "The X-Factor" has also failed to produce a hit single as of yet but several contestants have released albums that charted in the top 40.

Part of the success of "American Idol" is probably because it was the first show of this type (at least in this century), but more likely it is the fact that it provides a more complete development package for its contestants. Even while they are still competing, contestants will have singles sold on iTunes and played on radio. After the season is over finalists will go on tour, performing both as a group and solo, on as many as forty dates during the summer and mostly in smaller markets where fewer shows may be available.

Fitting In

If your band and your music are way out there, bleeding edge kind of stuff, then you should not expect your buddy who just got a job at a major record label to be able to get you a deal. If you and your music are a risk then only somebody at the label with an established record of success is likely to persuade the higher-ups that you are worth a chance. And remember, the A&R person is taking a big risk even promoting you to their bosses. Too many misses and they may be coming back to you looking for a job as part of your road crew! Regardless of where they are on the scale of musical tastes, artists have to capture the attention of A&R through their music, personality, and work ethic. Labels have limited budgets and they will invest in the artists that they feel are most likely to lead to a commercially successful recording. In the glory days of the business, labels might have signed an artist or two every month. In these more conservative times, the label probably won't sign more than two or three artists in a year.

Labels also must consider how each additional artist signed impacts the existing roster of artists. If the label already has several solo female acts in the early stages of their career they are not likely to sign another. The label can only support a limited number of artists before they begin to cannibalize their own sales. There have been cases where an artist had to wait years before a label decided it was the "right time" to sign them (Catino, 2014). Once signed, most contracts don't guarantee the label will ever actually release an album. Sadly, some artists have been signed and kept on label rosters for years, then cut without ever releasing an album.

Repertoire: A&R After the Signing

The ultimate job of the A&R department is to acquire masters (recorded songs) for the label to market. To do so, labels obtain masters in several ways. As in any business, labels need to manage risk. Repackaging and remastering previous recordings is the safest way to produce a master. By creating compilations of known artists with successful sales histories, a master is produced. How it is marketed eventually will determine its success, but the label knows what it has from the start. A successful example of repackaging would be the Beatles #1 reissue (still the number one selling artist on the EMI/Universal roster).

A&R is not a traditional marketing function but, depending on the level of involvement and input after the signing, A&R has a lot of impact on the marketer's job. The A&R person that brought the act to the label may heavily influence song choices and branding.

A&R's primary objective in this development process is to find the best possible songs for the artist—to develop their repertoire. Songs, like artists, may be discovered anywhere, but the most common sources are the artist themselves or a publisher. Songs have also been discovered from other artists (covers), producers, and unsigned writers. Sometimes that means that the artist-writer, the one with the record contract, has to be professional enough to accept that somebody else's song may be the better choice for their album than their own song.

The riskiest of masters is that of signing an unknown artist. Record labels must feel that the talent warrants a contract, with an ear in the creation of the master. Labels take many approaches to this process. But each project has similar determining qualities, such as which songs should be recorded, how many songs will be included, and who will produce the sessions. Along with the artist, the A&R department designates a producer or producers for a project. Trust is placed in the producer, who needs to be compatible with the artist and hold the same vision that the record label has conceived for the act. The success of a project can pivot on who produces. Some producers are very "hands on" with regards to the creative input, by sharing the cowriting role or playing on the master itself. These producers tend to lend credence to the project and assist its marketability. One example is Alicia Keyes and Ludicris debut releases, both produced by Kanye West, who became a notable producer on Jay-Z's albums, and who has become an artist in his own right.

Other artists may need less of a heavy-handed producer, but more like subtle guidance from both the A&R department, as well as producer. Songwriting artists have a unique quality since both the content, as well as the creation is open to direction. And these artists can have strong ideas as to their music and how they want it produced. But A&R representatives can take very active roles in the direction of the project, by aiding the songwriting process and nurturing the recording through the entire timeline. It all depends on what the label is looking for in the final product.

ARTIST DEVELOPMENT

Once a contract is signed, an artist usually assigned an AD representative, 12 months prior to street date.

CREATIVE SERVICES

As the A&R process evolves, a photo shoot is needed to represent the content of the recording, six months prior to street date.

ARTIST DEVELOPMENT/RELATIONS

Sometimes called the product development department, this department manages the artist through the maize of the record company and its needs from the artist. Madelyn Scarpulla describes them as, "the product manager in effect is your manager within the label" (Scarpulla, 2010). Artist development specialists hold the hand of the artist, helping them clarify their niche within the company. Artist development usually develops strong relationships with the artist and artist manager, with other departments in the company looking to artist development as a clearing house, helping to prioritize individual department needs with the artist.

Artist development not only manages the artist through the process such as the delivery of the recording, photo and video shoots, and promotional activities, but looks for additional marketing opportunities that maximize the unique attributes of the act. And in some labels, the AD can assist in the actual grooming of the talent—taking the act from lounge singer to star quality. But the role consists of being a "hub of the wheel" just as the rubber meets the road and the artist gains traction inside the record label company. AD is also responsible for the scheduling and logistics of all the players and keeping the calendar for the act while carving out their place within the company.

CREATIVE SERVICES

Depending on the company, the creative services department can wear many hats. Artist imaging begins with creative services assisting in the development of style and how that style is projected into the marketplace. Special care is taken to help the artist physically reflect their artistry. Image consultants are often hired to assist in the process. "Glam teams" are employed to polish the artist, especially for high profile events such as photo and video shoots, as well as personal appearances.

The creative services department often manages photo and video shoots, setting the arrangements and collaborating on design ideas and concepts with the artist. Once complete, images are selected to be the visual theme of the records and the design process begins. In some cases, creative services contain a full design team that is "in house" and a part of label personnel. Such "in house" teams can insure quality and consistency in imaging of the artist—that the album cover art is the image used on promotional flyers, sales book copy, and advertisements. When there isn't an in-house design crew, design of album cover art and support materials is farmed out to outside designers. Interestingly, the use of subcontractors can enhance unique design qualities beyond the scope of in-house designers, but there can be a lack of cohesive marketing tools if not managed properly.

PUBLICITY

Creating a press kit for advance awareness occurs shortly after photo shoot, five months prior to street date.

PROMOTION

Depending on the genre, the first radio single needs to released to garner airplay and create demand. Artist visits with radio can enhance airplay, three months prior to street date.

PUBLICITY

The priority of the publicity department is to secure media exposure for the artists that it represents. The publicity department is set into motion once an artist is signed. The biography of the artist via an interview is created. Other tools such as photos from the current photo shoot, articles and reviews, discography, awards and other credits, are collected into one folder, creating a press kit for each artist. These press kits are tools that are used by the publicity department to aid in securing exposure of their artists and are often sent to both trade and consumer outlets.

Pitching an artist to different media outlets can be a challenge. As an artist tours, the ideal scenario would be that the local paper would review the album and promote the show. Additional activities would be to obtain interviews with the artist in magazines and newspapers that can also be used as incremental content for web sites by these same entities. Booking television shows and other media outlets falls to the publicity department as well. On occasion, artists will hire their publicists to assist in creating higher profile events for the act. These publicists try to work with label publicists to enhance in-house efforts and build on relationships already established.

As artists become more established, many acts hire their own independent publicist to enhance their profile. This additional media punch is usually coordinated between label and indie publicist so that redundancy is avoided, as well as efficiency of best efforts among staff and hired guns.

RADIO PROMOTION

In most record companies, the number one agenda for the promotion department is to secure radio airplay. Although the Internet has created a new way for consumers to find new music, surveys continue to show that listeners still learn about new music and artists via the radio. Typically, radio promotion staffs divide up the country into regions, and each promotion representative is responsible for calling on specified radio stations in that region, based on format. Influencing the music director and radio programmer is key in securing a slot on the rotation list of songs played. These communications often take place on the phone, but routinely, radio promotion staffs visit stations, sometimes with the artist in tow, to help introduce new music and secure airplay.

With the consolidation of radio stations continuing, developing an influential relationship with individual stations is getting harder and harder. But there still exists a level of autonomy within each station to create its own playlist as it reflects their listenership. To strengthen the probability of radio airplay, promotion staffs conceive and execute radio-specific marketing activities such as contests, on-air interviews with artists, listener appreciation shows, and much more.

SALES

By visiting retail buyers, artists can assist in the set-up and sell-through of their record. Solicitation process begins two months prior to street date.

SALES AND MARKETING

The sales and marketing department sell product into retail, and create visibility of the product at the consumer level. If all the other departments have done their job well, selling the music to retail should be easy, since there will be a pent-up demand for the release. But to insure sales success, the sales and marketing department must create awareness with the gatekeepers.

Many sales departments look to their distributor as an extension of the label brand and are the first line of customers to educate regarding new releases. The distribution company needs to be well informed as to new releases and the marketing plan that goes with them so that they can represent the product to retailers, both physical and digital. To do so, sales departments educate their distributor by sharing with them detailed marketing plans for upcoming releases.

The second line of customers is retail. Sales departments continue the education process by informing retailers as to new releases and marketing plans. In tandem, record label sales reps, along with the distribution representative, will visit a specific retailer together, and on occasion, may bring an artist by to visit with the retail buyer. In this visit, the amount of music that is to be purchased by the retailer, along with any specific deal and discount information, is discussed. Co-op advertising is usually secured at this time as well, with pricing and positioning vehicles along with other in-store marketing efforts concluded.

This department also secures sales initiatives with "interactive" digital portals such as Amazon and iTunes plus exclusive offers through online engagement such as VEVO, YouTube, iHeartMedia, AOL, and MySpace— to name a few. These activities can be marketing awareness campaigns that include single give-aways, pre-sales, and contests that create excitement about new music from targeted artists.

PARTNERSHIPS AND BRANDING

This department wears a big hat that covers lots of territory. Partnerships can lead to strong leveraging positions that allow for accessing the deep pockets of corporate sponsors to help place an artist in front of millions of targeted consumers. The Rolling Stone tour sponsored by Citi Visa Card gave sizzle to the credit card by allowing special access to a "cool" artist while offsetting expenses for the tour. This sort of partnership leads to ongoing branding opportunities that reach worldwide where all participants win in the equation.

SOCIAL MEDIA

The social media department has their finger in every cookie jar! From receiving clearances via the contract to help with artist websites, to the creation of "widgets" and contextual banner advertising for Internet marketing, which is in coordination with the marketing and sales department, as well as the development of tools that may be used as content for online media through magazine and newspaper sites, the social media department can wear various hats. This department can also be responsible for aligning new business arrangements with key digital partners including telecom entities such as Verizon or AT&T, in creating additional exposure for new releases using downloads as the bait. Depending on the company, this department can be a "catch all" for all things "digital."

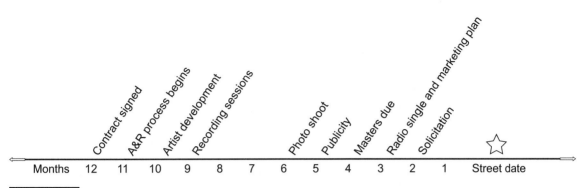

FIGURE 8.1 *Major Record Label Timeline*

INDEPENDENT LABELS

In an era when album sales continue to decline, the Internet has fragmented the market, and major record labels have tried to maintain their prominence as "creators of superstars," consumers have been looking elsewhere to satiate their burgeoning musical tastes—and they're eating from the independent labels' table. Historically, independent labels have been relatively small in sales stature and were genre- or regionally-specific, tending to react quickly to marketplace trends. Some independent labels strike deals of distribution with the major conglomerates where others find their way to consumers through independent distributors—but all are looking to sell records, either literally or virtually to an ever-fractionalizing marketplace. And this new age of consumership has brought with it an entirely new dimension for "indie" labels—one that can include the surprise of big artists with big sales along with even bigger opportunities.

Fueled By Ramen emerged as one of the most successful independent labels so far, with several of their initial acts going on to major label status such as Jimmy Eat World and Fall Out Boy. Youngster John Janick started his fledgling label from his college dorm room at the University of Florida. Since 1996, he has ridden the viral marketing wave by utilizing the Internet as his label's mouthpiece to cascade popularity of one band with another. His label has been attributed in creating the "360 Deal" where artists are signed to label, tour support, web promotion, and merchandising deals where Fueled By Ramen and its website acts as the portal for all revenue streams. Panic At the Disco, Gym Class Heroes, and Paramore have all risen to top 200 *Billboard* chart status via this model, with FBR continuing to evolutionize its marketing strategies with its popular YouTube Channel, as well as sales of merchandise through teen retailer Hot Topic and placement of recordings on the ever-popular video game, Rock Band (Levine, 2008). With what looks like only a dozen staffers, this organization has three folks dedicated to social media with offices in Tampa, Chicago, and New York City.

To look at their organizational chart, Big Machine Records in Nashville, TN is structured similarly to that of any major record label, including their size, which is 85 strong. But what makes them unique is that they are truly independent, meaning that they are not owned by a major conglomerate but act in their own interest and are in charge of their own destiny—and they seem to know where they're going. With virtually five imprints, Big Machine, Vallory, Republic Nashville, Dot, and Icon, each label seems to have their own "flavor" and artist emphasis with Taylor Swift being the flagship act the drives the entire entity. Taylor Swift was one of four acts in 2014 to sell over over 1 million records and her label controversially pulled her album along with all catalog from interactive streaming services such as Spotify a week after her *1989* streeted (the delay helped her leap over the 1 million mark for first week sales.) The label, distributed through UMGD, negotiated "free agent" status and can sell directly to consumers. Big Machine works on branding itself as well, with the hopes that music consumers will believe in the imprint when the label introduces new artists into the marketplace. As an example, the label leverages its artists by featuring them in a music company-wide campaign called "Outnumber Hunger" that partners with General Mills in feeding hungry Americans.

David Macias, President and founder of Thirty Tigers, states that his company is the "Home of the 60 Deal"—meaning: artists in the fold keep ownership of their masters and are under a record contract only. Thirty Tiger artists net over 70% of the proceeds of their recordings and are not expected to share profits from other revenue sources such as publishing, touring, or film as is outlined in many 360 deals. And this arrangement with the 75 artists that Thirty Tigers currently represent in a marketing and distribution deal are doing well. In an era where labels are struggling to find relevance in the food chain and keep the lights on, Thirty Tigers has increased their bottom line by 75% from 2008 to 2013. By looking for artists who know what they want and are "fully formed" artistically, the label goes to work to create a compelling narrative that will resonate with a specific market. Thirty Tigers looks for artists with a "sheen of authenticity" and finds consumers who attract to it.

Thirty Tigers

As an example, the company's P.R. and Artist Manager, Traci Thomas, wrote a story that was exclusive for *The New York Times* which shared the compelling story of Thirty Tiger's artist Jason Isbell and his coming out of rehab, pulling back from the edge—and authored a masterpiece that spoke to the readers. The timing of the story was two Sundays prior to Jason's latest release, giving many other media outlets a chance to interview him and his first week sales went through the roof. Jason Isbell's Southeastern release debuted top 10 on the *Billboard* top 200 and has sold almost 150,000 units to date, a huge feat for an independent artist. Starting in 2002 with two employees out of David's guest bedroom to running a multi-million dollar company with 22 employees, Thirty Tigers' business model of marketing, distribution, and artist management works for them because of their unique relationships with their artists, the understanding of outcomes desired by their acts, and their passion to reach a specific consumer in a shifting market that demands the creative approach that Thirty Tigers delivers (Macias, 2014).

So what is considered "success" in the music business today? Some indies have hit it big and compete with majors head on. But there are many smaller independent labels that are making a go at not just surviving, but creating a new model of business where incremental sales on a smaller level are enough to not only keep the lights on, but serve as a creative outlet for many more artists who would otherwise not have a voice to sing. If selling 50,000 records at $10 bucks a pop makes a half a million dollars, keeps an artist on the road, and his small indie label in business for another year—is this success? To the consumer looking for the coffee-house songster whose record may never be heard on mainstream radio—you bet.

APPENDIX: A&R ADVICE FOR THE UNSIGNED ARTIST

PERFORM, PERFORM, PERFORM. THEN PERFORM SOME MORE.

One thing is guaranteed in the music business: if you don't put yourself out there, if you never perform in public (live or recorded), you will never get discovered. You can tell everyone how good you are but until you show us, we will remain unconvinced. So, play at every opportunity. Play your songs; play somebody else's songs; perform the national anthem at the local high school game; but perform. Play for free until you don't have to. Just get yourself out there and create a buzz about how good you and your songs are. Work your social media and when the bars and clubs start filling up to see you then you no longer have to play for free. That is the time to shift gears and give away some music for free and get paid for your live shows.

GIVE AWAY FREE MUSIC

Giving away music free online gives the artist "social currency" and may attract more fans as friends of friends like their Facebook page or follow them on Twitter. It can also build a fan base that rewards the artist's generosity by buying concert tickets or merchandise. Sure, some people will take advantage of the situation and never give the artist a dime. Sometimes it is because they are using the free download as a trail and they are never converted to a fan. In other cases, they are just freeloaders. Take Radiohead's *In Rainbows* as an example. The band gave away the entire album in a "name your own price" deal and fans, after a minimal service fee, paid whatever they wanted. According to Comscore.com, 62% of downloaders paid nothing beyond the small access fee, but the average payment was $2.26 and the promotion grossed $1.36 million in three weeks (Comscore Insights, 2007). Piracy of the album was rampant and the number of units actually sold is disputed (SoundScan list the number at 972,926 as of October 26, 2014), but one thing is clear: the publicity from the promotion was followed by a world tour of mostly sold out shows.

Derek Webb, founder of NoiseTrade.com, did this in 2006. Webb gave away an album online in exchange for a name, email address and a zip code. "In three month's time Webb gave away over 80,000 full downloads of his album and collected valuable information [from] as many new fans.

As a direct result, Derek saw many sold-out shows, increased merchandise and album sales, including a curious spike in sales of the very album that was given [away] for free" (NoiseTrade.com/info/questions). Other artists that have posted songs on Noise Trade include the Lumineers, Fun, and Civil Wars (Hollabaugh, 2013). Alternatives to Noise Trade include Sound Cloud, Topspin, and Bandcamp. Established artists have also packaged free music along with concert tickets as Prince did with Musicology in 2004, Madonna during her 2012 MDNA Tour and Tom Petty in 2014 (Crenshaw, 2012; Lawrence, 2014). So, now that you're selling the experience of seeing you perform live it is the time to start inviting your friends in the industry and the A&R people to come see your show.

BUILD A SOCIAL MEDIA PRESENCE

Be the Bieb! Or Katy Perry. Both have been on the *Billboard* Social 50 chart for over 200 weeks. Building a social media presence is now an important part of making yourself attractive to a label or publisher, too. They really do pay attention to those sorts of things—how many Facebook friends you have; how many followers you have on Twitter—and how you interact with your fans on social media. Web presence alone is not going to get you a record deal—some labels only look at your sites after they have developed an interest in you because of your live performances. But the label is going to look at how you are trending on the net—whether you are adding fans at a faster rate or if your fame has stagnated.

MAKE PROFESSIONAL CONNECTIONS

Most publishers and record labels stopped taking unsolicited submissions because of copyright infringement lawsuits, so it helps to get to know somebody on the inside, or somebody who knows somebody on the inside. This is not as hard as it might seem. If playing a lot of shows has not caught the attention of the A&R department at your dream label then hiring a lawyer, publisher or manager may be the ticket in.

As noted above, an established entertainment attorney with connections in the industry or a publisher with whom you have already developed a relationship may be willing to take on an unsigned artist in a "shopping deal." Join organizations of other writers and artists and build a network of connections. You never know where that big break is going to come from.

REFERENCES

Brabec, J., and Brabec, T. *Music, Money and Success* (7th edition). New York: Schirmer Trade Books. 2011.

Bronson, F. "Ten Years of 'American Idol' Dominance: Clarkson, Underwood, Daughtry, Fantasia, More." *Billboard*, June 11, 2012. Accessed November 2, 2014. http://www.billboard.com/biz/articles/news/1093819/ten-years-of-american-idol-chart-dominance-clarkson-underwood-daughtry.

Catino, J. Personal interview. October 15, 2014.

Comscore Insights. *For Radiohead Fans, Does "Free" + "Download" = "Freeload"?* 2007. Accessed October 26, 2014.

Crenshaw, A. "5 Shrewd Ways Musicians Give Away Free Music." Houston Press Blogs. *Houston Press*, November 29, 2012. Accessed October 26, 2014.

Lipsman, A. (ed.). "For Radiohead Fans, Does 'Free' + 'Download' = 'Freeload'?" *Comscore*, November 5, 2007. Accessed. October 26, 2014. http://www.comscore.com/Insights/Press-Releases/2007/11/Radiohead-Downloads.

Hollabaugh, L. "NoiseTrade Exchanges Downloads For Fan Information." *Music Row*, November 18, 2013. Accessed November 6, 2014. MusicRow.com.

John Hammond Biography. Rock and Roll Hall of Fame, n.d. Accessed October 25, 2014. https://rockhall.com/inductees/john-hammond/bio/.

Krasilovsky, M. W., and Shemel, S. *This Business of Music* (10th edition). New York: Billboard Books. 2007.

Lawrence, J. "Tom Petty Announces Tour Dates, Will Include Free New Album With Each Ticket Purchase." *Forbes*, May 21, 2014. Accessed October 26, 2014. Forbes.com.

Levine, R. "An Alternative Approach to Marketing Rock Bands." *The New York Times*, May 5, 2008. Accessed November 16, 2014.

Macias, D. Personal Interview, October 17, 2014.

NoiseTrade.com/info/questions. November 4, 2014.

Passman, D. S. *All You Need to Know about the Music Business* (8th edition). New York: Free Press. 2012.

Scarpulla, M. "Artist Development or Tour Marketing." *TomHutchison.com*. Ed. T. E. Hutchison. N.p., July 22, 2010. Accessed June 24, 2015.

Yano, S. All Music Guide. All Media Network, LLC, n.d. Accessed October 25, 2014.

CHAPTER 9

Record Label Finances

PROFITABILITY OF A PROJECT . . . OR THE LACK THEREOF!

Record companies use profit and loss (P&L) statements to both predict the success of a record prior to its release, as well as analyze a project as it nears the end of its lifecycle. To understand the "math," let's look at all the components of a P&L, recognizing the financial significance of every line item. To appreciate the impact of each element, read through the actual statement in the back of this chapter and realize the impact of each line item as it affect the bottom line.

SRLP: SUGGESTED RETAIL LIST PRICE

The suggested retail list price (SRLP) or manufacturers suggested list (MSL) is set by the record label and is based on cost of the recording project, the artist's status, genre of music, competitive landscape, and what the market will bear. Although the royalty models are changing, many labels still pay artist royalties based on the SRLP. The SRLP has a correlating wholesale price, which is usually structured by the distribution company.

CARD PRICE

This line item is the wholesale price. The *card price* is the dollar figure that is set by the distribution company that sells the product. For fair trade practices, wholesale card prices are published entities that are the basis of further financial negotiations between seller and buyer. The chapter on retail shows an actual published rate card of wholesale prices.

DISCOUNT

Music product regularly receives a discount, which is set by the label and is administered via the distributor. (In some situations, it can be specified or capped by the artist's contract.) In mainstream music, the discount is applied to the wholesale price. In some genres of music such as Christian music, discounts are applied to the retail price, and negotiations are then based on retail prices. But the majority of music is sold on wholesale pricing strategies.

Discounts are based on many variables. If a label has a really hot artist that is a big seller, and demand at the consumer level is high, a discount may not be offered, since most retailers will buy the product at full price. To entice retailers to purchase a new artist, record labels will offer discounts, which will increase the margin and profitability at the store level. Fair trade practices require that labels and their distributors offer the same discount on the same release to all retail purchasers. But the discount can be changed based on the marketing elements that the retailer may offer.

For example, a label has a new artist and is offering a 5% discount. This discount will increase the potential profitability of the retailer. If the retailer agrees to include this new artist on a "new artist" endcap for an additional 5%, then the discount will be increased to 10%, adding to the potential profitability of the retailer.

This same discount does not apply to the life of a project, but can be offered at different times at different levels. So when evaluating a project's profitability, music companies use an "average" discount that would apply to the life of the project—be it a forecasted P&L or an "actual" average when taking a look backwards and assessing the record's overall success.

GROSS SALES

Gross sales refers to the wholesale price (value) of the product with the reduction of the discount included. It usually reflects the number of records shipped minus returns. Remember that music retailing is basically a consignment business and that stores can return product back to the distributor and label and receive a credit for this unsold product.

DISTRIBUTION FEE

Depending on the relationship between the label and distributor, the distribution fee is based on sales after the discount. Meaning, it is in the distributor's best interest to keep discounts as low as possible to help increase their profitability. This fee is a percentage of sales and varies greatly by the distributor. The conglomerates that own both the labels and distributor often charge between 10–12%. Independent distributors structure deals with indie labels and artists that range between 18%–30%, depending on the services being offered.

GROSS SALES AFTER FEE

Once again, the gross sales figure is determined on shipped minus returned product, being *NET*. Net units are multiplied by the per unit price, after discounts and distribution fees are deducted.

RETURN PROVISION

The industry averages about 20% returns, meaning that for every 100 records in the marketplace, 20 will be returned. To protect their business, record labels insure against returns by "reserving" a percentage of sales. Record labels "reserve" 20% of sales by pocketing these funds in an escrow account. Not until the life of the record has run the majority of its cycle will the reserve be adjusted. In some cases, a record may only have a 5% return in its lifecycle. The record company will then adjust the profitability statement to reflect such a low return, and royalties will then be distributed. In other cases, a record may have a 30% return, which adversely affects the company's overall profitability, since they expected they had sold 10% more of a particular release then previously accounted.

NET SALES AFTER RETURN RESERVE

Basically, net sales after return reserve is the computed net sales minus the 20% reserve deduction. This sales number reflects a record company's hedge against potential returns in the future. If returns do not occur, the reserve is returned as profit and appropriate royalties are distributed as noted in the next line item.

RETURNS RESERVE OPP/(RISK)

Labels incur returns on records shipped into the marketplace. For accounting purposes, the profit and loss statement reflects these returns within the overall equation. The returns reserve line item is usually computed at the end of the lifecycle of the release. A label will calculate what actual returns occurred and plug this adjusted number into the P&L, determining the ultimate profitability of the project. But labels will use the P&L or pro forma as predictor equations to determine potential profitability of a future project and will plug return reserve standards into the overall equation, helping to evaluate a project's future. Note that return reserves only apply to physical sales and do not apply to digital sales.

GROSS PHYSICAL SALES

The gross physical sale is the adjusted sales number deducting discounts, distribution fees, and returns of actual product in the marketplace.

GROSS DIGITAL UNITS/ALBUMS AND SINGLES

Digital sales continue to grow as the significant contributor to the bottom line; 2013 data reflects that 57% of albums were still in CD format but the lion share of music purchased is in digital configuration. Music companies are continually working on financial models that reflect consumer behaviors so that this line item can better serve the pro forma equation, helping to predict project potential. Album and single per unit dollar figures include royalties that have yet to be distributed.

GROSS STREAMING REVENUE: SOUNDEXCHANGE (NON-INTERACTIVE)

Based on the Digital Performance Right in Sound Recordings Act of 1995, non-interactive Internet transmissions such as Pandora are required to pay a statutory license that is established by the Copyright Board and is distributed by SoundExchange (http://en.wikipedia.org/wiki/Copyright_Royalty_Board). The current 2014 exchange for streaming on the Internet is $.0023/stream for the label with the other $.0023 going to the artist directly) with the collective 1,500 streams being the equivalent of a wholesale album (1,500 streams × ($.0023+ $.0023) = $7.50 Wholesale) This value is approximate.

SEA = Streaming Equivalent Album

GROSS STREAMING REVENUE: SPOTIFY (INTERACTIVE)

Interactive Internet transmission services such as Spotify (where consumer calls up specific artist and songs) are required to negotiate a license agreement with the copyright holder directly and can vary by company and sound recording. In the included P&L, the revenue generated reflects a streaming equivalent of $.0035/stream. But again, these rates are highly negotiable and vary by label and service.

TOTAL NET SALES

The total net sales is the adjusted sales number deducting discounts, distribution fees, and returns of both physical and digital sales.

COST OF SALES

The cost of sales are costs associated with making the actual product, including the pressing of the disc, printing of paper inserts, marketing stickers on the outside of the product, all-in royalties and mechanical royalties, and inventory obsolescence. In accounting terms, these are the variable costs since the amount varies based on the number of units produced.

Many of these costs are negotiable, such as the mechanical rate. Labels often receive a reduced mechanical rate on artists that are also the songwriter. Older songs, as well as reissued material, can often receive lower mechanical rates based on the age and inactivity of the copyrights.

Dealing with the inventory of product that does not sell into the marketplace costs money. So, labels build into the cost of goods an amount that will fund the management of returned and obsolete inventory. It takes manpower and resources to "scrap" a pile of CDs. This includes moving the inventory off the warehouse floor, pulling the inserts and CDs from the jewel cases (which are recycled for new releases), and breaking/melting the actual CDs into pellets, which can then be used to make new CDs.

The one-time fee of preparing the digital masters for online retailers must also be accounted for, as included in the cost of sales.

GROSS MARGIN BEFORE RECORDING COSTS

Gross margin before recording costs deducts the cost of sales from the total net sales. Know that the recording costs are initially funded by the record label. But built into most artist contracts, recording costs are recoupable, meaning that once the release starts to make money, the record company will pay itself back prior to the artist receiving royalties.

Also included in recording costs are advances. Advances are monies fronted to the artist to assist them with living expenses. It takes time to record a record, which doesn't allow an artist to make money elsewhere. Advances are recoupable.

GROSS MARGIN

Gross margin reflects gross sales minus discounts, distribution fees, returns, cost of sales, and recording costs. These are revenues made prior to marketing expenses.

MARKETING COSTS

To launch an artist's career in today's climate, it takes a lot of money. How a company manages its marketing costs can determine the success of an album . . . and artist. Beyond the making of the record, marketing costs include the imaging of an artist, advertising at both trade and consumer publications, publicity, radio promotion, and retail positioning in the stores.

Account Advertising

An expensive marketing cost is the positioning of a record in the retail environment, known as *co-op advertising*. In the grocery business, this is called a *slotting allowance*. Ever notice that premium, name-brand items are placed in the most prominent positions within a grocery store? The same occurs in record retail. The pricing and positioning (P&P) of a record, meaning a "sale" or reduced price along with prime real estate placement in retail stores, can cost record labels hundreds of thousands of dollars. Other types of account advertising include print advertising via the store's Sunday circulars, end-cap positioning, listening stations, event marketing such as artist in-store visits, point of purchase materials placement guarantees, and more. But the practice of co-op advertising in the physical retail environment is becoming less common as the digital marketplace continues to grow as the place of commerce for music purchases.

Advertising

Most schools of advertising include a lesson on internal versus external advertising. Trade advertising is considered an internal promotional activity. Trade advertising creates awareness of a new product to the decision makers of that industry by using strong imaging of the artist and release along with relevant facts such as sales and radio success, tour information, sales data, and upcoming press events.

In the record business, decision makers include music buyers for retail stores, radio stations and their programmers, talent bookers for television shows, reviewers for newspaper and consumer magazines, and talent buyers for venues. Some prominent trade magazines for the music industry include *Billboard, Amusement Business, Pollstar,* and *Variety,* to name a few.

Consumer advertising is considered external advertising since it is targeting the "end consumer." Consumer advertising also creates awareness, but to end consumers who will purchase music. The most popular consumer advertising today is Internet. This tool allows for "test driving" the music along with an endless array of imaging with immediate click 'n' buy opportunities. Although there are several online marketing strategies via banner placements on key websites, the most popular consumer advertising has come through funding the preparation of turn-key "widgets" that allow for fans to imbed and forward them on their personal website through social networking.

Video Production

Negotiated into the recording contract, artists are usually responsible for 50% of the cost of video production. Although the record company will fund the video shoot, the record company will expect to receive recoupable pay of 50% of the overall costs from record sales, and 100% recoupable from video/DVD sales.

Artist Promotion

This line item usually covers the costs associated with introducing and promoting the artist to radio. Still a primary source for learning about new music, labels often take artists to visit with radio stations, including on-air interviews, dinners with music programmers, and Listener Appreciation events. The record company incurs this cost and is usually not recoupable.

Independent Promotion and Publicity

Depending on the artist's stature and current competitive climate, record companies will hire the services of independent promotion companies and independent publicists. In addition to the label's efforts, these independent agents should enhance the label's strategy by assisting in creating exposure for their artists via additional radio airplay and media. This line item is not recoupable.

Media Travel

Record labels will fund the costs associated with travel for an artist who is doing a media event. Media travel is usually an isolated event where an artist is doing a television talk show or an awards event. Again, an expense to the label and not recoupable.

Album Art

The imaging of an artist can take time . . . and money. Most artists receive some type of grooming, if not just a polishing of what the artist already represents. Professionals trained at artist imaging are hired to create a "look" that is unique and defining. Such professionals as hair and make-up artists, clothing specialists, even movement/dance professionals are often required to give an artist a specific shine. As a part of the team, photographers are hired to shoot cover artwork, as well as press images for publicity use. And then a designer is hired to create the overall album art concept, from artist image use, album title treatment, booklet layout, and so forth. All of these efforts are expenses to the label and not usually recoupable.

No Charge Records

A marketing tool often used by labels is the actual CD. No charge records are those CDs that are used for promotional use such as giveaways on the radio, or in-store play copies for record retailers. The value of these records and their use must be accounted for, and are not recoupable. This practice is not used as readily in age when a download is just a click away.

CONTRIBUTION TO OVERHEAD

"Contribution to overhead" is another phrase for profit! From the gross margin, a label would subtract the marketing costs including account advertising, trade, and consumer advertising, video production, artist promotion, independent promotion and publicity, media travel, album art, and no charge records to determine the release's contribution to a record company's overhead.

Note the percentages given within the spreadsheet. The total net sales percentage reflects 100% of money generated by the sale of the project. Each line item subheading also reflects a percentage, causing each department to consider how much is being spent as a reflection of the project as a whole. Although the contribution to overhead may be a large number, its percentage of the whole determines how effective and efficient the project was managed.

Contribution to Overhead: Example of Two Different Projects

Example	Project A	Project B
Total Net Sales	$3,500,000	$5,000,000
Cost of Sales	$2,150,000	$3,500,000
Recording Costs		
Manufacturing		
Marketing		
Contribution to Overhead (Profit)	$1,350,000	$1,500,000
Percentage	38.6%	30.0%

Looking at the numbers, Project B made more money for the company, but as for efficiency, Project A was a more effective release since the company did a better job at managing its expenses.

WAYS OF LOOKING AT PROJECT EFFICIENCY AND EFFECTIVENESS

Record labels look at the "numbers" in many ways to determine how well they are performing. By "spinning" these figures, a company should analyze where spending is less effective, thus causing the overall project to be less profitable.

Return Percentage

Gross Ships − Cumulative Returns = Net Shipment
Example: 500,000 units − 35,000 returns = 465,000 units Net Shipments

Cumulative Returns/Gross Ships = Return Percentage
35,000 units / 500,000 units = 7% return

Again, industry standards reflect an approximate 20% return percentage, which is usually built in to the P&L spreadsheet. Generally, a record that returns more than 20% is not performing at the market average. A couple of decisions could have affected this percentage. The record company over-sold the project and caused the returns, or the record company did not promote and have a hit with the project, thus causing returns. In any case, the record company needs to determine what occurred to ensure that it doesn't happen again.

If a project's return percentage is below the industry average such as a lifetime return average of 7%, (as noted above), the project could be considered a great success, minimizing return costs, as well as manufacturing costs. But the company should also be sensitive to the fact that they may have undersold the project, thus not realizing the full sales potential of the project too. Again, further analysis is vital to the overall success of future projects and the company.

SOUNDSCAN AND SELL OFF

SoundScan is an information system that tracks sales of music and music video product throughout the United States and Canada. Sales data using UPC bar codes from point-of-sale cash registers is collected weekly from over 14,000 retail, mass merchant and non-traditional (online store, venues, etc.) outlets. Weekly data is compiled and made available every Wednesday. SoundScan has been considered the gold standard regarding consumer behavior and sales since 1993. But with the advent of social media and the evaluation of streaming as part of the "success" equation, new models are being considered. Still, traditional valuation of inventory management and manufacturing models have to be considered, and sell-off data is relative to a record label's core business.

To evaluate inventory, record companies can use a simple equation to know how many units remain in the marketplace:

Net Shipments − SoundScan Sales = Remaining Inventory
SoundScan Sales/Net Shipments = Sell Off Percentage

Example:
465,000 Net Ships − 437,000 units SoundScanned = 28,000 units
437,000 SoundScan/465,000 Net Ships = 94% Sell Off

If this project is a steady seller, quietly moving units every week, a record company could use SoundScan weekly sales to determine how many weeks of inventory are left in the marketplace. If this project sold 2,000 each week:

28,000 remaining inventory / 2,000 SoundScan each week = 14 weeks of inventory left

Knowing inventory levels and keeping aware of sales is critical to the success of a project. A record company does not want to run out of records, which is called *can't fill*. If it's not on the retailer's shelves, it cannot be purchased.

Percentage of Marketing Costs

Isolating marketing costs and analyzing their effectiveness in selling records is best reflected in the following equations:

Percentage of Marketing Cost to Gross:
Marketing Costs/Gross Sales × 100 = %
Percentage of Marketing Cost to Net:
Marketing Costs/Net Sales × 100 = %

Example:
$800,000 Marketing Costs/$4,238,000 Gross Sales = 18.9%
$800,000 Marketing Costs/$3,402,000 Net Sales = 23.5%

The lower the percentage reflects, the better performing project. Keeping marketing costs in check and knowing when to stop "fueling the fire" is usually a great determiner of seasoned record labels.

Marketing Costs per Gross Unit:
Marketing Costs/Gross Shipments = Cost per unit
Marketing Costs per Net Unit:
Marketing Costs/Net Shipments = Cost per unit

Example:
$800,000/500,000 units Gross Shipped = $1.60
$800,000/465,000 units Net Shipped = $1.72

In all of these equations, the "real" picture is best drawn when using net shipments, since that is the ultimate number of units in the marketplace. Using digital sales in the mix, both TEA and SEA, these evaluations are driven down even lower than when calculating with only full album sales exclusively.

As noted in the example P&L in the back of this chapter, marketing costs/gross reflect $.84 and marketing costs/net show $.91 and reflect the inclusion of track equivalent album sales and streaming equivalent album sales, increasing the bottom line by over 200,000 album units. These single sales and streams do add up.

AS A PREDICTOR EQUATION

Record companies can use the profit and loss statement as a predictor of success. Often, labels will "run the numbers" to see how profitable, or not, a potential release could be. By using the spreadsheet and plugging in forecasted numbers, including shipments, cost of sales, recording costs (or the acquisition of a master), and marketing costs, the equation will help a label evaluate and determine whether a project is worth releasing.

Small considerations can dramatically affect the contribution to overhead. List price, discounts, number of pages in the CD booklet, royalties—both artist and mechanical, and the various marketing line items can either make a project profitable or not.

BREAK-EVEN POINT

When does a record "break even" in covering the costs that it took to make the project, and when does it start to turn a profit? Depending on the equation, record companies look at this value in several ways.

Without marketing costs, a number can be derived simply by dividing the total fixed costs by price, using wholesale dollars.

Break-even point

$$\text{Break-even point (units)} = \frac{\textit{Total fixed costs}}{\textit{Price} - \textit{(total variable costs/units)}}$$

Break even with costs listed

$$BE = \frac{\textit{Recording Costs} + \textit{Advances}}{\textit{Price} - \textit{Variable costs}}$$

Break-even example

Example:

$$BE = \frac{\$250{,}000 \text{ recording costs} + \$75{,}000 \text{ advance}}{\$7.50^a - (\$3.20^b + \$.88^c)}$$

[a] *(card price of SRLP $11.98)*
[b] *(based on total cost of sales)*
[c] *(Distr. Fee)*

BE = 95,029 units without marketing costs

List price

List Price (Retail)	11.98	
Card Price (Wholesale)	7.50	
Distribution Fee	.90	@ 12.0%

Cost of Sales		
Manufacturing (per unit shipped)	0.77	7%
Royalties (Artist/Producer)	1.125	15%
AFM/AFTRA	0.11	1%
Copyright (.091 per track)	0.91	8%
Other (Obsolescence, Sticker of Product, etc.)	0.21	2%
Returns Fee (1.75% of Returns)		
Digital Mastering Prep for Online Retailers		
TOTAL COST OF SALES	3.125	28.4%

Percentages in last column reflect value based on retail

Adding marketing costs changes the outcome of the equation dramatically:

A label has to derive predicted/budgeted marketing costs to add to the equation. Most seasoned labels have an idea as to how much each activity may cost to launch a record. Using the following dollars, check out the break even analysis:

Marketing costs

Marketing Costs	
Account Advertising	100,000
Trade Advertising	40,000
Consumer/Other Advertising	100,000
Video Production	125,000
Artist Promotion (Radio)	134,000
Indy Promotion/Publicity	10,000
Media Travel	25,000
Album Art/Imaging	25,000
No Charge Records	10,000
TOTAL MARKETING COSTS	569,000

Break-even example with marketing costs

Example:			
BE =	$250,000 recording costs + $75,000 advance + $569,000 mkt costs		
	$7.50[a] − ($3.20[b] + $.88[c])		
	[a] (card price of SRLP $11.98)	[b] (based on total cost of sales)	[c] (Distr. Fee)

Break-even point in units

$$BE = \frac{\$894,000}{\$3.42} =$$ 261,404 units just to recover the costs of making and marketing this record. This does not take into account overhead costs and salaries of the employees getting this job done.

Clearly, the job of the profit and loss statement is multifaceted. It can be used as a predictor of success (or not), as well as an evaluation tool of existing projects. Not a part of the equation is the overhead that it takes to operate the business, such as building expenses, salaries, supplies, and so on. But lessons learned from this type of analysis should aid record labels as to the better allocation of funds and resources.

A LOOK AT REAL P&L STATEMENTS

This statement was based on real-life scenario where the new album release was only a month old, but the single had been out for 12 weeks. As you can see, the single has sold over 1.3 million units and the album has sold over 275,000 units, with 118,000 of them being digital. Also recognize the revenue generated through streaming. The single was a viral hit, with nearly 90 million hits, converting to 40,000 in album sales to the bottom line. All these sales figures indicate of a youth buyer, with over 60% of the album sales being digital, if digital single sales are converted to album sales via TEA. Make note that the single has generated more revenue than the digital album sales, so far. And that the revenue generated from 85 million streaming views converts to nearly a quarter of a million dollars to the bottom line.

Check out the discount at 8%. This applies to physical sales only at this time.

The recording costs were completely recouped, so this line item cancels itself out.

Marketing costs are also shifting as consumers have taken on the role of word-of-mouth marketing via social networks. Not long ago, account advertising for a release this size would have cost nearly twice as much. This is a pop artist—note the expensive video costs and radio promotion costs since more than one format is being worked.

Even though this release is early in its lifecycle, it has made over $1.7 million dollars.

Change one item, and see what happens to the PROFIT line (contribution to overhead.) In Figure 9.2, the discount has been reduced to 0%, meaning that retailers did not receive a discount. Those 8% points would mean that this

Profit and Loss Statement

8% Discount
20% Returns Provision

	Amt	%	Units	Dollars	
			Gross Units		
CD List Price	11.98		325,250		
CD Card Price (Wholesale Price)	7.50				
Std. Discount	0.60	8.0%			
			Net Units (Actualized)		
Adj. Price to Accounts	6.90		276,463		
Gross Sales				$2,244,225	
Returns Provision (after Fee)				−$448,845	20.0%
Sales After Reserve (Net Revenue)	6.90			$1,795,380	
Distribution Fee	0.83	12.0%		$215,446	12.0%
Net Sales After Fee	6.07			$1,579,934	
Sales Reserve Opp/(Risk)			16,263	$112,211	5.0%
Gross PHYSICAL Sales				$1,692,146	
Gross Digital Units/Albums	7.25		118,412	$858,487	
Gross Digital Units/Singles	0.80		1,313,816	$1,051,053	
Gross DIGITAL Sales				$1,909,540	
Gross Streaming Revenue					
Non-Interactive					
Sound/Exchange: Gross Streaming Unit/Single (2014)	0.0023		45,150,356	$103,846	
SEA Streaming Equivalent Albums (based on 1500/album)			13,850		
Interactive					
Streaming Services: Gross Streaming Unit/Single	0.0035		40,456,254	$141,597	
SEA Streaming Equivalent Albums (based on 1500/album)			26,971		
TOTAL NET SALES (Physical and Digital)				**$3,847,128**	
Cost of Sales					
Manufacturing (per unit shipped)	0.77	12.7%		$250,443	14.8%
Royalties (Artist/Producer)	1.125	18.5%		$610,333	36.1%
AFM/AFTRA	0.11	1.8%		$81,550	4.8%
Copyright (.091 per track)	0.91	15.0%		$541,585	32.0%
Other (Obsolescence, Sticker of Product, etc.)	0.21	3.5%		$68,303	4.0%
Returns Fee (1.75% of Returns)				$854	0.1%
Digital Mastering Prep for Online Retailers				$1,000	0.1%
TOTAL COST OF SALES	**3.125**	**51.5%**		**$1,554,067**	**40.4%**
Gross Margin (Before Recording Costs)	**2.95**	**48.5%**		**$2,293,061**	**59.6%**
Recording Costs					
Recording & Advances				$625,000	16.2%
Recording Recoupment				−$625,000	−16.2%
NET RECORDING COST				$0	0.0%
Gross Margin – also known as Gross Profit				**$2,293,061**	
Marketing Costs					
Account Advertising				$100,000	2.6%
Trade Advertising				$40,000	1.0%
Consumer/Other Advertising				$100,000	2.6%
Video Production				$125,000	3.2%
Artist Promotion (Radio)				$134,000	3.5%
Indy Promotion/Publicity				$10,000	0.3%
Media Travel				$25,000	0.6%
Album Art/Imaging				$25,000	0.6%
No Charge Records				$10,000	0.3%
TOTAL MARKETING COSTS				$569,000	14.8%
Contribution to Overhead (Profit)				**$1,724,061**	**44.8%**
Gross Shipped to date				325,250	
Cumm Return to date				48,788	
Return Percentage				15.0%	
Mkg Cost per Gross Unit				$0.92	
Mktg Cost per Net Unit				$1.00	
SoundScan / Physical Product only				187,366	
Sell of Percentage				67.8%	

FIGURE 9.1

Profit and Loss Statement

	Amt	%	Units	Dollars	
8% Discount					
20% Returns Percentage			Units	Dollars	
			Gross Units		
CD List Price	11.98		325,250		
CD Card Price (Wholesale Price)	7.50				
Std. Discount	0.00	0.0%			
			Net Units (Actualized)		
Adj. Price to Accounts	7.50		276,463		
Gross Sales				$2,439,375	
Returns Provision (after Fee)				–$487,875	20.0%
Sales After Reserve (Net Revenue)	7.50			$1,951,500	
Distribution Fee	0.90	12.0%		$234,180	12.0%
Net Sales After Fee	6.60			$1,717,320	
Sales Reserve Opp/(Risk)			16,263	$112,969	5.0%
Gross PHYSICAL Sales				$1,839,289	
Gross Digital Units/Albums	7.25		118,412	$858,487	
Gross Digital Units/Singles	0.80		1,313,816	$1,051,053	
Gross DIGITAL Sales				$1,909,540	
Gross Streaming Revenue					
Non-Interactive					
Sound/Exchange: Gross Streaming Unit/Single (2014)	0.0023		45,150,356	$103,846	
SEA Streaming Equivalent Albums (based on 1500/album)			13,850		
Interactive					
Streaming Services: Gross Streaming Unit/Single	0.0035		40,456,254	$141,597	
SEA Streaming Equivalent Albums (based on 1500/album)			26,971		
TOTAL NET SALES (Physical and Digital)				**$3,994,271**	
Cost of Sales					
Manufacturing (per unit shipped)	0.77	11.7%		$250,443	13.6%
Royalties (Artist/Producer)	1.125	17.0%		$610,333	33.2%
AFM/AFTRA	0.11	1.7%		$81,550	4.4%
Copyright (.091 per track)	0.91	13.8%		$541,585	29.4%
Other (Obsolescence, Sticker of Product, etc.)	0.21	3.2%		$68,303	3.7%
Returns Fee (1.75% of Returns)				$854	0.0%
Digital Mastering Prep for Online Retailers				$1,000	0.1%
TOTAL COST OF SALES	3.125	47.3%		$1,554,067	38.9%
Gross Margin (Before Recording Costs)	**3.48**	**52.7%**		**$2,440,204**	**61.1%**
Recording Costs					
Recording & Advances				$625,000	15.6%
Recording Recoupment				–$625,000	–15.6%
NET RECORDING COST				$0	0.0%
Gross Margin – also known as Gross Profit				**$2,440,204**	
Marketing Costs					
Account Advertising				$100,000	2.5%
Trade Advertising				$40,000	1.0%
Consumer/Other Advertising				$100,000	2.5%
Video Production				$125,000	3.1%
Artist Promotion (Radio)				$134,000	3.4%
Indy Promotion/Publicity				$10,000	0.3%
Media Travel				$25,000	0.6%
Album Art/Imaging				$25,000	0.6%
No Charge Records				$10,000	0.3%
TOTAL MARKETING COSTS				$569,000	14.2%
Contribution to Overhead (Profit)				**$1,871,204**	**46.8%**
Gross Shipped to date			325,250		
Cumm Return to date			48,788		
Return Percentage			15.0%		
Mkg Cost per Gross Unit				$0.92	
Mktg Cost per Net Unit				$1.00	
SoundScan / Physical Product only				187,366	
Sell of Percentage				67.8%	

FIGURE 9.2

Profit and Loss Statement
8% Discount
20% Returns Provision
20% Artist Royalty

	Amt	%	Units	Dollars	
			Gross Units		
CD List Price	11.98		325,250		
CD Card Price (Wholesale Price)	7.50				
Std. Discount	0.60	8.0%			
			Net Units (Actualized)		
Adj. Price to Accounts	6.90		276,463		
Gross Sales				$2,244,225	
Returns Provision (after Fee)				−$448,845	20.0%
Sales After Reserve (Net Revenue)	6.90			$1,795,380	
Distribution Fee	0.83	12.0%		$215,446	12.0%
Net Sales After Fee	6.07			$1,579,934	
Sales Reserve Opp/(Risk)			16,263	$112,211	5.0%
Gross PHYSICAL Sales				$1,692,146	
Gross Digital Units/Albums	7.25		118,412	$858,487	
Gross Digital Units/Singles	0.80		1,313,816	$1,051,053	
Gross DIGITAL Sales				$1,909,540	
Gross Streaming Revenue					
Non-Interactive					
Sound/Exchange: Gross Streaming Unit/Single (2014)	0.0023		45,150,356	$103,846	
SEA Streaming Equivalent Albums (based on 1500/album)			13,850		
Interactive					
Streaming Services: Gross Streaming Unit/Single	0.0035		40,456,254	$141,597	
SEA Streaming Equivalent Albums (based on 1500/album)			26,971		
TOTAL NET SALES (Physical and Digital)				$3,847,128	
Cost of Sales					
Manufacturing (per unit shipped)	0.77	12.7%		$250,443	14.8%
Royalties (Artist/Producer)	1.5	24.7%		$813,778	48.1%
AFM/AFTRA	0.11	1.8%		$87,649	5.2%
Copyright (.091 per track)	0.91	15.0%		$547,683	32.4%
Other (Obsolescence, Sticker of Product, etc.)	0.21	3.5%		$68,303	4.0%
Returns Fee (1.75% of Returns)				$854	0.1%
Digital Mastering Prep for Online Retailers				$1,000	0.1%
TOTAL COST OF SALES	3.5	57.6%		$1,769,709	46.0%
Gross Margin (Before Recording Costs)	2.57	42.4%		$2,077,420	54.0%
Recording Costs					
Recording & Advances				$625,000	16.2%
Recording Recoupment				−$625,000	−16.2%
NET RECORDING COST				$0	0.0%
Gross Margin – also known as Gross Profit				$2,077,420	
Marketing Costs					
Account Advertising				$100,000	2.6%
Trade Advertising				$40,000	1.0%
Consumer/Other Advertising				$100,000	2.6%
Video Production				$125,000	3.2%
Artist Promotion (Radio)				$134,000	3.5%
Indy Promotion/Publicity				$10,000	0.3%
Media Travel				$25,000	0.6%
Album Art/Imaging				$25,000	0.6%
No Charge Records				$10,000	0.3%
TOTAL MARKETING COSTS				$569,000	14.8%
Contribution to Overhead (Profit)				$1,508,420	39.2%
Gross Shipped to date				325,250	
Cumm Return to date				48,788	
Return Percentage				15.0%	
Mkg Cost per Gross Unit				$0.92	
Mktg Cost per Net Unit				$1.00	
SoundScan / Physical Product only				187,366	
Sell of Percentage				67.8%	

FIGURE 9.3

same project would yield nearly $150,000 MORE to the label—if they did NOT discount the product.

Another change is shown in Figure 9.3. Again, change the royalty by increasing by $.38 to $1.50 and again, the bottom line is greatly impacted. This time, the label loses nearly $216,000. The negotiation of every line item matters—down to the penny and the percentage point.

CHAPTER 10

Publicity

> Without publicity a terrible thing happens—nothing.
> —the great showman P. T. Barnum (1810–1891)

INTRODUCTION

Publicity is arguably the most important part of any marketing plan. It lays the foundation on which every other part of the plan is built. Labels often handle publicity for the artist's recording career, as well as for news and press releases about the label itself through a publicity department. Sometimes an artist will hire a personal publicist to handle other areas of their life and career. By creating awareness of the artist, publicity makes all other aspects of the marketing plan more effective and promotion and sales efforts easier. By definition, publicity is earned media, promotion whose placement is not directly paid for (advertising) or owned (websites), and therefore, the most accessible part of any marketing effort regardless of whether you are an independent artist or a major label act at the pinnacle of your career. Since both the publicity department and the publicist do the same job we will refer to them both as simply the publicist.

PUBLICITY DEFINED

In traditional marketing, **publicity** is part of the public relations function that includes media relations, creating **press kits** and press releases, and

CONTENTS

CONTENTS

lobbying. The AMA Committee on Definitions once defined publicity as "non-personal stimulation of demand for a product, service or idea . . . not paid for directly by the sponsor" (Dommermuth, 1984). Nowadays, the term has fallen out of favor with marketing academics and publicity is referred to as part of the broader category of public relations. Around 2009, practitioners began to distinguish promotion by its ownership (Figure 10.1). Owned media are the promotion channels the artist or label controls, like their website or Twitter accounts. Paid media is advertising or other promotions where the label pays to have their message strategically placed. Earned media, formerly known as publicity, is when a magazine or television show shares the story as content (as opposed to advertising). At its extreme, earned media includes word of mouth, word of mouse or viral marketing. The news is shared directly between individuals via email, text or social media. This view is artist or business centric—the artist owns, buys, or earns the media.

Media Type	Definition	Examples	The Role	Benefits	Challenges
Owned Media	Channel the brand controls	Web Sites Mobile Site Blog Twitter Account	Build for longer-term relationships with existing potential and earned media	▪ Control ▪ Cost efficiency ▪ Longevity ▪ Versatility ▪ Niche Audiences	▪ No guarantees ▪ Company communication not trusted ▪ Takes time to scale
Paid Media	Brand pays to leverage a channel	Display Ads Paid Search Sponsorships	Shift from foundation to a catalyst that feeds owned and creates earned media	▪ In demand ▪ Immediacy ▪ Scale ▪ Control	▪ Clutter ▪ Declining response rates ▪ Poor credibility
Earned Media	When customers become the channel	Word of Mouth Buzz "Viral"	Listen and respond – Earned media is often the result of well-coordinated owned and paid media	▪ Most credible ▪ Key role in most sales ▪ Transparent and lives on	▪ No control ▪ Can be negative ▪ Scale ▪ Hard to measure

Source: Forrester Research, Inc.

FIGURE 10.1 *Types of Promotion*

In 2013, Forrester Research introduced another way of looking at publicity and social media that focuses on the consumer's perspective. They called this new model Marketing RaDaR (Nail and Elliott, 2014). The model will be discussed in more detail below.

Publicity is distinguished from other forms of promotion by its low cost, but that low cost comes with a sacrifice of control. Because you are not paying for space in a magazine or time on a television network there is no guarantee your message will get out and, even if it does, there is no assurance that your message will be communicated the way you intended.

The purpose of label publicity is to place nonpaid promotional messages into the media on behalf of the artist's recorded music project. That can range from a short paragraph in *Rolling Stone* to a mention in a music blog, to an appearance on "Saturday Night Live." Mentions and appearances in the new and traditional media contribute to the success of the marketing of the artist and their music. Earned media on behalf of a recording artist has a certain credibility that paid advertising does not. While an advertisement can be bought, a feature article or review gets published only because a journalist thought the artist or their music was interesting enough to write the article and a publisher thought it was interesting enough to make the space available to present the story or run the review.

An online article, or one in a newspaper or magazine, suggests to the reader that there is something more to the label's artist than just selling commercial music. Published articles and TV magazine-style stories (for example, "E! News") give credibility to the artist in a way that paid advertising cannot.

There are key differences between publicity from the label and the advertising placed by the label. Label publicists generally create and promote messages to the media that are informative in nature and do not have a hard "sell" to them. Consumers are resistant to paid media (advertising) messages, but more receptive to the subtle persuasion of the publicity effort that takes the form of an interesting story or review. Theoretically, the more impressions consumers receive about a recording, the more likely they are to seek additional information about the recording, and to purchase it. Advertising planners and sophisticated publicists use the term "reach and frequency" as they compile a strategy and its related budget. This means they plan an affordable campaign that can "reach" sufficient numbers of their target market with the "frequency" necessary for them to remember the message and act by purchasing. Publicity becomes the foundation, or at least a nice complement, to the advertising strategy without the direct costs of paid advertising.

HISTORY

The earliest music promoters were in the publicity business at the beginning of the 20th century, primarily helping to sell sheet music that was heard on recording playback devices or at public performances. Those who worked in the publicity profession in the early 1900s relied primarily on newspapers and magazines to promote the sale of music. In 1922, the federal government authorized the licensing of several hundred commercial radio stations, and those in the recorded music business found their companies struggling as a result. People stopped buying as much music because radio was now providing it for "free," and newspapers and magazines were no longer the only way the public got its news. Radio became the entertainer and the informer. But publicists found themselves with a new medium and a new way to promote, and quickly adapted to it, much in the way they did in 1948 with the advent of television as a news and entertainment form.

History has a way of repeating itself, only this time the new technology was not a terrestrial broadcast medium, but the Internet. The work of the label publicist today involves servicing not only traditional media outlets but online music and entertainment blogs and websites as well. Some of these online outlets are Internet extensions of magazines, newspapers, and video channels, which did not exist 10 years ago (and may not exist 10 years from now) while others are brand new with no traditional media presence. The label publicist also works with print media for feature articles, and they work with television program talent bookers to arrange live performances.

LABEL VS. INDIE PUBLICIST

Large record labels usually have one or more people on their staff responsible for publicizing their activities and those of their artists, but with the contraction of the industry much of the work has been outsourced to **independent publicists** on an "as needed" basis. This allows the label or the artist to hire the best person for the job rather than use an in-house publicist. If the in house publicist lacks the contacts or expertise needed for a particular job, a label will have to hire an independent publicist anyway. Depending on the artist's status, an indie publicist can easily cost several thousand dollars a month and expect a six-month contract. Good publicity is clearly not free in that case.

TOOLS OF THE PUBLICIST

> The only thing worse than being talked about is not being talked about.
> —Irish writer and poet Oscar Wilde (1854–1900)

The traditional tools of the label publicity department or publicist have been the press kit, the press release, an artist bio, a couple of 8 × 10 glossy photos and a Rolodex full of contacts. The Rolodex has been replaced by the computer and the smartphone with a database of contacts.

A publicist is only as good as their contact list. Many come from media backgrounds and are sought after because their relationships with key people can get a story placed in a particular magazine or get an artist an appearance on a particular show. If you had the contacts (and the time) you could do the publicist's job yourself! So, before you hire a publicist ask about their contacts and recent successes.

The publicist creates and distributes communications on a regular basis, so the maintenance of a quality, up-to-date contact list is critical to the success of that communication effort. Publicists may maintain their own contact lists or rely entirely on subscription database providers, but usually it is a combination of the two. The value of maintaining a quality database is that the information enables the publicist to accurately target the appropriate media outlet, writer or producer.

An example of a subscription, or "pay" service, for media database management is Bacon's MediaSource. Bacon's updates its online database daily with full contact information on media outlets and subjects on which they report. Services like Bacon's can literally keep a publicity department on target. Most labels and their independent publicist partners maintain keyword, searchable lists within their databases to assure they are reaching the appropriate target audience with each new message. These lists are used to distribute press releases, press kits, promotional copies of the CD for reviews, and complimentary press passes to live performances.

Bacon's parent company, Cision, offers an array of services including in-depth information on media writers regarding their personal preferences and peeves as journalists. Journalists frequently change jobs, so having a service like Cision can make database maintenance easier for the publicist. Publicist also use paid wire services such as PR Wire and Business Wire for news release distribution for major stories.

The most effective way to reach media outlets is through email, because "it's inexpensive, efficient, and a great way to get information out very quickly" (Stark, 2004). A few media outlets still prefer regular mail or through expedited delivery services, but the immediacy of the information is lost. An effective publicist learns the preferred form of communication for each media contact. Bacon's MediaSource and Cision provide some of that information but it is always best to check with the journalist.

Internet distribution of press information from a label requires the latest software that will be friendly to spam filters at companies that are serviced with news releases. The most reliable way to assure news releases and other mass-distributed information are received by a media contact is to ask them to add you to their email contacts. Some companies use services like Emma (www.myemma.com) or MailChimp (www.mailchimp.com) to track whether a news release was received and whether the receiver opened the email containing the news release. It becomes an effective way to be sure news releases are accurately targeted to interested journalists and to be sure they were able to get through spam filters.

THE PRESS KIT AND EPK

The term "press kit" comes from the package of materials that traditionally were prepared, usually in a folder of some sort, to give to the press or the media as a way of introduction of a new artist. As stated above, the press kit would contain a brief **biography** of the artist, a picture or two and a press release to go along with a copy of the artist's latest CD. If possible, the publicist would include album or concert reviews.

Physical press kits are used less frequently these days. A good publicist will know whether the recipient prefers a physical or electronic version of the press kit, call an **EPK (electronic press kit)**. The EPK offers several obvious advantages, not the least of which is the ease of distribution. Once created, an EPK can be sent to hundreds of outlets with the click of a few buttons.

Another major advantage of the EPK is the additional content that it can contain. Because printing costs are eliminated, the variety of high resolution photos that can be made available so that a local newspaper doesn't have to use the same old publicity photo that every other publication has used, is limitless. Videos of the artist performing live or in the studio can be included making it possible for the music supervisor of a late night television talk show (or a local promoter) to actually see that artist perform before making a decision whether or not to have them on the show.

The Internet is slowly but surely replacing the DVD and CD as the preferred method of delivering the EPK content. Many labels and artist managers post their artists' EPKs on a password protected website accessible to the media and music supervisors for film and television. The reason for limiting access is to be sure that the high-resolution pictures and copies of music videos are not misused. Services that provide EPK templates and hosting include Sonicbids.com, presskitz.com, and powerpresskits.com. Bandcamp and Reverbnation are also good sites for EPK-like information without the password protection.

PHOTOS AND VIDEOS

Publicity photos and videos are not without cost. A full-blown photo shoot with a big name photographer (think Annie Liebowitz or Randee St. Nicholas) can cost thousands of dollars, like $10,000–$12.000! If you are going to spend that kind of money you want to not only get some good head shots for the press kit, but photos that can be used in the CD booklet and on the artist's and label's websites.

Why is a photo shoot so expensive? The largest expense is usually the photographer herself, but like so many things you often get what you pay for. And using a big name photographer creates its own publicity angle. Imagine the publicity your artist would get if you hired Anne Geddes to do their photo shoot! In addition to the photographer you will need to hire a makeup artist, a hair stylist, and maybe a wardrobe stylist or costume designer. If your artist is a band, then your expenses can easily double as each member may require their own hair, makeup, and costumer.

Speaking of clothes, will the artist bring their own or will they be rented or purchased? If they are a big enough act you may be able to score some free outfits as the exposure is mutually beneficial for the designer and the artist. If you want to shoot photos on somebody's private property you may have to pay for that privilege. The more locations, the more expenses, including transportation. Sometimes, location is used to help portray the artist's identity, but studio shots are easier to control (Knab, 2001). Most photographers will do the shoot using digital equipment that will save both time and money; however, some photographers still prefer to use film because of its unique qualities, kind of like how some producers prefer analog over digital recording.

Behind the scenes, there is catering and probably a film crew shooting video of the photo shoot for behind the scenes footage that can be used in the EPK or for additional publicity. Since this army of crew and artists may be on location for a day or two and having to travel to a restaurant is disruptive and time consuming, food will need to be provided at the location of the photo shoot.

All these decisions—food, clothing, locations, hair styles—should be determined *before* the day of the shoot by the marketing team, not left up to the artist and the photographer. Remember, these photos will determine or reinforce the artist's image or brand maybe for their entire career and, therefore, need to be consistent with that brand.

The music video is an important part of the EPK, and the marketing department's effort to promote the album as consecutive singles are released. Production of the video is usually overseen by the label's **creative department** and may be promoted by same people who work the single to radio or by an outside agency like Nashville's Aristomedia. The artist's music video is also a valuable tool used by publicists in securing live and taped appearances in television programs. (Other uses of record label videos are discussed in chapter 14.)

PRESS RELEASE

A **press release** is a brief, written communication sent to the news media for the purpose of sharing something that is newsworthy. It frames an event or a story into several paragraphs with the hope that media outlets will find it interesting enough to use as a basis for a story they will create. The key concept in the definition of the press release is that the item be newsworthy. Hard core fans may want to know every detail of the artist's day but the media and the average fan will not. The publicist and the marketing team must find the balance between too much information too often and too little too infrequently. Some events obviously call for an announcement to the press—the signing of a new record deal, the launch of a tour, the nomination or winning of an award or contest, a marriage, the birth of baby. Best to save the more mundane, day-to-day stuff for the artist's blog or their Twitter followers.

These days, it is not enough to send the press release just to the traditional news media because most of the tastemakers in entertainment are outside the mainstream media, like Perez Hilton, for example. Bloggers are a key target for press releases and a key source of information for traditional media. They are also likely to be more receptive to your story than traditional media, especially if you artist is not a star already. A good publicist will target the right media outlets for each story based on the appropriateness and likelihood that they, and more importantly, their readers or viewers, will find the story interesting.

Press releases may be harder to write than you think. There are certain formats to be followed and the need to be concise and still tell the full story.

WHAT SHOULD BE IN A PRESS RELEASE

- Attention-grabbing headline and subhead
- Who (name of artist/band)
- What (style of music/gig/recording being promoted)
- Where (location of event/where recordings are available)
- When (is event taking place)
- Why/How (is show a benefit/for who/why people should be interested/advance tickets/ where to find the venue)
- Quick recap

—Source: (Stewart, 2010a and b)

HOW TO WRITE A PRESS RELEASE

- Get right to the point
- Lead—5 W's and 1 H—Who, What, When, Where, Why, and How
- Make clear the news that it is announcing
- Body: Explain lead and details of the facts concisely
- The writing should be crisp and informative
- Engage the reader and draw them into the news item
- End: Information of least importance should come in last paragraph
- Tell the full story so that an editor could use the press release "as is."
- Include the appropriate contact information so the recipient can follow up if they want to

Source: (Stewart, 2010b)

The press release should be written with the important information at the beginning. Today's busy journalists don't have time to dig through a press release to determine what it is about. They want to quickly scan the document to determine whether this is something that will appeal to their target audience. A well-written headline will contribute greatly to this end. It is important to include links to relevant photos and videos when the press release is going to Internet based media so that they can quickly and easily post your story.

THE ANATOMY OF A PRESS RELEASE

The press release needs to have a **slug** line (headline) that is short, attention-grabbing, and precise. The purpose or topic should be presented in the slug line. It is suggested that a sub-heading be placed under the slug line that supports the point of the press release. The release should be dated and include contact information, phone number, email, and links to the artist's social media. The body of text should be double-spaced.

The lead paragraph should answer the five W's and the H (who, what, where, when, why, and how). Begin with the most important information; no unnecessary information should be included in the lead paragraph (Knab, 2003a). In the body, information should be written in the inverse pyramid form: in descending order of importance. Tell the story in such a way that the article or post is all but written for the journalist or blogger. Minimizing their work improves the chances that your press release will get used.

For the electronic news release sent via email, it should include embedded links where appropriate. Some links that should be considered to be included in the text of the news release should take the reader to the artist's website, the label's website, or to any other site that contributes to the journalist's understanding of the importance of the story. The added benefits of embedding links in electronic news releases is that they often become featured in blogs and music websites which can improve page rankings by search engines on subjects relating to the artist, as well as helping fans find their ways to sites maintained for the artist's benefit.

THE BIOGRAPHY

Writing a **bio**, like writing a press release, takes practice and a special talent. While anyone can put down the facts in some ordered and organized way, a good biographer makes the story compelling and easy to read. A bio for a press kit needs to be short, usually two pages, maybe three for an established artist. Before writing the bio, research is done on the artist's background, accomplishments, goals and interests to find interesting and unique features that will set the artist apart from others. (Knab, 2003b) Keep in mind the target readership of the bio. The bio should be succinct and interesting to read (Hyatt, 2004), and create an introduction that clearly defines the artist and the genre or style of music. The hardest decision for the biographer may be deciding what to leave out. The point is, you should hire a professional to write the artist's bio if you can afford it and if you can't afford it plan to write and rewrite the bio several times before you get the story just the way it needs to be.

An interesting new trend in biographies aimed at more visually oriented consumers is the bio and infographic. Created mostly by third party sites, infographics are a creative, colorful way to convey otherwise humdrum facts. Biographical infographics have been created for celebrities as diverse as director David Fincher (www.hark.com), Conan O'Brien (sdrscreative.com) and Michael Jackson (www.biography.com).

PRESS CLIPPINGS

Nothing succeeds like success! Every press kit, whether physical or electronic, should contain examples of positive press the artist or their music has received, called **press clippings**. Newspaper articles should be reduced to $8\frac{1}{2} \times 11$ inch paper and the print kept large enough to be readable.

One need not include every article, but enough recent articles to let the recipient of the press kit know that others are excited about the artist and their music. Finally, it is important to include links to the press clippings on the artist's website.

PUBLICITY AND BRANDING

As noted previously, a major downside of earned media is neither the label nor the artist can control it, or at least not all of it. It takes only one bad decision by the artist, one momentary lapse in judgment, to undo years of work by the publicist and the marketing team building a strong brand. When an artist is signed to a label or embarks on a career with an independent marketing team, the major parties, artist, manager and label marketing, need to get together and talk about who the artist is, what they want their public reputation to be, and how they will work together to achieve that goal. This is a time for a frank and honest discussion. Pushing the artist to be something they are not will be extremely difficult to sustain in the long run. At their core, the artist has to be who they are. Marketing and management can put some polish and shine on them with media training, wardrobe and make up, but in the end the person on the inside will find his or her way to the surface. Don't try to make an Amy Winehouse into a Taylor Swift or vice versa. It just won't work.

Once the team has a grasp on the image for the artist, everyone must set about the business and building and constantly reinforcing that brand. Of course, the most important contributor to this effort are the artists themselves. Every outfit they wear, every song they sing, every video they make, and every word that comes out of their mouth should reinforce the brand. Some labels hire hair stylists and clothing and costume consultants, some will pay for dental work, and some are rumored to pay for cosmetic surgery in order to polish the artist's image to prepare them for their expanded public career (Levy, 2004). A media consultant may be hired to train the artist to handle themselves in public interviews and other non-musical performance occasions. Working with the artist, the consultant prepares the artist for interviews by taking the unfamiliar and making it familiar to them, teaching them what to expect, and giving them the basic tools to conduct themselves well in a media interview. "In the long-gone, golden, olden days . . . before TMZ, texting, Twitter, cell phone video, and YouTube, image consultants were better able to protect both their clients and the egos of journalists, who privately agreed to clear-cut parameters in exchange for celebrity access" (Carol Ames, 2011–2012) Today's "journalists" may lack the professional training and ethics of the traditional media. More importantly, with cameras built into almost every cell phone, every public act, and some private ones, are likely to end up as a thread on Reddit or a video on YouTube or somewhere else on the Internet.

Recording contracts which include comprehensive and multiple rights over career management of their newest artists create a number of things that impact the image of the artist around which the publicist must work. They include:

- The name chosen by the artist
- Physical appearance of the artist
- Their recording style and sound
- Choices of material and songwriting style
- Their style of dress
- The physical appearance of others who share the stage
- The kind of interviews done on radio and TV
- Appearance and behavior when not on stage

(Frascogna and Hetherington, 2004)

When an artist is ready to remake their image, the transition, whether intended to be temporary or permanent, must be handled with extreme care and forethought. The history of music is littered with stories of attempts to rebrand artists. Some failed (e.g., Garth Brooks' alter ego Chris Gaines) and some succeeded (Katy Hudson the Christian artist became Katy Perry, the pop star; Darius Rucker morphed from pop to country and Rod Stewart shifted from Rock to singing standards with a big band).

THE PUBLICITY PLAN

The publicity plan is designed to coordinate all aspects of getting nonpaid press coverage, and is timed to maximize artist exposure and record sales. The plan is usually put into play weeks before the release of an album. In the case of music magazines, the plan begins months in advance due to the long lead-time necessary to meet their deadlines for publication.

Once the genre of music and the target audience for the artist have been determined, publicity planning begins by coordinating with the label's marketing plan and linking the plan's timetable with the marketing calendar. Remember, publicity is just one part of the overall marketing plan for the artist, all of which is coordinated by the marketing department or a marketing director. The media marketplace is then researched, and media vehicles targeted that have audiences, readers or viewers that align with the artist's target audience. One of the major advantages of an Internet world is that it brings together people from all over the world. There is an outlet for almost every interest on the Internet and the publicist must seek out those outlets that align with the artist's target market. Next, materials are developed and the pitching to journalists and **talent bookers** begins. Lead-time is the amount of time in advance of the publication that a journalist or editor needs to prepare materials for inclusion in their publication. A schedule is created to ensure that materials are created far enough in advance that they can be provided in a timely manner to make publication deadlines. Long-lead publications, mostly print magazines, are particularly problematic for the publicist as they need to have materials prepared months in advance of the release date, and sometimes those materials are not yet available. If an artist suddenly breaks in the marketplace, it is too late to secure a last-minute cover photo on most monthly publications. Fortunately, today's print media almost always have an online version that can respond quickly to breaking news.

Before the materials are sent out, pitch letters are then sent to targeted media requesting publication or other media exposure. The pitch letter is a carefully thought-out and crafted document specifically designed to grab the interest of a busy, often distracted journalist, TV producer, blogger, or online website editor. It is never emailed to a bulk list, but is specifically tailored to each media outlet being contacted (D'Vari, 2003). The pitch letter should begin with a few words presenting the publicist's request, and then quickly communicate why the media vehicle being contacted should be interested in the artist or press material. In other words, the publicist will point out why the media's audience will be interested in this particular artist. Prep sheets are also developed and sent to radio programmers and their consultants so that DJs can discuss the artist as they prepare to play the music on the air. Retail and radio are given the first "heads up" about 16 weeks prior to street date for the album. This may be little more than telling a buyer or program director that the artist has an album scheduled to come out on a particular date, but it serves to create initial awareness and, hopefully, a buzz about the forthcoming release. Serious planning with radio and retail begins about 10 weeks prior

Table 10.1	The Publicity Plan

The Publicity Plan

- Identify Target Market
- Set Publicity Goals
- Identify Target Media
- Create Materials
- Set up Timetable with Deadlines
- Pitch to Media
- Provide Materials to Media
- Evaluate

to street. New release materials, including one-sheets which summarize the information about the new release, are mailed to retailers and media six to eight weeks prior to street in order to make deadlines. Interviews may be done during this time as well so that articles will be ready for magazines that will hit newsstands the same time as the album hits store shelves.

Review copies are mailed or made available electronically to magazines two months in advance. Major newspapers, having shorter **lead times**, get theirs about three weeks in advance of street date. The artist will begin self promotion (calling stations, broadcast media interviews, etc.) the week before street date. As the publicity plan unfolds, its success can be evaluated through clipping services and search engines to see how many "hits" the effort has resulted in.

Table 10.2	Example Publicity Timeline for a Major Label
Time Frame	**Publicity Task**
Upon signing the artist	Schedule meetings with artist
	Press release announcing signing
During the recording	In-studio photos
	Invite key media people to studio
Also during this time period	Schedule media training if needed
	Select media photos
	Determine media message
When masters are ready	Hire bio writer
	Create advance copies for reviews
	Create visual promo items
When advance music is ready–Ideally four months out	Send advances to long-lead publications
	Send advances w/bio and photo to VIPs—magazines, TV bookers, and syndicators
	Begin pitch calls to secure month-of-release reviews
	Start servicing newsworthy bits on the artist on a weekly basis to all media
Advance Music—one month out	Send advances/press kits to key newspapers, key blogs, and TV outlets
	Begin pitch calls to guarantee week-of-release reviews
One week out	Service final packaged CDs to all media outlets
	Continue follow up calls and creative pitching
After release	Continue securing coverage and providing materials to all media outlets

Source: *Amy Willis, Media Coordinator, Sony Music Nashville.*

BUDGETS FOR MONEY AND TIME

"Time, energy, and talent can be more important than budget"
—charity founder Scott Harrison

A budget for the publicity campaign is developed based on the objectives of the project, the expectations of the label for the part publicity will play in stimulating interest in the artist's music, and the degree to which the label is managing the artist's career. If this is the first album for the artist, the development of new support materials may be necessary, such as current photos and a bio. If the new artist is working under a multiple rights contract, tour press support will be necessary. If it is an established artist, budgets could be considerably higher, in part because of the expectations of the artist to receive priority attention from the director of publicity.

Publicity costs include the expense of developing and reproducing materials such as press kits, photos, bios, video, and so forth, communication costs (postage and telephone bills, maintenance of contact lists), and staffing costs. The minimum cost for an indie label would run about $8,000, with $3,000 of that for developing press kits and $2,800 for postage. Adding an outside consultant to the project would add another $1,500 or more per month. For major label projects, an outside publicist can be hired for six months to provide full support to a single, and album, and tour publicity for $25,000, which includes out-of-pocket costs such as postage, press kits, website maintenance, and anything else the label requires to support the publicity effort.

An equally important part of the plan is to budget adequate time to support the album based on when it will be released during the annual business cycle of the label. If the in-house staffing is adequate, given the timing of the project, the plan can be executed without additional help. If, however, the publicity department is overloaded, the director may consider hiring an independent company to handle publicity for the project. This seemingly removes the burden from the director, but it adds oversight duties since the director must be sure the outside company is working the plan according to expectations. The ultimate success (and failure) is still the responsibility of the director of publicity.

PUBLICITY STRATEGY AND OUTLETS

According to Forrester's Marketing RaDaR, consumers go through four stages in the way they interact with a company, brand or artist. First, they *discover* a product or service. If they have additional interest they will *explore* for more detail and information. The next stage of the process is *buying* or consumption of the music, song, album or a live performance. The final stage in the Forrester model is *engagement*. For entertainment these stages may not be clear cut and distinct and the later stages may be repeated (for any product) over and over. In the engagement stage consumers interact with the company (artist) and other consumers to share the experience. Each stage requires different types of media channels to build "Reach and Depth and Relationships" (RaDaR) and thinking about your marketing strategy from this perspective places the emphasis on the right channel at the right time rather than ownership of the channel.

FORRESTER RESEARCH | MARKETING LEADERSHIP PROFESSIONALS

Effective Marketing Programs Must Support The Customer Life Cycle

Mix Art And Science For Marketing Success

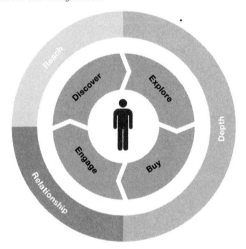

86562   Source: Forrester Research, Inc. Unauthorized reproduction or distribution prohibited.

FIGURE 10.2

Depth channels tell the artist's story. It will be made up mostly of owned media, but could include earned media in the form or magazine articles and public appearances.

Relationship channels are about staying in touch with the artist's existing fans and would include email, Twitter, and other social media that the artist controls.

Reach channels are the earned media, plus advertising and point of purchase merchandising, used to encourage existing or would-be fans to learn more about the artist (Nail and Elliott, 2014).

Some of the existing publicity tools are discussed below.

The Internet—One of the negatives about the Internet is how it has fragmented the music market and made it so much harder to have a multi-platinum seller. One of the positives about the Internet is how it has brought the fragmented markets together in one place. No matter how obscure your interests you can probably find someone else, a blog, an e-magazine, a website that shares your interest. You will want to scour the Internet for the sites appropriate for your artist and target them for press releases, interviews and reviews.

Blogs—Bloggers play a critical role in entertainment publicity because they often have greater credibility with young consumers who are actively searching for new music. A recent study found that traditional media is using blogs as sources of information for their own stories more and more often (Messner and Distaso, 2008) Hypebot.com recommends identifying bloggers and writers specifically, rather than sites or blogs, because they have their own musical preferences and they will do the actual writing of the story. (Five Tips for Identifying & connecting with bloggers, 2012). Pitchfork, Consequences of Sound, Tiny Mix Tapes, and Stereo Gum are some of the most influential music blogs.

E-zines—Electronic or online magazines offer a good way to introduce a label's artist to the target market. MarketingTerms.com defines *e-zines* as an electronic magazine, whether posted via a website or sent as an email newsletter. Some are electronic versions of existing print magazines complete with magazine style formatting, whereas others exist exclusively online or digitally. The web-posted versions usually contain a stylized mixture of content including photos, articles, ads, links, and headlines, formatted much like a print equivalent. Smaller versions may be emailed to subscribers as a pdf file. Most e-zines are advertiser-supported, but a few charge a subscription fee.

Many established music e-zines are genre specific or have particular subject areas dedicated to genres. They may feature music news, concert and album reviews, interviews, blogs, photos, tour information, and release dates. As a result, their readers are predisposed to be receptive to new and unfamiliar artists and their music, provided that the artist is within the genre that the e-zine represents. A study of the readers of the Americana music magazine, *No Depression* found that 90% of their readers learned about new music from an article published either in a print or in an online version of a magazine. It is probably no coincidence that *No Depression*, which started out as a print magazine in 1995, added an online version in 1997 on its way to becoming exclusively online in 2008.

What to Send

E-zines are mostly interested in feature articles and press releases pertaining to some newsworthy item (such as an album release or a tour schedule announcement). Label publicists write the article with an assumption that it will appear unedited in the online publication, in the inverted pyramid style. Articles like this also include an attached publicity photo or two along with the article for submission. As with any other publication, label publicists should take care not to send a news release to an e-zine if there is nothing that is considered newsworthy. Like all journalists, those at e-zines will look at the news release to determine its relevance and timeliness for their readership, and an irrelevant news release strains the relationship the publicist has with their contacts.

Where to Send It

The Ezine Directory has a listing of many of the better-known music e-zines, along with descriptions and ratings of each (www.ezine-dir.com/Music). The goal, like any target of publicity, is to find those with the correct target market and submit articles, music, and photographs to the editor, encouraging him or her to include a link to the artist's web site. Some e-zines have submission forms available on their website, whereas others are not as specific about their submission policy. When an article does appear online, the publicist will link the artist's and sometimes the label's websites to it.

Resources for E-zines and Distribution of Press Releases:

The Ultimate Band List: www.ubl.com
Rapping Duck Press: rappingduck.com
The Ezine Directory: www.ezine-dir.com
Music Industry News Network: www.mi2n.com
MusicDish: www.musicdish.com (try the open review and "submit your article")
PRWeb: www.prweb.com

Amazon.com and Wikipedia as resources for basic information.

Labels supply and offer product through Amazon.com not because of the sales, but because people use it as a resource when looking for information. According to Yahoo!, the number two search engine, "those searching for artist information are selecting the Wikipedia entry link over that of artists' MySpace pages by a factor of more than 2-to-1. The Wikipedia entries are also more popular than artists' Web sites" (Bruno, 2008a).

Wikipedia.com has over 4.5 million articles in 287 languages and receives over 500 million unique visits per month. "While anyone can contribute to a given article, they must first past muster from a team of volunteer editors with a particular passion about the subject before the text appears live." Photos and clips of audio and video can be added to the page (not full songs or videos). This is not the place for breaking news because the editing process takes too long, but it is a good place to post the artist's bio and historically-oriented factual information, including the artist's discography and other facts about the artist and their career (Bruno, 2008b). Avoid hyperbole and unsubstantiated claims ("she is the world's greatest singer") or your submission may be edited or rejected.

Broadcast—Getting publicity on broadcast outlets will be more of a challenge for a new or unsigned artist because of the limited time, so being persistent (without being annoying) and targeting the appropriate outlet with the right information is important. Many factors go into these booking decisions, but mostly the publicist needs to do their homework to make sure that the artist and their story are a good match for the show.

TELEVISION APPEARANCES

News Shows

Major entertainment television news shows, including syndicated news shows on major network affiliates and cable channels, are most often interested in major acts. Their viewers want to know the latest information about their favorite recording artists. Stars with the highest **Q factor**, that is those celebrities that are easily recognizable, are most often sought for their entertainment news stories because they draw a lot of interest and big audiences. With major artists filling prime interview opportunities, it becomes a genuine challenge for the record label that is trying to publicize a new artist. In order to compete with the superstars who can easily get airtime, a new act must have an interesting connection with consumers that goes beyond the music. There are more artists looking for publicity than slots on talk shows, forcing the label publicists to be as creative as they can to get the attention of producers and music bookers for their new acts.

Television interviews with new artists require a backstory that sets them apart from every other "new artist with a great voice" who is seeking the media spotlight. Shows look for that added dimension to a new artist that makes them interesting to the viewers, and they often look for the nontraditional setting in which to present the story. Though at times it is overdone, connecting an artist with their charity work becomes an interesting angle for television.

The challenge to the label publicity department is to find those key personal differences that make their recording artists interesting beyond their music. Label publicists are sometimes criticized for citing regional radio airplay, chart position, or YouTube views as the only positives that make their newest artists stand out. Those in the media say they look for that something special, different, and newsworthy that gives an angle for them to talk or write about. In marketing terms, the media is looking for strong brands that are uniquely positioned against the competition. In that light, it puts the responsibility on the label publicist to find several different angles to offer to different media outlets to generate the interest needed to get a story placed. Writers for major media want their own angle on an artist when possible because it demonstrates to media management that an independent, standout story has been developed, making them different from their competition. Sometimes, though, the story angle about an artist is different enough that it stands on its own and most media will see the value it has for their audiences. Entertainment writers and producers are often self-described storytellers, and delivering that unique story to them is a continuing challenge to the successful label publicist.

Talk-Entertainment Shows

Label publicists are often the facilitator of an artist's appearance on popular talk and entertainment shows, often with the result of introducing an artist to an audience that is not actively seeking new music. This would include shows like Jimmy Kimmel Live, The Late Show, and daytime shows like "Ellen DeGeneres Show" and the "Queen Latifah Show."

According to Tompkins (2010) "Saturday Night Live" has the biggest impact on sales after an appearance. This is probably because there are no competing shows and the band gets to play two songs. Shows like "The Late Show" (on NBC) and the "Tonight Show" (on CBS) typically limit the artist to one song and it is at the end of the show when many viewers may have already tuned out or fallen asleep. The effectiveness of an appearance on various television shows will be discussed in greater detail in the video chapter.

The payoff to the label for an artist's appearances on television shows comes in the form of sales. For example, after spending a week appearing on an array of television shows like "The View" and "Today" the week of February 24, 2007, Josh Groban's album, *Awake* rose in the *Billboard* 200 sales chart from number 33 to number 17 (Paoletta, 2007). The British rock band, The Heavy saw sales of their album, *The House That Dirt Built* jump 537% in the week after their appearance on the "Late Show with David Letterman" (Tompkins 2010).

Bookings on programs like these are handled by the publicist based upon their relationships with talent bookers on these shows. It is not uncommon for a publicist to precede a pitch for an artist to appear on one of these shows by sending a big fruit basket. However, the success of placing the label's new artist on one of these shows is also based on the ability of the publicist to build a compelling story for the artist that will interest the booker. Often the publicist will offer another major artist for a later appearance in exchange for accepting the new artist now.

Major labels have the benefit of their high profile roster of artists and the financial resources to promote live performances to major shows. Independent labels, with their much smaller promotion budgets must, by necessity, approach a pitch for a live performance keeping those limited dollars in mind. Cole Wilson was the music booker for "The Late Late Show With Craig Ferguson" and offers these points to the indie publicist seeking a performance on the show.

- The artist needs an online presence where the booker can see performances and read the comments left by fans.
- Talent bookers for late night shows in New York and Los Angeles frequently spend time visiting live entertainment venues in those cities, providing the bookers an opportunity to see a prospective artist.
- If an artist is "different from the norm" it gives the booker an opportunity to present something fresh to the audience. She says an artist who sits on a stool and sings for three minutes lacks visual appeal.
- The publicist should remember that it isn't just the talent/music booker who must be convinced the artist should be invited to the show. Often it is a committee who will want to view the artist from as many perspectives as the publicist can present.
- The artist should remember that an appearance on television does not mean that they can make "outrageous demands" from the show. (Donahue, 2008)

Melissa Lonner, who served eight years as the "Today" show senior producer and entertainment booker gives the following advice to label publicists pitching an artist for an appearance on the NBC morning show.

- Keep your pitches short over the phone and/or email.
- Don't pitch on voicemail.
- Send a CD of music with selective press clippings.
- Send an email to follow up and recommend a track.
- Don't send the deluxe press clippings collection.
- Don't say why the artist should be on "Today."
- Provide the music, stats and the facts—not the hype.
- Don't say that the artist is the next "——."
- Follow up on pitches via email or phone.
- Be kind, calm, and honest.
- Don't stalk, threaten, or demand.

(Paoletta, 2007)

Award Shows

The value of having an artist perform on an award show is obvious—it provides tremendous artist exposure and sells recordings and concert tickets. These slots are coveted by all the record labels, and lobbying efforts may pay off in a big way. Many awards shows are showing modest declines in viewership, artists who are nominated or who perform on music award shows can see spikes as high as 700% in the sales of their music. For example, the week after the 2011 Grammy Awards show, Mumford & Sons sales went up 99%. Digital track sales went up 93% from 52,000 to 100,500 and the album went to #2 on the *Billboard* 200 chart (Grammys Spike Sales, 2014). Two years later, their appearance on the Grammy Awards show boosted sales of "Babel" by 242% (Grammy Awards, 2013). In 2012 Adele's 21 saw a post-show jump in sales of 493,000 units, a 207% increase over the previous week (Molanphy, 2013).

Comparing Publicity and Record Promotion

> The savvy labels recognize how important publicity is to the mix—it's almost as important as record promotion.
> —Phyllis Stark, former Nashville editor, *Billboard*

The effort of radio promotion has great impact on the marketing of a recorded music project. Table 10.3 is a look at the relationship that publicity has with its counterpart in the overall promotional effort for an album.

Table 10.3	Comparison of Promotion and Publicity Departments	
Record promotion	**Publicity department**
Develops and maintains relationships with key radio programmers (gatekeepers).	Develops and maintains relationships with key writers, news program producers, and key talent bookers for network and cable channel TV shows.
Tells radio programmers that a new single or album is about to be released and to prepare for "add" date; sends promo singles and albums.	Prepares and sends a press kit to journalists announcing the new single or album project.
Schedules the new artist for tours of key radio stations for interviews and meet 'n' greets with station personnel.	Schedules the artist and sometimes the album producer for interviews with both the trade press and consumer press.
Employs independent radio promotion people who have key relationship with important radio programmers.	Employs independent or freelance publicists who have key relationships with important media outlets.
Effectiveness of their work is measured by the number of "adds" they receive on the airplay charts of major trades.	Effectiveness is measured by the number of "gets" they receive, meaning the number of articles placed, number of TV news shows in which stories run, the number of talk/entertainment shows on which the artist performs (Phyllis Stark).
Gets local radio publicity and airplay for new artist based on the promise of an established artist making a local appearance sometime in the future.	Gets new artists booked on major talk/ entertainment shows based on the promise of making an established artist available to the show sometime in the future. Supports local press during touring.

CHARITIES AND PUBLIC SERVICE

People expect successful artists to give back to society. After all, the world, the fans, has given them so much. But don't wait until your artist wins her first Grammy to do some charity work. Being proactive and building good public relations can go a long ways towards minimizing negative publicity when and if it occurs.

Having an artist associated with a respected charity is always good publicity. If the cause is personal, even better. We once worked on a marketing plan for a band whose members had lost their mother to breast cancer. They played every cancer charity gig they were offered. It was personal, not just something they were doing for publicity. As the artist's star rises, the association with a specific charity becomes even more of a win-win, giving both the artist and charity greater exposure. Some artists have become synonymous with their charities: Elton John and the Elton John AIDS Foundation; Jars of Clay and Blood Water Mission; and Bono and Band Aid, for example.

Another way for artists to get involved in charities is through organizations like Global Citizen. This organization fights poverty by getting people to volunteer in order to receive free concert tickets. "The goal is to reward volunteer work with live music. Fans can take part in various social actions, ranging from signing partitions to calling their representatives to earn points they can use to win free concert tickets" (Waddell, 2013). Artists donating tickets to the cause range from the relatively unknowns to One Direction, Tim McGraw, and Bruno Mars.

BAD PUBLICITY

> You can't shame or humiliate modern celebrities. What used to be called shame and humiliation is now called publicity.
> —political satirist, journalist, writer, and author P. J. O'Rourke (b. 1947)

Is all publicity good publicity? What comes to mind when you hear these names: Ty Herndon, Janet Jackson, Bono, Ozzie Osbourne, Kanye West, Amy Grant, Paul Simon, Ian Watkins? All have received "negative" publicity at some point in their musical careers and most, but not all, have survived in the business. So the question is "Why?" Why do some artist bounce back or even thrive despite arrests or drug addiction and others have their careers ruined? The answer is brand image—or more precisely, consistent brand image. When Kid Rock was arrested (Billboard.com, 2005) for punching a strip club DJ, did he lose any fans? Probably not. It hardly made the news. Why? Because the behavior was not inconsistent with his image, his brand. But imagine what would happen if Christian artist, Steven Curtis Chapman had been arrested for punching out a strip club DJ? He didn't, but if he had that would have made the news and probably ended his career because it is not consistent with his image, his brand. For a real life example, one need only look at Steven Page's career trajectory. Page was a founding member of the Barenaked Ladies, a Canadian pop-rock band known for its witty and mostly lighthearted songs. But his clean-cut image has destroyed when he was arrested in July 2008 for drug possession. Perhaps, had the band not just released an album targeted at children the offense would have been overlooked, but his behavior didn't fit with the band's new target audience; it didn't fit the band's brand. Page officially left the band early the next year. His career has yet to recover.

What to Do When Negative Publicity Occurs

Every situation is unique and will call for professional judgment. The following is meant to be a general guideline. There is no "one size fits all" solution to negative publicity.

Chances are the first person the artist will call when something bad happens is their personal manager. Managers are usually easier to face than an angry spouse and may have more experience with bailing people out of jail. Then the manager will call the publicist. Before anyone makes a public statement, get the facts about what happened. Talk to your artist and anyone else who was in their entourage when the "event" occurred. Make sure you are getting the truth and not just the artist's version of what happened. You don't want to be surprised later when contradicted by other witnesses. Every cellphone has a camera in it. Look for videos of the event to show up on YouTube.com, TMZ, and the evening news.

Once you have the facts you can decide how you and the artist should respond. You have two basic choices: Ignore it or respond to it. You can safely ignore the situation, that is not have an official response, if the event simply reinforces the artist's reputation. If your blues artist was seen smoking an illegal substance, but not arrested, and the story makes it to the tabloids, fine. The behavior is not unexpected or contrary to his image. On the other hand, if your artist is involved in an accident or hurts somebody you will need to respond publicly. Be direct and as open and honest as possible. If the artist has a drug problem or other reoccurring behavioral problem then management will probably want to direct them to a rehab program. Fans are usually sympathetic when the problem rises to the level of disease or addiction, and sometimes managers use this to their advantage, even if the artist is really just a badly behaved spoiled brat! The rehab program will take the artist out of the public view for weeks or even months, allowing the furor to subside and for them to get their act together. And with any luck, they may even come out a better person and be inspired to write a hit song about it.

CONCLUSION

Publicity, or earned media, is one of the most powerful tools that a record label marketer has at their disposal. Unfortunately, much of it is also out of their control. For publicity to be successful, the label needs the cooperation and assistance of the traditional and Internet media. The most successful publicity will go viral and spread directly from fan to fan. This will be discussed further in chapter 11.

APPENDIX

STRUCTURE FOR A PRESS RELEASE

FOR IMMEDIATE RELEASE or FOR RELEASE DATE
FOR MORE INFORMATION: Name, phone, email
City/State, Date

Headline

- Considered to be the most critical part of the release
- You have 20 seconds to grab the reader's attention
- Often the only part of the press release that the media reads
- Use short, clear and hard-hitting one-line summaries to identify what you are promoting. Don't be afraid to be dramatic.
- Sub-headings are also used to attract attention and provide information.

First Paragraph

- Purpose of the first paragraph is intended to alert the media and to inform them of what you are promoting—who, what, when, where, why, how
- Make sure that the first paragraph has no more than three or four sentences
- Needs to set forth all of the main points covered in the release

Body of Press Release

- This is an opportunity to provide the details of the story
- Press release should be written in the third person
- Don't weigh the piece down with extraneous details

End Paragraph

- Should summarize the story
- #### or—30-

About Us
Contact Information
Source: (Stewart, 2008)

GLOSSARY

Bio—Short for **biography**. A brief description of an artist's life and/or music history that appears in a press kit or other publicity material.

Creative department—This is a department or division at a record label that handles design, graphics, and imaging for a recorded music project. Also called creative services.

Clippings—Stories cut from newspapers or magazine.

Discography—A bibliography of music recordings.

Electronic press kit—An electronic version of a standard artist press kit that includes digital images, documents, audio and video files and PDF versions of all documents and news clippings. Some may contain video clips that can be used on-air and magazine-quality images for reprint purposes.

Independent publicist—This is someone or a company that performs the work of a label publicist on a contract or retainer basis.

Lead time—Elapsed time between acquisition of a manuscript by an editor and its publication.

Press kit—Collection of printed materials detailing various aspects of an organization, presented to members of the media to provide comprehensive information or background about the artist.

Press release—A formal printed announcement by a company about its activities that is written in the form of a news article and given to the media to generate or encourage publicity.

Q Factor—A term used to indicate the overall public appeal of an artist in the media. A high Q factor means an artist is able to draw large television audiences.

Slug—A short phrase or word that identifies an article as it goes through the production process; usually placed at the top corner of submitted copy.

Talent bookers—These are people who work for producers of television shows whose job it is to seek appropriate artists to perform on the program.

REFERENCES

Ames, C. "Popular Culture's Image of the PR Image Consultant: The Celebrity in Crisis," *Image of Journalist in Popular Culture Journal* (2011–2012), pp. 90–106.

Bruno, A. "It's High Time You Edited—Or Perhaps Created—Your Entry." *Billboard*, Vol. 120, No. 13 (March 29, 2008a).

Bruno, A. "Music fans prefer Wikipedia to MySpace," *Reuters* (March 22, 2008b). Accessed July 31, 2014. http://www.reuters.com/article/2008/03/22/us-wikipedia-idUSN2148195720080322.

Dommeruth, W. *Promotion: Analysis, Creativity, and Strategy,* Belmont, CA: Wadsworth, Inc. 1984.

Donahue, A. "The Indies Issue: How To Get On A Late Night Show," *Billboard*, Vol. 120, No. 20, (June 28, 2008), p. 27.

D'Vari, M. "How to Create a Pitch Letter." 2003. http://www.publishingcentral.com/articles/20030301–17–6b33.html.

"Five Tips for Identifying & Connecting with Bloggers" 2012. http://hypebot.com/hypebot/2012/11/5-tips-for-identifying-connecting-with-music-bloggers.html.

Frascogna, X., and Hetherington, L. *This Business of Artist Management*, New York: Billboard Books. 2004.

"Grammy Awards 2013: Sales soar for Mumford & Sons, Fun., and Goyte in Wake of Awards Show Glory," May 1, 2014. www.nydailynews.com/entertainment/music-arts/mumford-fun-gotye-enjoy-post-grammy-spike-article-1.1268860.

"Grammys Spike Sales". May 1, 2014. http://www.grammy.com/blogs/grammys-spike-sales.

Hyatt, A. "How to be Your Own Publicist." 2004. http://arielpublicity.com.

"Kid Rock Arrested on Assault Charge." 2005. Accessed May 1, 2014. www.billboard.com/articles/news/64079/kid-rock-arrested-on-assault-charge.

Knab, C. 2001. Promo Kit Photos.http://www.musicbizacademy.com/knab/articles/.

Knab, C. "How to Write a Music-Related Press Release." November 2003a. http://www.musicbizacademy,com/knab/articles/pressrelease.htm.

Knab, C. 2003b. http://www.musicbizacademy.com/knab/articles.

Levy, S. *CMA's music business 101.* 2004. Unpublished.

Messner, M., and Watson Distaso, M. "The Source Cycle." *Journalism Studies*, Vol. 9, No. 3, 2008.

Molanphy, C. "A Brief History of the Grammy Sales Bump." *The Record: Music News from NPR.* NPR, February 15, 2013. Accessed June 27, 2015.

Nail, J., and Elliott, N. *Mix Art and Science for Marketing Success.* November 21, 2014. https://www.forrester.com/Mix+Art+And+Science+For+Marketing+Success/fulltext/-/E-RES86562

Paoletta, M. "As Seen On TV," *Billboard*, Vol. 119, No. 16, (April 21, 2007), p. 27.

Stark, P. Personal interview. April 21, 2007.

Stewart, S.M. "Artist-generated Publicity: Part I." *Fringe Magazine* Vol. 1, No. 5, (March/April 2010a). Web. fringemagazine.com.

Stewart, S.M. Artist-generated Publicity: Part II." *Fringe Magazine*, Vol. 1, No. 5, (May/June 2010b). fringemagazine.com.

Stewart, S.M. "How to Write a Press Release." 2008. Course handout, ts.

Tompkins, T. "The Impact of Late Night Television Musical Performances on the Sale of Recorded Music," *MEIEA Journal*, Vol. 10, No. 1, (2010).

Waddell, R. "Global Cause." *Billboard*, May 11, 2013.

CHAPTER 11

Social Media

Contributed by Ariel Hyatt

This chapter is contributed by leading music publicist and social media strategist Ariel Hyatt. Hyatt's company, Cyber PR®, specializes in branding, social media, and public relations. The New York based firm concentrates on connecting artists with fans via social media, blogs, podcasts, Internet radio, and the myriad of messaging that takes place via cyber space and beyond. The firm has represented over 1700 musician of all genres, including George Clinton, John Popper and the Nitty Gritty Dirt Band. This is her social media advice for artists and labels.

THE CONCEPT OF TRIBES

When Seth Godin first came out with his book *Tribes* in 2008 it felt like a revelation. The reason was someone had finally articulated the way forward in building community and a pathway to success in the music business.

Here are the exact words from his famous Ted talk on *Tribes*:

Tribes . . . is a very simple concept that goes back 50,000 years. It's about leading and connecting people and ideas. And it's something that people have wanted forever. Lots of people are used to having a spiritual tribe, or a church tribe, having a work tribe, having a community tribe. But now, thanks to the Internet, thanks to the explosion of mass media, thanks to a lot of other things that are bubbling through our society around the world, tribes are everywhere.

(Godin, 2008)

As you are building your tribe, there is one thing to keep in mind: Social media creates the appearance that each of your fans holds the same weight, be it one 'like,' one 'follow,' or one 'friend.' This couldn't be further from the truth.

Your fans are all different. The fact is that you will run into a wide range of fans, some of whom are passively connected to you online but may not have actually heard you; meanwhile others will be dedicated super fans who actively evangelize your music to others. Of course, most of your fans will fall somewhere in between these two extremes.

However, no matter how small the percentage of your fan base that could be considered super fans, these are your true moneymakers and thus should be the focal point of a majority of your time and attention.

Super fans are the ones who will not just evangelize your music, but will spend the most money—on downloads, physical albums, tickets, and mercy.

So what makes super fans so special? An emotional connection has been established.

These fans, or tribe members, more than just *like* your music. They have a *connection* to you, your music, and/or even the fan base that is so strong that it is a part of them.

The more emotionally connected fans you have, the more money you will make both in the short-term and the long-term.

Take this into careful consideration: the Internet is a wide-open place full of countless opportunities. The more focused you are, the easier it will be to connect and build your audience swiftly. The golden keys that unlock the door are the niches. Once you have connected to a niche you will be able to swiftly build your tribe. One very effective way to do this is through focusing on the blogosphere and establishing yourself as credible in the areas you are most connected to.

With 2,000 new releases every single week and hipster, indie rock music blogs taking up the lion's share of "music writing" these days, the options are limited if you do not fall into a specific niche.

EXAMPLE BLOG NICHES

Humanitarian Charity/Causes

Pets/Animal Rescue

Religion

Your Community

Gardening

Vegan/Gluten-Free

Travel

Cancer/Survival

Politics

Positivity

Your City/State

Crafting

Fashion

Yoga

Teen/Tween

Cooking/Recipes

Eco/Green

Health Books/Reading Parenting

Guerilla and ambush marketing tactics

Artist and label websites

Facebook

YouTube

Twitter

Instagram

Pinterest

Internet marketing

Mobile marketing

App development

Indie labels and unsigned artists

International implications

INTERNET MARKETING: ARTIST AND LABEL WEBSITES

Your Website Is Your Front Door

You can't enter a house without a front door. Keeping up with all of the developments and strategies in the online space can be a tough job. As soon as you master a social platform or technique that works, it may change (hello, Facebook!). One thing that doesn't change and never will is this: Your entire online presence starts at your website. Social media is where a lot of action and conversation takes place but your website is the part of your online house that you built and own. Social sites may come and go, but your website is yours (as long as you pay your hosting fees!). This is why it is crucial to keep your website updated.

How to Set Your Artist on the Right Path

First, you must have a domain name. To register a domain name go somewhere like godaddy.com (U.S.) or crazydomains.com.au (Australia).

Register the domain that you would like to use. Get a dot com (.com) if you can, with no dashes, dots, or underscores (e.g., www.artistname.com). And make sure that the corresponding YouTube, Twitter and Facebook page names are also available. It is important to make sure your social media sites match.

Free!

A free site owned by AOL called About.me is one of the best ways to start your first footprints on the Web. Using widely accepted search optimization formatting, it allows you to build and design your own webpage attached to your own choice of URL.

Pay-as-You-Go

A pay-as-you-go option with a website builder can get you up and running very quickly and you won't need a designer to build for you. Here are four sites that host musician and band websites. All have excellent call-in customer service to help ease the confusion. All also have tons of fabulous templates to choose from and many social media options and plug-ins to bring social media into your site layout.

Bandzoogle—http://bandzoogle.com
Bombplates—http://www.bombplates.com
Hostbaby—http://www.hostbaby.com
Wordpress—http://www.wordpress.com

Don't pay more than $500 for a basic WordPress site. If you do desire a designer, check out http://www.crowdspring.com or http://www.linkedin.com for finding affordable WordPress designers. Make sure you read the designer's reviews and see examples of his or her work before you hire him or her so you don't get any unpleasant surprises.

NINE STEPS TO A GREAT HOME PAGE

1) The entire website should be easy to navigate with a navigation bar (nav bar) across the very top of each page or down the left hand side so visitors can see it, not buried where they have to scroll down.

2) Create a unified brand with the look, the colors, and the logo (if your artist has a logo) and of course a stunning photo of the artist or the band. Your social media sites should all match your website colors.

3) Your artist's social media sites should feature your artist's name and pitch (or specifically what your act sounds like in a few words). If you feel weird creating a "pitch," use one killer press quote or fan quote that sums up the way you sound.

4) Feature a free MP3 in exchange for an email address.

Use ReverbNation, PledgeMusic, Topspin, and NoiseTrade widgets. These "widgets" are imbedded into the site and allows for a site visitor to be able to download a song for free, pass along the song to others, or imbed the song onto their own sites.

ReverbNation—http://www.bit.ly/reverbfreebribe

PledgeMusic—http://www.bit.ly/pledgefreebribe

NoiseTrade—http://www.noisetrade.com

TopSpin—http://www.topspinmedia.com

5) Link your artist's home page to your social media: Facebook, Twitter, YouTube, Pinterest, ReverbNation, Sonicbids, Last.fm, and anywhere else you maintain an active profile.

6) Include a Facebook "like" widget.

7) Include a Twitter stream or a group Tweet stream that updates in real time.

8) A blog feed or news feed, or new shows updating onto the page via widgets. Again, using widgets similar to those listed above, show listings can be shared among interested visitors.

9) If you like sharing photos, use a Flickr or Instagram stream, which imports over to your blog!

Navigation Bar Elements and Tabs

It is crucial to think through how you present your artist and your brand to people who may want to write about them, feature them, or share their music. More and more, we see online tastemakers influencing their friends; having elements and assets that are easy for visitors to grab and share will make this a win-win for you both. Navigation bars should include:

1. Bio or Press Kit. For the press kit, use Sonicbids or ReverbNation. Make sure your photos really capture artist's essence. Make sure they have clear instructions on how they can be downloaded.

2. A link to buy music—iTunes or a storefront

3. A tour or live performances tab linked to an up to date list and locations

4. The blog: Linked to Tumblr, Wordpress, Blogger or whatever resource used to create the artist's blog

5. The artist's contact information

Historically, it's been very difficult to find simple press components on many artists' websites. Here are three critical components that should be included on the press page of your artist's website. These components show music writers and calendar editors that you care about making their lives easier. Editors need access to artist information quickly, because they are constantly under a deadline. If you do not make it easy for them to get the artist's information from their site, they may move onto another one of the 50 artists that are playing their market that same week.

1. Music—Album or Live Tracks

Make sure you have some music available on the website or a very obvious link to your Facebook where people can hear the music instantly. Many newspapers are now including MP3s of artists coming to town in the online versions of their papers, so make it easy for them to download the tracks to add to their own sites; this is additional excellent exposure for your artist.

2. Biography—Including the Artist Pitch

Make sure you have a short, succinct bio that can be easily located on the site in addition to the long-form bio. I suggest having three bios:

1. Long form

2. 50 words or less

3. 130 characters (a tweet plus a username)

Make sure this bio can be easily cut and pasted so writers can drop it into a preview or a column. Also include a short summary (less than 10 words)—**the pitch**—that sums up the artist's sound for calendar editors.

You can also include the blogs and all the opinions from each band member. But remember that while these extras are fun for the fans, they are not necessarily useful for music writers.

Website Takeaways

- It is important that you identify your niches before you take on rebuilding or creating the website.
- Assess what you need to do for your artist's website, whether a touch-up or full rebuild. Which plan will you choose? Free? Pay as you go? Work with a designer?
- Make sure your site is easily updatable by you (and that you are not beholden to anyone but you!).
- Create a personalized system for site updates that you can follow so you have a system that your team or interns can follow as well!

NEWSLETTERS

Email is still king when it comes to generating revenue, specifically for musicians. Artists build relationships with fans on your social networks, and turn them into customers by driving them to sign up to their newsletter.

Are you including content in your artist's newsletters that is interesting and amusing? If you are just talking about their next show, or their next release, then you are missing the mark.

The WIIFM Principle

People are constantly making decisions based on what I call the WIIFM (What's in it for Me) Principle. Whether they perceive it consciously or not, people are always thinking with this principle in mind.

The artist needs to provide new, potential fans with an incentive to sign up on their newsletter list. Unless they already adore the artist, they will not grant you permission to market to them unless you give them some sort of content in exchange for their email address.

Be Consistent—Send Newsletters at Least Once a Month

You should send your newsletters out consistently—and statistics prove that people are most likely to open their emails on Tuesdays and Wednesdays, so send on one of those two days, at around 12 pm. The reason for this is people often check mail from work, and they are swamped on Mondays (digging out from the weekend), and they are checked-out on Fridays. A recent case study also suggests sending on Saturdays because people have time on weekends. The one downfall of this is that you may see a higher unsubscribe rate on a Saturday, but your open rates may indeed increase.

Choose Your "From" Field Carefully

According to Doubleclick surveys, 60% of your readers determine whether to open your email based entirely on the "from" field. So choose this field carefully, and keep it consistent every month. Something generic may seem impersonal, like Newsletter@YourBand.com, so try putting your first name or first initial and last name there to personalize the address.

Keep Your Subject Line to 55 Characters

Most email programs cut off the subject line after 55 characters, so keep your subject line short and sweet, and to the point: five to six words max.

Shorter Is Sweeter

Remember, people have short attention spans. Keep the length of each of your newsletters short and sweet: four to five paragraphs max. If you have more content than that, link them to the artist's blog via a hotlink, so interested people can read more.

The Artist Doesn't Have to Have an Upcoming Show to Send a Newsletter!

How about inviting everyone on the newsletter out for drinks for an evening, or to join the artist for a show of another artist they love? Or share something fun that they've done recently; maybe they just purchased a new album and they love it, and they want to talk about it.

> If you have shows to promote, keep your gig list short. If you have a long list of upcoming gigs, you don't want them to take up a large portion of the newsletter. Showcase a few, and then link fans to your site or Facebook Event page to find out more information.

Give Away a Free Gift or Make a Special Offer

For example, the first three people to respond get something: a free exclusive MP3, something that they've done recently, like a special poster to one of the artist's gigs, or a T-shirt. One of our artists, Pete Miser, actually held a contest for a pair of his old Adidas shoes! It was a great idea—he actually got his entire mailing list engaged, and one of his fans won a pair of his old sneakers. It was fun, and it was effective!

Mailing Address and Unsubscribe Link

By law, you need to put your mailing address and an unsubscribe link at the bottom of each of your newsletters. If you are uncomfortable adding your home address, then open up a P.O. Box and use that. This is why using a newsletter provider is critical.

HTML Is Better Than Text

Studies have shown that HTML emails get a better response than regular text emails. However, if you're dedicated to using all-text formats, keep your lines to 65 characters each. This way they will be formatted properly.

HOW TO WRITE AN ENGAGING NEWSLETTER

Greeting: Make it Personal

Share something non-music related in the greeting. Pull people in on a human level. Make them care about the artist as a person, not just as a musician.

Some ideas:

- Vacation

- Family time (doesn't need to be too specific)

- Whatever they are reading or listening to

- TV and movies they are into and why they liked them

Post photos of these personal touches on Instagram, Facebook, Pinterest, their blog, and so on.

Guts: The Body of the Newsletter

What are they up to as an artist? Are they in the studio? Are they touring? Writing new tracks? Remember, people love and connect to stories, so TELL STORIES!

Getting: Putting Readers into Action

This call to action is the part of the newsletter that gets the fans to take action and it is the most critical part of the newsletter:

Ask them to join the artist on socials

Ask them to vote for the artist in an online contest

Ask them to review the artist's Music on CD Baby, iTunes, or Amazon

Give them a survey or a contest to participate in

Gift them a free download: A gift makes your artist memorable

Invite them to an upcoming show

There should only be one call to action per newsletter. Fans will get confused and choose nothing if they have more than one choice.

Newsletter Takeaways

- Your newsletter doesn't have to be perfect the first time just send it out and improve upon it each time

- Don't just fill your newsletter with news and information about the band's activities—include "fun" information that will get the fans excited

- The key to the success starts with the size of its list; start building your list!

- Add people to your list carefully; don't spam—make sure fans have opted in to receive your newsletter.

- Read your newsletter and think about how you would respond if it came to your inbox every week?

- Have fun with it! The fans will notice.

FACEBOOK

My Cyber PR® team and I have written so many "how to" articles on Facebook and the site gets easier to use as it develops, while the articles get shorter and easier to craft. That being said, Facebook has changed so many times it's hard to keep up.I'm betting Facebook will change yet again even after this book gets released, so please bear that in mind as you read this.

Facebook Questions That Always Gets Asked!

Do I really need to have a Facebook personal profile as well as a Facebook fan page?
   The answer is YES! Here's why:

1. You need to have a personal profile in order to administer your fan page.

2. Personal pages max out at 5,000 friends. This means if you friend more than 5,000 people, bands, and brands on a personal profile, you will have to unfriend people in order to add new ones, or start another personal profile (which Facebook does not consider "legal"). So, having a fan page is crucial, because it allows the artist have to have more than 5,000 friends.

3. You cannot track Facebook insights (analytics) from a personal profile. Knowing the effectiveness of your marketing is vital to improving your reach and your marketing. You cannot measure this from a personal profile.

4. Some things really are personal. Facebook is now an important tool for many of our non-musical, self-promoting friends. Some people in your life may want to share with you and your family, tag photos of you in personal settings, etc. These friends may not care so much about your Fan Page updates, song giveaways, and tour announcements. There is now a way to separate what is truly private more effectively than ever before.

How Your Artist Arrived at This Problem

Your artist may have started promoting herself on Facebook before Facebook Fan Pages existed, so their Facebook fans and their "real" friends were muddled together.
   I'm not implying that their fans are not their "real" friends; I'm just saying there is probably a separation that they make, and what they share with each group may be different.

This can be problematic because now, all of the sudden, they want to add their music information to their personal page. While they are also cross-posting information on their fan page, they find themselves doubling up on the amount of work, while possibly annoying their Aunt Matilda and Uncle Bob.

Facebook's New Reality

Facebook makes it ALMOST impossible to make any sort of organic growth happen.

In other words, it's a pay-to-play platform.

Facebook makes most of their enormous profit from advertising.

At the end of the day, Facebook is catering to their customers. Believe it or not, Facebook's customers are not artists or the labels. They are the advertisers. They are the people willing to spend money to be connected with others (others you bring to the site), and this algorithm was created to ensure that this happens.

Facebook has created an option for those who are NOT full-time advertisers, and for better or worse, gives them the opportunity to "gain access to fish in the sea" more quickly and effectively. This is the dreaded "promoted post" function that Facebook introduced. By paying even as little as $15, you are FAR more likely to see true engagement happen on your posts, simply because Facebook is ALLOWING this to happen (because you've paid for it!). As ridiculous as this seems, this option does present an artist or label with a good opportunity to jumpstart the engagement of a new page by promoting select posts that nurture strong engagement with your audience.

Pay Attention to Analytics—Facebook Insights

Facebook Insight is the analytics tool built into the back-end of Facebook. It's free and easy to use. It won't track personal pages, only fan pages (yet another reason to have a fan page). If you are the admin to your fan page—which you should be—you can gain access to Insight.

Facebook's Insights give you a detailed look at whom your fan base is, where they live, and most importantly, what content they are most willing to engage with. Your content strategy never needs to be a static thing—it should be fluid! It should shape-shift as you find out more about who your fans are and what their needs are. Using Facebook Insights is critical to a strong Facebook fan page that holds well in Facebook's algorithm.

Of course, using Facebook Insights are only helpful if you know what the average metrics on Facebook are, so that you can compare your efforts to the standard.

First off, you have to understand the average number of fans on a Facebook page . . . this will help you establish a realistic goal to work for.

Six Shocking Reasons Facebook Fans Lack Engagement

A vast majority of social media users are unaware of just how difficult growth of a Facebook fan page can actually be. This fact is not meant to scare you away from building your community on Facebook; the purpose is to shine a light on the harsh reality that is Facebook-centric community building. I sincerely hope that you use the following information for good: to set more realistic goals, put more effective strategies in place, and build stronger fan communities.

1) Average Life of a Facebook Post

One common complaint about the Internet in general by new users (or non-users) is that they don't like the idea that something published online "lives forever." Well, fret not . . . because this couldn't be further from the truth.

Yes, your content will technically be online, but the average lifespan of a Facebook post is just a short three hours. This means that after the average three-hour timespan, your Facebook posts will no longer show up in any of your fans' news feeds.

Although three hours is actually a far higher number than, say, Twitter (see: http://moz.com/blog/when-is-my-tweets-prime-of-life), this does mean that you MUST understand when your fans are online and most likely to engage with your posts so that each day you post in your most effective three-hour window.

2) Average % of Total Fan Base Willing to Engage with a Brand Page

A recent study by Napkin Labs exposed the shocking statistic that only 6% of a total fan base will ever actually engage with a brand page. While this does fuel the fire of the argument that you need more fans—A LOT more fans—in order to build any sort of community on your Facebook page, this study actually contributes to the idea of having 1,000 True Fans.

The fact that 6% of a total fan base will engage with a brand page is partially because of Facebook's algorithm . . . Facebook's news feed algorithm is a system that ranks and displays only the most "relevant" and "important" content on your news feed from your friends and pages you have liked. But the issue of your fan base lacking engagement also has a lot to do with the fact that the overwhelming majority of the average fan base is made up of "Ambient" or even "engaged" fans, and not the kind of "super fans" needed to truly build a consistently engaged community.

3) Average Engagement Rate

And as scary as the "6%" stat above is, this one gets even more frightening: There have been several recent studies done on the actual average engagement rate of a page (the "People Talking About This Page" number) and it is shown that the average engagement rate of a fan page is only .96% (yes, that's less than 1%). This means anything above a 1% engagement is considered strong, results-wise.

The difference between this stat and the one above (see #2) is that the number above reflects the percentage of fans who are EVER willing to engage with a page over the lifetime of the relationship. In other words, for the average fan page, 94% of your fans never have nor will they ever engage. Meanwhile, this stat reflects the true engagement of a fan page on Facebook's rolling seven-day scale. In other words, within seven days (on average), the number of 'People Talking About This Page'/the number of 'Total Fans' will equal around 1%.

4) Posting Less Will Garner Stronger Results

After seeing these stats, the natural reaction is to want to publish content even more often to do everything you can to build this number higher than a measly 1%, right?

Well . . . sorry, but that won't work either.

Further studies from several sources have shown that the sweet spot of one or two posts per day on Facebook fan page will garner the strongest results. Anything more than that and you will actually start to see increasingly diminishing results.

Jeff Bullas research shows how meaningful the difference between one or two posts vs. three or more posts can actually be: By posting only one or two times per day, your page is likely to receive 32% more likes and 73% more comments when compared to the engagement you're likely to receive from posting three or more times per day.

5) You Will Be Published For Being Efficient

Because consistency is key to a strong social media strategy, it is critical that you publish content every single day. No breaks. Even though you should only post one (or possibly two) posts per day, this can get overwhelming quickly. But thankfully there are several apps, such as Hootsuite, that allow you to schedule content for Facebook ahead of time.

PHEW . . . Great news!

Don't get too excited just yet. Let's see what Facebook's news feed algorithm has to say about this . . .

Unfortunately, a study by Hubspot has shown that publishing content from any 3rd party (yes, including Hootsuite) will average a whopping 67% lower engagement rate than content published directly to Facebook. The good news: Facebook has introduced a way to publish content ahead of time using their own platform. To do so, go to your fan page, click the Status Update box, and click the little clock icon and add in the date and time you'd like to publish each post.

6) Likelihood of Posts going "Viral" of Facebook

We have all seen those photos, memes or videos that have gone viral in the past. And understandably so, this idea of "going viral" is why many people started using social media to promote themselves or their brand in the first place.

Well, we're sorry to say that it look's like Facebook's news feed algorithm has reared its ugly head once again.

The news feed algorithm is a system that ranks and displays only the most "relevant" and "important" content on your news feed from your friends and pages you have liked . . .

Not that it has EVER been easy to "go viral" online, and the fact remains that most viral videos, photos and memes do so because of chance. There IS NO science behind viral content. But with that said, Facebook doesn't make it easy for this type of viral reaction to ever happen to content published from a fan page.

In the suite of Facebook Insights data (Facebook's native analytics tools given to all Facebook fan page admins) there is a stat called "virality" which is the percentage of fans who have shared your content on their own timelines for their own friends, family, and fans to see.

A recent study from Edge Rank Checker shows that as of March 4, 2013, the average "virality" rate of a post on Facebook fan page only 1.5%. This means that of your 100 fans, only 1.5 of them (on average) will share the content with their communities.

Quickstart Guide to Get Your Facebook Fan Page Set Up Correctly

1) Cover photo—The most noticeable change on Facebook fan pages is the giant cover photo that artists can now implement. This was a cool feature for personal profiles and allows users to show off their creativity (or that amazing photo you took of the sunset on vacation), but it is a really powerful branding tool for artists. Uploading an eye-catching photo will help draw fans into your page and solidify your image. Branding has always been

a little tricky on Facebook fan pages as you had to use certain apps to effectively do this, but this left the wall essentially "unbranded." Cover photos solve this issue and bring a whole new life to the "wall."

2) Pinning—Facebook now allows you to "pin" certain posts to the top of your Timeline. I'm particularly fond of this feature because it allows artists to keep promotional posts at the top of their page instead of endlessly posting the same promotion over and over again. Instead of continuing to post updates about an upcoming show, bands can focus on providing more interesting and engageable content to pull fans toward their page. Keeping the show's promotion pinned to the top of the page ensures all visitors will see that post and will be reminded of the show. It's a beautiful thing.

3) Highlight—You can "highlight" certain posts to make them stretch across the entire width of the timeline. As fans scroll through your timeline, these highlighted posts will jump out and will be sure to catch attention. Highlighting videos is a great way to get more fans to watch them, even as they fall down on the timeline. This feature also helps draw attention to significant events that have happened throughout the band's history.

4) Milestones—The whole point of timeline is to allow users to scroll through someone's Facebook history. This is a very powerful tool for artists. Coldplay has documented the band's entire history on their timeline. Going so far as to show a picture of the band's first rehearsal, first show, first EP, first NYC show, etc. I'm not a huge Coldplay fan (If you are, I hope you'll still keep reading!), and even I found this very interesting. Imagine how much actual Coldplay fans love this! For new, up-and-coming bands, this is a great opportunity to let fans grow with you and celebrate all your accomplishments as they happen.

5) Unique fan experience—Facebook has ensured that each fan's experience on an artist's timeline will be unique. What does this mean? It means that when I visit Coldplay's Timeline, I see what my friends are saying about Coldplay, whether or not they actually posted on Coldplay's timeline. For example, Cyber PR® team member, Jon Ostrow, posted about Coldplay back in August, and because Jon and I are friends and interact frequently, his post shows up high on Coldplay's timeline for me. This feature will intensify the sense of community your fans experience on your fan page.

6) Messages—One of the most exciting features of timeline is messaging. This has long been a desire for fan page admins, and Facebook has found a way to deliver without allowing fan pages to spam their fans. Fan pages can now receive messages from personal profiles and then respond.

7) Admin panel—It's not all about the fan's experience; Facebook has made it easier for admins with the newest version of its "admin" panel. In the admin panel, you can easily see new notifications, new likes and a

basic insight graph. The "build audience" tab makes it easy for admins to grow the number of likes. You can use the age-old "invite friends" technique (obviously not a new feature, but now easier to find). And now you can also send out invites from your email contacts. This is a great feature that can be extremely helpful for artists with a large email list and for new artists who have a large personal email list.

8) Landing page—This isn't really a feature; instead it's more of an omission. Facebook has, unfortunately, taken away the ability to assign different landing pages on timeline. But, there is a way to work around this. The various artist pages have unique URLs. You can now give fans this URL and send them directly to where you keep your music and email sign up form. While this isn't as good as a default landing page, the added benefits of timeline and the new layouts of most of the artist pages more than make up for this (in my humble opinion).

Since the initial launch of timeline for fan pages, brands have seen a significant increase in engagement (as much as 46%). This stat is huge and it's vital for artists to take advantage of this increased engagement while it lasts. However, the new timeline won't do the work for you—you have to get in there and make it happen.

It's now more important than ever that artists and their labels push out quality and engaging content. Know what you want your "brand image" to be, and post content accordingly.

How to Make your Artist's Facebook Timeline Pop!

There is a fabulous feature that will help you highlight the things that happen throughout your artist's life and career that you can post onto your Facebook fan page. It is the event, milestone button. This is a phenomenal tool for going back in time and recording important things in the history of your artist's personal life, their band life, or anything else you might like to have highlighted. For artists that have histories with other bands, this is doubly amazing because you can go back and create milestones for practically anything, and really build your story.

Here Are a Few Things You Can Add

Past tour dates
    Past album release dates (with album cover image)
    Past press placements and radio ad dates
    The day you saw an amazing concert that inspired you
    The day you were signed to a label
    The day you were dropped from a label

The day you got your publishing deal
The day you recorded your first song
The day you entered the studio
The day you had your first vocal lesson

FACEBOOK IS ALWAYS CHANGING

Eight Ideas for Increasing Artist's Reach on Facebook

1) Like Other People's Stuff. But don't stop there! Leave real comments that are thoughtful and noteworthy.

2) Ask Questions. Ask easy questions that people can weigh in on like, "What's your favorite music to clean your house to?" or, "What's the most annoying jingle of all time?" or, "What's your favorite movie for crying?" etc. People can easily answer things that they have emotional connections to. This is a smart way to get fans involved.

3) Wish People a Happy Birthday. (This is from your personal profile, but it's a considerate way to appreciate people, especially fans.)

   AH Tip #4: Create a great Happy Birthday video of you/your band singing and post that on every single person's timeline who has a birthday.

4) Photo Themes. How about a church of the week, plate of French fries of the week or a thing that makes you LOL each week? Having a theme that people can recognize and comment on can be a way to have people expect great content from you.

5) Photo Contests. Don't just stop at posting and sharing photos. Why not hold a contest of favorite photos from your hometown, or of your pet, or of your favorite food, etc. We love a service called One Kontest, and for a fee, they can help you run a fantastic photo contest with your fans.

6) Video Curation. Pick something that goes with your aesthetics—or a great video from the 1970s, 1980s, or 1990s, commercials from your childhood, moments in sports history you remember—anything that is meaningful to you may be meaningful to your fans as well.

7) Spotify Playlists. Create playlists everyone can contribute to and share them on your fan page. Songs about tennis, songs about broken hearts or songs to play while driving with the top down—there are a million themes.

8) This Day In History. Are you into art or theatre or great rock and roll history? Post photos and videos!

Facebook Takeaways

- Yes, you need to have a fan page! Remember that average unpaid (organic) engagement on a fan page is 1% (this is why having a fully integrated social media house plan is crucial).

- Photos are the most shared item on Facebook, so share lots of them.
- Take the time to pimp out your page properly; branding is everything.
- Lists are fantastic when you are grouping people, and they will help you tremendously with targeting on Facebook.
- Supplement your page with acts that show your other favorite social media streams and activities.
- Analytics are your friend; understand how to use Facebook Insights.

YOUTUBE

YouTube is a social network, not just a video site, it is a critical social media site to include in your arsenal of social media and online marketing techniques. There is no way to refute the power of video, or of the site: YouTube is the second largest online search engine. It has hundreds of millions of users from around the world and is localized in 53 countries across 61 languages. In 2006, Google bought YouTube for $1.65 billion dollars. So, YouTube now operates as a subsidiary of Google (which is, of course, the largest online search engine). It's important to understand how they relate to each other in order to maximize your presence.

YouTube is an interactive social network just like Facebook or Twitter, and you are encouraged to make friends with other users. Making friends, leaving comments, and "thumbing up" as many videos as you can are key moves on the YouTube platform.

If you start looking around online, you will see people leading you back to their YouTube channels all over the Internet. If you see links to YouTube on people's Facebook, Twitter, websites, or blogs: head on over, subscribe to their channels, watch videos, leave comments, and make friends. That's how to be social!

AH Tip #5 Being active on YouTube is a great newsletter-building strategy. Once the artist (or somebody acting on behalf of the artist) engages someone on YouTube and gets into a conversation with them, they can ask if they can add them to your newsletter.

Be Consistent Across Platforms

Make sure the channel name matches the artist's website and user name on Facebook and Twitter. Ideally, you should make the user name the band's or artist's name. If that name is unavailable, add something like "music" or "official" onto the user name.

Watch What is Already Working, and Follow!

Before you upload a well-produced video of the artist performing or pay for an expensive camera shoot, take a look at what is "going viral" on You-Tube. They make this easy by posting the current most-viewed videos here: http://www.youtube.com/charts. I suggest you watch the most viewed videos in the music categories for several weeks and see if you can decipher a trend or come up with an idea.

Think about Your Audience

Keep in mind who your audience is, and where they watch. Many people watch videos while they are at work. And people pass along videos that are funny, amusing, or interesting to them in some form or fashion.

Always Include a Call to Action

Remember, in order to get people to take action, you need to tell them what to do. Add a Call to Action (CTA) to every video you post. A CTA could be "Follow us on Twitter," "Like us on Facebook," or "Visit Our Site." Only include one CTA in each video you post so you don't confuse your audience. Without a CTA, you are leaving potential traffic on the table. The whole point of videos is to drive traffic back to your site.

It Doesn't Have to be Fancy or Expensive to Go Viral!

Look at the charts. There are a lot of videos on YouTube that are lo-fi, not "produced," and they get millions of views.

Remember the Purpose of Videos

It is important to remember the reason you make videos in the first place! It is not just about getting a lot of views (that's just the beginning); it is to get people to take an action. Whether that action is coming back to the artist's website or following them on Facebook or Twitter, it's critical to remember why you made the video. Keep your eye on the end goal and make sure you constantly measure your videos' effectiveness.

Know What People Are Searching for

Caution: YouTube is a search engine. This means people are searching for things that they already know and love on YouTube. So, including content that people already know is crucial, because they are already looking for it!

Share Video on Social Media Sites: Twitter, Facebook, and Blog

One of the best parts of YouTube is that it's easy to share videos using the embed codes and share buttons that YouTube provides. And a solid marketing strategy involves cross posting your artist's videos on their website, on their blog, and on their social media sites. According to YouTube, a link to the video in a tweet "results in six new YouTube.com sessions on average." So, once you post your artist's videos on YouTube, post them on Twitter as well!

Optimizing Your YouTube Channel

1. Titles are Key

I've been reading a lot of YouTube studies, and it turns out that the single biggest contributing factor to your success on a video click through will be directly related to how you title your video.

> AH Tip #6: Make sure that the title of each of your videos includes the artist or band name, song title, and any other relevant information. If it's not a straight-ahead music video, create a title that will make the viewer want to watch the video. Something captivating and catchy is key. (However, remember that it must relate back to the actual video.)

2. The Description Box Is Crucial

The description box is critical to optimization of your channel. Always start with your URL at the very beginning of the description box and don't forget to include http:// or else it will not show up as a hyperlink. And it's not a bad idea to add this link (or a link to your Facebook, Twitter, etc. at the end of the description as well).

3. Select a Proper Video Category

This will more than likely be "music." But it could be "humor" or "education" too. Remember to think about which best fits your content!

4. Tag Thoroughly

Google ranks tags, so always start with the tag that is the most important. If you are adding a title that is more than one word, put it in quotes—i.e., "Sympathy For The Devil"—or else the words will become individual tags. Don't overdo the number of tags. (Google frowns upon over-tagging, so keep it to seven or eight at the most). Some tag ideas: artist or band name, song name, any related artists names (especially if you add a cover), similar artists (so that when people type in an artist they like, they will come across your video), genres of music, hometown, names of all band members, producer, themes in video, etc.

AH Tip #7: Don't overdo it! Your build-out will take some time. YouTube moderates user activity closely, much as Facebook moderates its own user activity. If you receive a notice to stop sending messages, adding friends, and so on, STOP, or else your account will be deleted. And you don't want that!

5. Subscribe to Other Channels

Subscribing to other channels can help you go a long way. Search for keywords that match yours. Start by subscribing to channels of similar artists or artists that influenced you or that you sound like. After you have subscribed to your favorite artist's channels, start subscribing to their fans' channels by going to the artist's channel and locating the "Subscribers" box. This will be a good place to start adding friends.

6. Add Videos as Favorites

Love a video on YouTube? Just click on the little heart. Keep in mind that these videos will be added to your "Favorites" section on your channel.

7. Comment on Videos

Respond to other people's comments, just as you would respond to others' comments on Facebook.

8. Rate the Comments

You can click on the "thumbs up" or "thumbs down." This process takes only about one second per rating.

The Official YouTube Playbook for Musicians

YouTube has just launched a new series of "Playbooks" for content creators. There are several topics covered, and one of them is specifically written for musicians. This free eBook details strategies and best practices on how to build a music channel, and includes techniques to help guide success of your artists' videos. Chapter themes include: better optimization, how to release an album or song, content frequency, and how to effectively engage with your fans. It's downloadable on YouTube's website, and is a great resource for anyone looking to grow their audience on this platform.

YouTube Takeaways

- Make sure that your artist's channel is skinned to match their branding, and all of the links and bio are visible and clear.
- In order to have a successful YouTube presence, you must post videos often.
- YouTube is a social network, and two-way engagement and following others is key.
- Just having an official music video and some live tracks probably won't cause your artist to have much of a following.
- Studying what is going viral really means studying the types of videos that have already gone viral.
- Understand how to select the proper video categories and thoroughly tag each of your artist's videos so that they can be found.
- Always include a clear call to action in each video you post.
- Don't be obsessed with big numbers; small numbers can be wonderful.
- Never buy views—it's just not worth it.

TWITTER

Many people resist Twitter because they don't understand the ins and outs of it, but I promise once your artists master it, they will come to realize it is a fabulous way to connect with new people and potential fans. All the while, they can give their current fans even more if they take at least five minutes a day to effectively use Twitter with either their mobile or computer.

Twitter is a great tool to share thoughts, ideas, tips, and advice. Even better, Twitter is great for generating awareness for a band, song, podcast, article, press release, review, and so on. Important Twitter terms are defined at the end of the chapter. You should read them (and so should your artists)!

Think About Your Artist's Handle (User Name) First

When you go to set up an account, or if you are helping someone rename their page, don't just pick a name you like. Use the name that matches their website, their Facebook profile and their other social media sites for consistency. And remember: Whatever name you choose on Twitter becomes Google-icious too.

Get a Mobile App For Twitter

There are many—choose one you like.

Follow Lots of People

Twitter does not work in a vacuum, so the key is to follow at least 250 people!

> AH Tip #8: In order to find people, search for them by keyword. To do this, you will go to the grey box on the top right of the page, type topics you are interested in, words about the music you play, and so on. When you find interesting people and follow them, Twitter will lead you to more people to follow.

Tweet Three Times a Day

Enough said.

> AH Tip #9: Don't Over-Hype Yourself. If all your tweets say things like, "Buy my album! Come to my show!" you are not going to build an audience who trusts you . . . or wants to hear from you!

@ People You Like! and RT

To comment on tweets you like or have a reaction to, or to connect directly with someone, just tweet @ and then their username. So if you want to say something directly to Derek Sivers, type @Sivers and then your message. This will turn up in the "Replies" section of Derek's Twitter dashboard, and he will see your comment. But so will everyone else! This is a public message that everyone on Twitter will see.

Connect Directly

To send someone a direct, private message, go to the icon on the top right of your Twitter home page and in the drop down menu choose "direct messages." Then choose the person whom you want to send a message from the pull-down menu at the top of the page. Direct messages are private messages. Only the user you choose can see these messages.

AH Tip #10: To send a DM, the person must be following you. If you want someone to follow you, simply @ them and ask them (9 times out of 10, they will follow you if you ask)!

TWITTER HOW TO: HOW TO CREATE A TWITTER LIST

1) From the dashboard on your Twitter home page, click the profile drop-down menu on the right-hand side. When you scroll down the menu, click "Lists." On the right side of your screen you should see the button, "Create a List," where a pop-up box will appear prompting you to give a list name and a description. Note: You can either make this list public, so others have access to the list, or private, so only you can see this list.

2) Once you have created the list, you can either use the search box to find people to add to the list, or click on "following" so you can see a list of those Twitter users you follow to add to the list.

3) Next to each name of the person you are following, there is a drop-down menu that allows you to add or remove each person to a list. Click on the list to which you would like to add them. Remember, you can always add more Twitter lists later.

The Cyber PR® Guide to Increasing Twitter Followers

Twitter is, in my humble opinion, the fastest way to grow your fan base. Here is a time tested strategy we have executed for multiple Cyber PR® clients. It takes some time and TLC, but it works!

To be safe, I do not recommend following more than 500 people every 24 hours, because if you are near 800 or 900, you could be flagged for "strange activity" (Twitter's phrase). Alternatively, if you follow between 100–300 new people each time, you should be okay, as far as steering clear of issues.

The fastest way to legitimately grow your artist's Twitter follower is by growing their mutual follow-back relationships. Here is how you do it:

1) Think of an artist who has similar fans to your artist.

2) Type in any band or solo artist that influenced your artist in the search box on the top right-hand side of your Twitter home page.

3) Click on any profile with a decent number of followers. Once you get to that profile, click on the "followers" link on the top right hand side of their home page.

You will now see a long list of profiles that you could follow. (Move down the list and start following by clicking the grey follow box with the little Twitter bird on the right of each profile.) I highly suggest that you don't get too nitpicky and read every profile—unless one looks fishy, or there is no bio or photo. Just follow—you can always unfollow later.

4) Repeat this process for several minutes until your artist is following around 300 people.

5) Wait at least two to three days to give the people who you followed ample time to follow you back. During those two to three days, you will notice some of the people that you followed will reciprocate by following your artist back and they will have many new followers.

Until your artist has over 2,000 followers of their own, you only have the ability to follow up to 2,000 profiles at one time.

6) Be sure to unfollow those who are NOT following your artist back. tweepi.com & manageflitter.com will help you swiftly identify and unfollow those who do not follow you back or who do not have profile pictures, etc.

After your artist has 2,000 followers, they will be able to follow more users, but that will be dependent on how many people are following them. So, after they break 2,000 people, it may be advantageous for you to only follow back the people who are very active.

Measuring Your Effectiveness on Twitter—Twitter Analytics

So, now that we've taken you through the tactical points of Facebook and Twitter, and you're on a roll with your new-found knowledge, it's time to dig into something that is critical: your analytics. Analytics are the key to understanding your marketing effectiveness and your artist's fan base. When you learn how to use them well, they will show you what is working effectively versus what is not. Once you can get a handle on these key differentiators, you can do more of what works and less of what doesn't, to capture and engage more fans.

USING ANALYTICS TO SUPERCHARGE YOUR TWITTER BASE

STEP 1: Go to Your "Home" Page

Go to your home page on Twitter (press the "Home" tab next to the search box at the top of your profile). Your Timeline (stream) will be displayed.

STEP 2: Look at Your @ Mentions

Click on @Mentions tab—Scroll down your @ Mentions and see if certain people stand out (people who have responded more than two times or people you know). Make sure you are following everyone who @'s you.

STEP 3: Create a List

Create a list for these followers so you can monitor them. And make SURE to interact with them regularly. Call your list whatever you like. For example: "People I Love."

Twitter Takeaways

- Branding your artist's profile is crucial—make sure you have a photo, bio, links and a background skin that matches your artist's brand— just choosing a default twitter background won't cut it.
- Understanding the basics will change your life on Twitter (@, RT, DM & Hashtags)
- It is virtually impossible to over tweet—six times a day is recommended.
- Mix it up—make sure you are tweeting photos, music, videos and articles as well as retweeting and @ replying people.
- Share links using bit.ly—it's a great way to shorten long links and you can track your results.
- Your Lists are key! Group great people together and make sure to interact.
- Analyze your Twitter account often, and interact with your strongest followers (the ones who have the most followers and tweet most often).

INSTAGRAM

Instagram is a social media channel and it's primarily used as a mobile phone app.

You probably know that Instagram was purchased for $1 billion by Facebook. And you probably know that it has become wildly popular. It is relevant because it super easy to use and makes editing and sharing photos a snap! If your artist is stressed out about having something to say they can let the pictures do the work for them.

Instagram has 15 different filter options and buttons to post directly to Twitter, Facebook, Flickr, Tumblr, Foursquare, or send them by email. Other Instagramers can follow users, "like" pictures, comment, and mark favorite images easily as well. The pictures don't show up directly in the Twitter stream, but they are one click away.

Like all social sites, the secret to great Instagram use is liking other's photos, commenting as often as you can, and of course using effective hashtags.

PINTEREST

Pinterest is a social network that allows users to visually share and discover new interests by "pinning" images or videos to their own or others' "pinboards" (i.e., a collection of "pins," usually with a common theme).

You can link your artist's Pinterest accounts to their Facebook and Twitter, so their pins can attract more eyes from different channels. And it is very easy to set up.

Pinterest Takeaways

- Pinterest is about developing relationships with other pinners. ·
- Diving in is the most important thing—start several boards, start pinning, and see what happens!
- Sharing music on Pinterest is still underexplored. See if your artist can build a group of pinners who will reshare their music.
- Set up analytics so that you can track clicks back to your artist's website or blog from Pinterest.
- Pinterest is not just for women, even though they are still the largest demographic that use the site.
- Find great artists' Pinterest sites and emulate them.
- Follow artists and others that your artist admires.
- Pin content from other artists in addition to original content.
- Work to find the niches where your artist will get the most traction with their pins.

BLOGS

Blogging is a crucial piece of your artist's social media strategy. Blogging also allows their fans to see more of them. However, it takes dedication and consistency to pull off a blogging strategy.

I believe that getting reviewed on blogs is critical for every musician because it helps create a bigger footprint for them online, builds awareness, and allows for a two-way conversation around their music.

All content on blogs is archived in Google and it can be a great portal for people to discover your artist when they blog consistently.

Dive In

According to the most recent statistics, there are currently over 180 million active bloggers. Some of these blogs are read by a few dozen people, but others are read by millions. And, as you likely know, blogs can be about any topic. The vast majority of bloggers create blogs for no financial gain whatsoever; in fact, it usually costs music bloggers money to host their files and maintain their blogs. A blog is usually a personal endeavor. Most bloggers create their blogs as an outlet where they can talk about things they are passionate about in their lives, their opinions, and the things that they like and dislike. A blog is basically an online diary.

Finding blogs that are right for your artist won't take long. Just dive in and start reading them. The ones that resonate will jump out at you.

TOP SEVEN REASONS EVERY MUSICIAN SHOULD BLOG

1) Blogging is a fabulous way of keeping your fans connected to you.

   Blogging goes much deeper than 140 characters on Twitter or images on Flickr. Blogging also gives your fans a platform to have in-depth, two-way conversations with you for the whole world to see.

2) Google loves blogs.

   If you set your blog up properly, you'll be indexed on Google for anything and everything you write about. This means that people who were searching for other topics can find you. For instance, let's say you blog about your dog. A person who is searching for "yellow lab" could come across your blog entry, discover your music, and become a fan!

3) Blogging puts you on a level playing field with other bloggers.

   Bloggers read other blogs, especially those pertaining to subjects they write about—like music. And a music blogger will trust you much more if you understand the whole world of blogging.

4) A blog allows you to invite your fans backstage and into your life so that they can see all sides of you . . . but only the sides you want to be seen. You are in control of your content. Fans can subscribe to your blog using an RSS reader and get new updates sent directly to them without having to visit your site over and over to check for new posts.

5) You can syndicate your blog posts all over the Internet.

   ReverbNation, Facebook, Twitter, and your own website are just a few places where your blog posts can show up so people can see them and engage with you.

6) Starting a blogroll adds to your credibility with other bloggers.

Add bloggers who acknowledge your blog onto your blogroll, which is a list of links to the other blogs you like or recommend. This is usually placed in the sidebar of your blog. In the blog world, it's critical to associate yourself with other blogs and communities of people with whom you would like to connect and with bloggers and communities that want to connect with you.

7) Blogging gets you community feedback fast.

Not sure about a song lyric, a photo shoot, which night to have a gig? Ask your fans to weigh in with their opinions!

Your artists don't have to only blog about their music; they can talk about their home life, their TV habits, their favorite foods, their day job, their fitness routine—anything! The key here is that they must post regularly and consistently. If they are a band, having each band member contribute one post a month is a great way to keep new content flowing.

AH Tip #11: Don't overthink. Just post! Do not treat this like a show or think you have to make every sentence perfect. The point of a blog is that it is an informal endeavor. Just get posting, don't stress about it, and tweak it to death (I would, however, recommend spell check).

Blogging Takeaways

- The key to your blogging strategy is consistency. If you commit to blogging, keep in mind that you should blog a minimum of twice a month.
- Bloggers respect other bloggers, so adding blogging to your strategy is a fantastic way of setting yourself up to get covered on other people's blogs.
- Commenting on blogs before you dive in and begin to write your own is a fantastic way of getting a feel for two-way conversations.
- Identify 10 blogs and read them regularly and become a known contributor to them. I recommend getting started with smaller blogs and not ones with huge amounts of traffic, if you want to be known in the community of that blog.
- If you do decide to blog, vary your content and don't only blog about music.

CONCLUSION

A good social media strategy is like a well balanced diet. Too much or too little of even the right stuff will lead to a less than healthy relationship with your artist's audience. Communicate with their fans in the proportions in the Cyber PR® Social Media Pyramid (Figure 11.2) and you'll enjoy a long and healthy relationship.

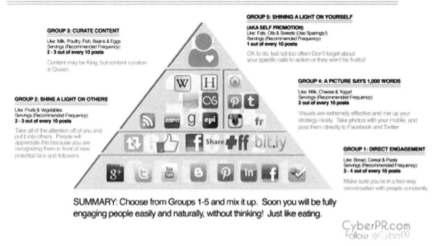

FIGURE 11.2

APPENDIX 1

TWITTER DEFINITIONS

Even if your artist thinks of his- or herself as an intermediate or advanced Twitter user, have them take a quick peek at these foundational definitions.

Twitter Address—This refers to the URL where your Twitter home page is. Ex: http://twitter.com/cyberpr

Twitstream/Tweetstream/Twitter Stream—The collective stream of Twitter messages (tweets) sent by the artist and the people that they're following. No two people's streams are exactly alike.

Twitter Handle—This refers to the artist's Twitter username and begins with an "@" symbol. My handle is @CyberPR.

Tweet—An individual Twitter message that is 140 characters or less. A tweet can:

1. Answer, "What's happening?" or just post an update

2. Allow the artist to share links to content (music, photos, blog posts, articles, or videos).

3. Help the artist engage with others by using @'s, #'s and RT's

Followers—The people who are following your Twitter messages.

Following—The people whose Twitter messages you follow.

RT (ReTweet)—Quoting or repeating someone else's tweet (Twitter message) in your tweet. This is also a significant measure of the value of a tweet; if something has been retweeted many times, it is ranked higher in the system.

@Reply—Also called "at reply," this is a tweet where your Twitter name is mentioned. This will create a public conversation between you and the other user.

DM (Direct Message)—Private tweet between you and another person. You can only send DMs to people who are following you. This is a private conversation, much like email.

Twitter Spam—Twitter messages with little or no conversational value.

Favorite—Tweet marked by you or others as favorite, as indicated by a little gold star. When you favorite something, you are adding it the list of Favorites that can be seen publicly on your profile. You are also giving feedback to the other user that their post (that you have favorited) is valuable to you in some way.

Trend—The most tweeted about words or phrases at the present time. This list is on Twitter's home page on the left. Trending topics often correspond with current events.

List—A group of users that any Twitter user can create, and then curate by trend, location, theme, and so on.

Who to Follow—Twitter will analyze the list of people that you are already following and will constantly make recommendations of other people to follow. You will find this on the left-hand side of your Twitter home page.

Listed—This section displays the lists within which you have been included by other Twitter users. You can follow entire lists here, or choose individuals to follow within these lists.

Hash Tag (#)—A hash tag is used on Twitter when people want to highlight a particular phrase, usually in reference to a topic or event. For example: #WorldCup, #MadMen, #SXSW.

REFERENCE

Godin, S. *Tribes*. New York: Penguin Group. 2008.

CHAPTER 12

The Business of Radio

The reports of my death have been greatly exaggerated.
—Mark Twain

RADIO

Despite the talk about the demise of traditional commercial radio on blogs and other online chatter, radio continues to be a major influence on music discovery. Nielsen Audio, formerly Arbitron, provides radio audience measurement services for the broadcast and advertising industries, and in its 2014 report, "State of the Media: Audio Today 2014", the company notes that 90% of people in the U.S. listen to radio every week (Radio Today, 2008). On average, these same consumers listen to 18 hours of radio each week with their time spent listening being down only 75 minutes over the previous six years; on a daily basis, they listen to one or more radio stations for more than two and one-half hours each day (Nielsen, 2014 B). Additionally, the Nielsen Music 360 Report (August 2012) finds that "radio is still the dominant way people discover music." Forty-eight percent of music buyers learn about new music through radio, according to the report and among teens, radio is second only to You-Tube for music listening. With the growth of smartphone apps like the one for iHeartRadio, the lines between radio and the Internet have been erased.

Anyone in the business of selling recordings is making a mistake by underestimating the reach and impact that radio has with consumers. Labels understand this, and it is why they continue to put considerable resources into influencing decisions by radio programmers and other radio

CONTENTS

gatekeepers to get their new music on the airwaves. To demonstrate the point, here are weekly audience sizes counted by Broadcast Data Services (BDS) for the week of August 4, 2014:

Table 12.1	Artist / Song Position with National Audience	
Artist	**Song/CHR Chart Position**	**National Audience**
Iggy Azelea	"Fancy" (#8)	56,445,000
Maroon 5	"Maps" (#6)	47,197,000
Pharrell Williams	"Happy" (*)	8,199,900
Florida Georgia Line	"This is How We Roll" (#25)	9,052,500

"Happy" peaked at #1 on Billboard's Hot 100 and had been out over 30 weeks when these counts were taken. This Is How We Roll was in its 26th week and had peaked at #15.

Online and other viral promotion of recorded music are important elements of any marketing plan, but a competitive international plan with a goal of selling more than 200,000 units must include promotion to commercial radio as key element of its success.

Veterans of the recording industry estimate that as many as 70% of consumer decisions to purchase recorded music can be traced directly back to exposure to the music by commercial radio. As much as consumers complain about the large number of commercials and repeated playing of the same music, radio still is the most important vehicle the recording industry has to showcase its product to the public. And despite our wireless devices, MP3 players, GPS units, dashboard video players, and other distractions, we still spend over two and a half hours per day time listening to radio (Nielsen, 2013).

Given the role radio plays in promoting recordings to consumers, it's important to have an understanding of radio and the people who make programming decisions at those stations. They are the gatekeepers to the radio station's airwaves. When the marketing and promotion staff at the label understand what radio needs, it becomes easier for them to find a way to get their new music programmed.

THE BUSINESS

One of the best adjectives to describe the relationship shared by the recording industry and commercial radio is "symbiotic." Though it is a term most often used in science, it means the two industries share a mutual dependence on each other for a mutual benefit. Radio depends on the recording industry to provide elements of its entertainment programming for its listeners, and the recording industry depends on radio to expose its product to consumers. No two other industries share a relationship as unique as this. However, the nature of the businesses of a record company and a radio station are very different. For a record company, it's easy to define the business: to sell recordings. Money moves from consumers to record companies when recordings are sold.

The business of radio is building an audience that it leases to advertisers. Radio uses music to attract listeners in order to attract advertising revenue. The larger the audience the station attracts, the more it can charge for its advertising. Radio, however, is not in the business of building recording careers, nor is it interested in selling recordings. The number of units that a recording is selling might be of interest to a radio programmer, but that information by itself does not necessarily affect programming decisions.

THE RADIO BROADCASTING INDUSTRY

The traditional over-the-air radio broadcast industry in the United States began consolidating when the Telecommunications Act of 1996 was signed into law. Prior to the new law, radio broadcasting companies were limited in ownership to no more than 40 stations (20 AM and 20 FM) nationwide (Williams, 2007). The law now allows companies to own unlimited numbers of radio stations nationally, but no more than eight stations per market and no more than five in the same service (AM or FM). The numbers decrease on a sliding scale for smaller markets and in no case may one owner control more than 50% of the stations in a market (Oxenford, 2011; Federal Communications Commission, 2014).

The relaxing of the radio ownership rules has created some of the largest media companies ever. In the first seven years after passage of the act the number of radio station owners decreased by 35% and the largest ownership group swelled to over 1,200 stations. The biggest radio companies in America are:

Table 12.2	Radio Broadcasting Companies
Radio Broadcasting Company	**Number of Stations**
iHeart Media	850
Cumulus Media (includes Citadel and ABC Radio)	525
CBS Radio (formerly Infinity Broadcasting)	126
Entercom Communications	106
Cox Radio	57

Station counts taken from company websites and reports in 2014. Numbers change often as stations are bought and sold for financial or regulatory reasons. There were a total of 11,349 commercial radio stations in the U.S. at the end of March 2014. There was an additional 4,057 educational stations. Sources and notes in Appendix B.

Revenue for the industry declined steadily from 2005 to 2009, with annual earnings hitting a low of $13.3 billion in 2009, but in 2010, aided by a stronger economy and the additional revenue from advertising online, radio advertising rebounded and is expected to continue to grow through 2017 (marketingcharts.com).

THE RADIO STATION STAFFING

Typical Radio Station

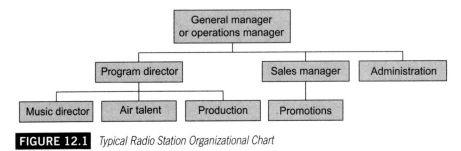

FIGURE 12.1 *Typical Radio Station Organizational Chart*

In order to see how decisions are made about music choices at a radio station, it is important to understand the organization within the station. The general manager or someone with a similar title, is responsible for the business success of the station. Reporting to the general manager is a manager of administration who has responsibilities such as accounting, commercial scheduling, and keeping up with regulatory matters. The sales manager has a staff of people who sell available commercial time to advertisers. Promotions are contests and other sales-oriented activities that are often the collaborative work of the sales team as well as the programming department. Some stations have added the position of marketing manager or marketing and promotions manager. Marketing managers may oversee sponsorships, publicity, events and research in addition to promotion duties (Denton, 2014).

From a record marketing standpoint, the key positions at a radio station are the program director, and to a lesser degree, the music director. The program director (PD) is genuinely the gatekeeper. Without the "okay" of the programmer, there is no chance that a recording will get on the air at most large radio stations. The PD is directly responsible to the general manager for creating programming that will satisfy the target market and build the existing audience base. The programmer decides what music is played, which announcers are hired, which network services to use, how commercials are produced, and every other aspect of the image the station has within the community it serves.

Critics of radio often say program directors have too much power because they can decide whether a recording is ever exposed to listeners. Large radio chains have group programmers who play an even larger role as a gatekeeper, recommending which music is appropriate for similarly programmed stations owned by the company across the country. As group owners of stations seek economies within their companies, group programmers—rather than a local program director—play a larger role than ever in the decisions regarding which music is played for the station's audience. Unknown to most of the radio audience, programming consultants armed with research about both the music and the local audience, assist the PD with decisions about music programming. To the record label, the consultant becomes another important gatekeeper.

Another criticism aimed at program directors is the decision to limit the size of their music playlists. Critics say that radio is serving as a filter for the massive amount of recorded music that is created every year and that they limit opportunities of newer artists. However, given the tens of thousands of new releases each year this may be more a case of sour grapes than fact. Theoretically, radio finds the most appropriate music for its audience and filters the music by choosing the best selections for the target audience. Program directors use advice, research, and their experience to find the best mix of music and information to retain and build their audiences for advertisers. Unfortunately, as the population ages, that mix has often included a larger proportion of sports and news, formats preferred by older males, and less music.

RADIO AUDIENCES

Nielsen Audio, the audience measurement company, publishes its annual "Radio Today" in which it provides an analysis of the makeup of audiences who use commercial radio in the U.S. The chart below is taken from their 2014 report, and shows the percentage of persons by age and gender who listen to radio during the average week. **The dark column represents males and the light column represents females**. Cume is a reference to the cumulative audience for the average week. For example, this charts shows that of people who are 12–17 years old (P12–17), 88.2% of males listen to radio each week whereas 92.4% of females listen to radio each week.

The size of the audience of a radio station is important to a label because it often determines how much time and other resources are put into promoting a song to the programming executive. Also important to the label is the time of day that the song is scheduled to be played. AQH is

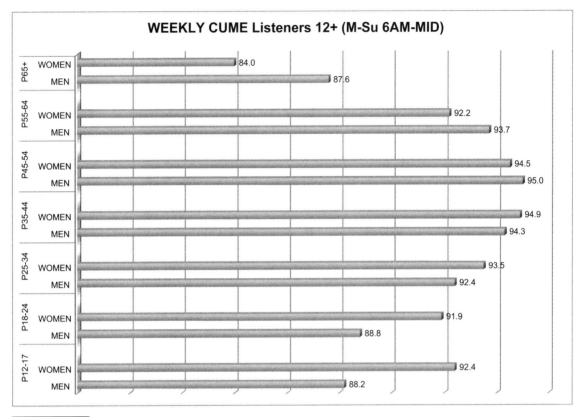

FIGURE 12.2 *Weekly Cume Rating (Source: Nielsen Audio Radio Today, 2014)*

a reference to the numerical size of the audience during the *average quarter hour* within the timeframe measured by Nielsen Audio. As you can tell by the Hour-by-Hour Listening chart, audiences are considerably larger during the week and during morning and evening rush hours. Radio refers to rush hour programming as "drive time" not only because the time corresponds to our morning and afternoon commutes but because listenership during that time drives the revenues of the station.

Radio listening peaks in the morning hours, known as morning drive time. Radio listening is divided into day parts of *morning drive, midday, afternoon drive, evening, and overnight*. From 6:00 a.m. until 9:00 a.m., listening is greatest as commuters wake-up to clock radios, and continue to listen as they drive to work. Listening picks up again around noon, declines slightly after lunch but remains relatively strong throughout the afternoon drive time, and then tapers off drastically throughout the evening and into the overnight period.

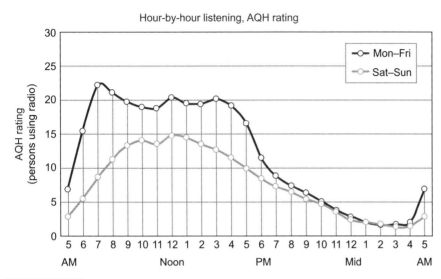

FIGURE 12.3 *Hour-By-Hour Listening (Source: Nielsen Audio Radio Today, 2014)*

Radio on the Go

About two-thirds of all radio listening happening away from home. Between 10:00 a.m. and 7:00 p.m. on weekdays, 72% of listening occurs outside the home. After 7:00 p.m. the majority of listening shifts to in-home.

TOP RATED DAYPARTS

	DAYPART AQH	P12+	P18-34	P25-54	P55+
W E E K D A Y S	AM DRIVE (6AM-10AM)	12.0	10.0	13.5	12.9
	MID DAY (10AM-3PM)	12.6	11.1	14.0	14.2
	PM DRIVE (3PM-7PM)	11.6	11.2	13.2	10.8
	EVENINGS 97PM-MID)	4.3	4.8	4.5	3.6
	WEEKENDS (6AM-MID)	7.2	6.7	7.4	7.5

FIGURE 12.4 *(Source: Nielsen: State of the Media: Audio Today 2014)*

Source: Nielsen National Regional Database, Spring 2013

RADIO FORMATS

Station owners choose radio formats by finding an underserved audience that is attractive to advertisers. When the format is chosen and developed, a programmer and staff are hired, and the audience develops. The chart

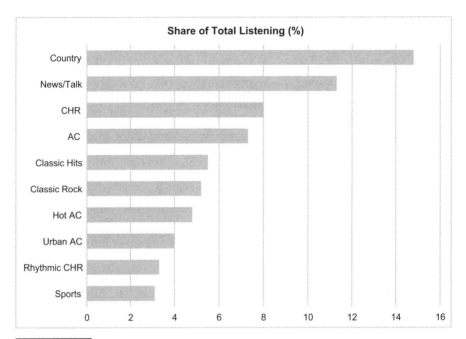

Share of Total Listening (%)

FIGURE 12.5 *Audience Format Trends Today (Source: Nielsen, 2014a)*

in Figure 12.5 shows the national radio audience sizes by format in 2014. The horizontal axis reflects the percentage of radio listeners who chose to listen to the radio formats represented in the chart.

The Nielsen Audio audience measurement service reports the national percentages of radio format shares in the chart. The top radio format is news/talk. This format has maintained its strength in recent years and continues to represent nearly eighteen per cent of all listeners.

Nielsen Audio is a subscription service, and is the only major company that measures the size and demographics of radio audiences. While the audience share chart shows the size of the national audience, Nielsen Audio measures the same information, radio market by radio market. The share and audience makeup of each individual commercial radio station is measured and reported to subscribing stations and advertising agencies. The size of the station's radio audience is directly related to the amount of money the station can charge for its advertising. The more listeners (or the larger its audience share), the more the station charges companies to access their audience through advertising. Nielsen Audio charges its clients hundreds of thousands of dollars for its audience measurement services. Since college and other noncommercial stations do not use traditional advertising, and therefore, don't have the revenues of the commercial stations, they are charged a greatly reduced fee (Buc, 2014).

With this in mind, a programmer is very careful in choosing music for airplay since the objective is to build its target audience. The program director is not inclined to experiment with an unproven recording that will turn an audience off. This will be discussed in more depth in a later chapter.

TARGETS OF RADIO FORMATS

The ability to obtain airplay can be a major factor in determining whether a recording will be released commercially. In order to be a commercial product, recorded music must find a target that is able and willing to buy it. Finding that target is the first step in the marketing process followed by the development of a strategy to reach the target. This table provides some broad definitions of music formats and their targets.

Table 12.3 Radio Formats and Targeted Demographics *CHR is for Contemporary Hit Radio. Artists listed in this chart are as they appear in the October 2014 charts for the MediaBase. America's Music Charts.

Format Name	Target Demographic	Artists in the Format
Adult Contemporary	Females 25–54	John Legend, Katy Perry, Justin Timberlake
Active Rock	Men 18–34	Sether, Godsmack, Pretty Reckless
Alternative	Persons 18–34	Black Keys, Arctic Monkeys, Cold Play
CHR/TOP 40	Persons 18–34	Maroon 5, Pharrell Williams, Magic!
CHR/Rhythmic	Persons 12–24	Nicki Ninaj, Iggy Azalea, Lil Wayne
Country	Persons 25–54	Blake Shelton, Keith Urban, Little Big Town
Hot Adult Contemporary	Females 18–24	Paramore, Nico & Vinz, Ariana Grande
Urban	Persons 18–34	Schoolboy Q, Chris Brown, Ca$h Out

The target market of a particular radio format is the logical consumer target for commercial recordings. The 2010 Broadcasting & Cable Yearbook lists over 35 musical radio formats ranging from AAA (Adult Album Alternative) to CHR to Variety (Diven, 2010). Music marketers carefully study the listeners of each format so that they can pitch their artist to the stations that reach their target market.

Radio station group owners further refine their format audience by gender and age. For example, many country radio stations specifically target females 35–44, while stations with CHR/Rhythmic formats target 12–24 males.

One of the key components of most of these radio markets is the 18–34 year-old. Young adults are a big consumer, setting up new households and making substantial purchases like appliances, furniture, or that first new car. Women in this age group heavily influence or actually make the purchase decisions for households, and advertisers highly value this demographic as a target for their messages.

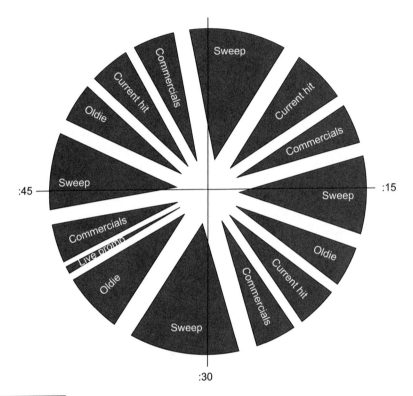

FIGURE 12.6

Programmers sometimes use a clock wheel to offer a visualization of how time is allotted to the various broadcast elements. It is a wheel indicating sequence or order of programming ingredients aired during one hour (Tarver, 2005). The clock face is divided into pie pieces, and each small section of time in an hour is prescribed a very specific item to be played on the air—from a song in a specific genre to a commercial, or news and weather. A sweep is a block of "non-stop" music, although station IDs and song introductions may be made. Notice that the sweeps in this clock all run through a quarter hour (15, 30, 45 minutes and the top of the hour). This is designed to improve the stations ratings in each quarter hour.

WHAT IS IMPORTANT TO PROGRAMMERS?

Convincing radio to play new music is "selling" in every sense of the word. And in order to sell someone anything, you must know what is important to them and what their needs are. High on that list of important things to radio is Nielsen Audio's measurement of radio audiences because it directly impacts the earnings of the station for its owners. Understanding concepts like this and their importance to programmers will help marketers of recorded music better relate to the needs of radio and its programming gatekeepers.

RATINGS RESEARCH AND TERMINOLOGY

The term P-1 listeners represents one of the prized numbers of radio programming. As Mike McVay with McVay Media puts it, "These first preference listeners . . . referred to in radio station boardrooms, focus groups, and inside the headquarters of Arbitron doing diary reviews," are "the most loyal of radio listeners," and every radio programmer courts this primary core of their radio audiences (www.mcvaymedia.com). The terms P-2 and P-3 refer to listeners with a lesser degree of connection and loyalty to a particular radio station.

Cume is the total number of unduplicated persons included in the audience of a station over a specified time period. This programming term that comes from the word cumulative, and it refers to the total of all different listeners who tune into a particular radio station, measured by Nielsen Audio in quarter-hour segments.

AQH, or average quarter-hour, refers to the number of people listening to a radio station for at least five minutes during a 15-minute period.

TSL means "time spent listening." It is an estimate of the amount of time the average listener spent listening to a station (or all stations) during a particular daypart. TSL is calculated by the following formula.

TSL = [Quarter-hours in a time period × AQH Persons]/Cume Persons

A radio station's **share** refers to the percentage of persons tuned to a station out of all the people listening to radio at the time.

Share = AQH Persons tuned to a specific station/AQH Persons in market currently listening to radio × 100

Rating refers to the percentage of persons tuned to a station out of the total market population.

Rating = AQH Persons tuned to a specific station/
Persons in market × 100

A complete list of radio rating terms and their definitions can be down-loaded from http://www.arbitron.com/downloads/terms_brochure.pdf.

FIGURE 12.7 *Personal People Meter (Courtesy of Nielsen Audio)*

Nielsen Audio manually measures and rates radio listener habits in about 290 markets in the US. Ratings are measured using the diary method in all but the top fifty-two markets. These latter markets, with the exception of Puerto Rico and the Hudson Valley metro area, are measured using electronic devices called the Portable People Meter shown in Figure 12.7 ("Market Survey," 2013).

For the diary method of tracking listening, Nielsen Audio selects households at random, and asks members' age 12 and above to carry the diary for one week and record their radio listening. Potential diary keepers are first contacted by telephone, and then diaries are sent to the household. Completed diaries are returned to Nielsen Audio and the data are entered into computers and analyzed on the following characteristics:

- Geographic survey area (metro or total survey area)
- Demographic group
- Daypart
- Each station's AQH: the estimated number of persons listening

- Each station's rating: the percent of listeners in the area of study during
- the daypart
- Each station's share: the percent of one station's total daypart estimated
- listening audience
- Cume: the total unduplicated audience during the daypart for an average week

For those using the Portable People Meter (PPM), it "is a unique audience measurement system that tracks what consumers listen to on the radio, and what consumers watch on broadcast television, cable and satellite TV. The portable People Meter is a pager-sized device that consumers wear throughout the day. It works by detecting identification codes that can be embedded in the audio portion of any transmission." (Arbitron, 2009) The meter is about the size of a pager, and is worn at all times during the day. At night, the People Meter is placed into a base unit and information is uploaded to a central database. Whereas the diary method of tracking radio listenership uses diaries targeted to a balanced yet random population sample, the PPM enlists families of people, pays them a fee, and seeks a two-year commitment for them to be part of the program.

Each Nielsen Audio Radio Market Report covers a 12-week period for the specified market and contains numerous pages like the example in Figure 12.8. At the top of each page, the target demographic is listed. Beneath that, the dayparts are laid out in columns. Then for each daypart, the AQH, the Cume, the AQH rating and AQH share are listed for each radio station in the area (listed in the left-hand column).

A station's ratings determine its advertising rates, and the example of a Nielsen Audio report of the Atlanta radio market makes the point that ratings mean money. The chart in Appendix A shows the call letters, the station format, the owner, and the percentage of listeners in the Atlanta market who choose each station. The table of ratings is shown beginning with the fall of 2003 as an indication of the trends of the ratings for each station in the Atlanta market. The column on the far right is the station rating for fall 2004. The station showing 9.7% of the market's radio audience can charge more for advertising than the station with only 0.4% of the audience because it has more listeners. Nielsen Audio rating points are the targets of the audience-building efforts of a radio programmer. For example, each one-tenth of a rating point, or 0.1%, is worth $1 million in advertising rates to radio stations in the Los Angeles radio market.

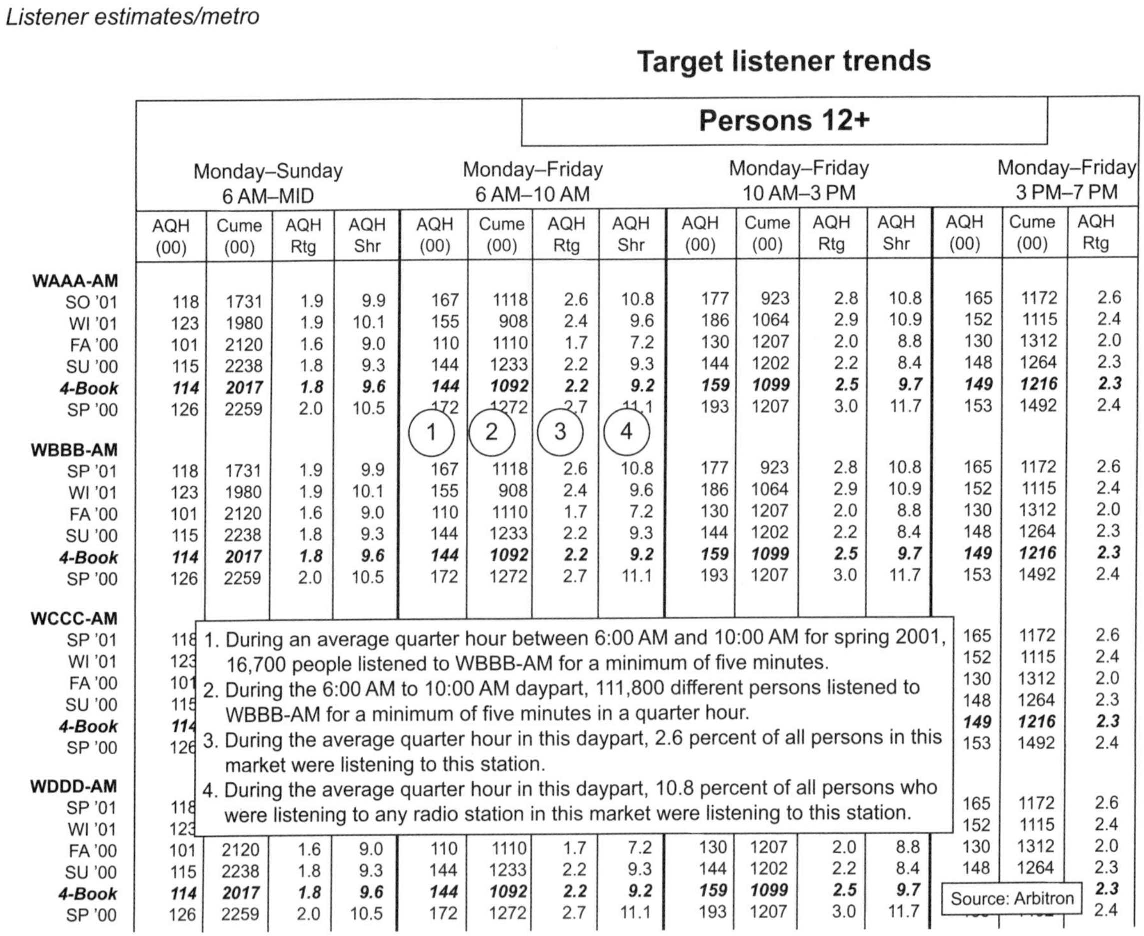

Listener estimates/metro

Target listener trends

	Monday–Sunday 6 AM–MID				Monday–Friday 6 AM–10 AM				Monday–Friday 10 AM–3 PM				Monday–Friday 3 PM–7 PM		
	AQH (00)	Cume (00)	AQH Rtg	AQH Shr	AQH (00)	Cume (00)	AQH Rtg	AQH Shr	AQH (00)	Cume (00)	AQH Rtg	AQH Shr	AQH (00)	Cume (00)	AQH Rtg
WAAA-AM															
SO '01	118	1731	1.9	9.9	167	1118	2.6	10.8	177	923	2.8	10.8	165	1172	2.6
WI '01	123	1980	1.9	10.1	155	908	2.4	9.6	186	1064	2.9	10.9	152	1115	2.4
FA '00	101	2120	1.6	9.0	110	1110	1.7	7.2	130	1207	2.0	8.8	130	1312	2.0
SU '00	115	2238	1.8	9.3	144	1233	2.2	9.3	144	1202	2.2	8.4	148	1264	2.3
4-Book	*114*	*2017*	*1.8*	*9.6*	*144*	*1092*	*2.2*	*9.2*	*159*	*1099*	*2.5*	*9.7*	*149*	*1216*	*2.3*
SP '00	126	2259	2.0	10.5	172	1272	2.7	11.1	193	1207	3.0	11.7	153	1492	2.4
					① ② ③ ④										
WBBB-AM															
SP '01	118	1731	1.9	9.9	167	1118	2.6	10.8	177	923	2.8	10.8	165	1172	2.6
WI '01	123	1980	1.9	10.1	155	908	2.4	9.6	186	1064	2.9	10.9	152	1115	2.4
FA '00	101	2120	1.6	9.0	110	1110	1.7	7.2	130	1207	2.0	8.8	130	1312	2.0
SU '00	115	2238	1.8	9.3	144	1233	2.2	9.3	144	1202	2.2	8.4	148	1264	2.3
4-Book	*114*	*2017*	*1.8*	*9.6*	*144*	*1092*	*2.2*	*9.2*	*159*	*1099*	*2.5*	*9.7*	*149*	*1216*	*2.3*
SP '00	126	2259	2.0	10.5	172	1272	2.7	11.1	193	1207	3.0	11.7	153	1492	2.4
WCCC-AM															
SP '01	118												165	1172	2.6
WI '01	123												152	1115	2.4
FA '00	101												130	1312	2.0
SU '00	115												148	1264	2.3
4-Book	*114*												*149*	*1216*	*2.3*
SP '00	126												153	1492	2.4
WDDD-AM															
SP '01	118												165	1172	2.6
WI '01	123												152	1115	2.4
FA '00	101	2120	1.6	9.0	110	1110	1.7	7.2	130	1207	2.0	8.8	130	1312	2.0
SU '00	115	2238	1.8	9.3	144	1233	2.2	9.3	144	1202	2.2	8.4	148	1264	2.3
4-Book	*114*	*2017*	*1.8*	*9.6*	*144*	*1092*	*2.2*	*9.2*	*159*	*1099*	*2.5*	*9.7*			*2.3*
SP '00	126	2259	2.0	10.5	172	1272	2.7	11.1	193	1207	3.0	11.7			2.4

1. During an average quarter hour between 6:00 AM and 10:00 AM for spring 2001, 16,700 people listened to WBBB-AM for a minimum of five minutes.
2. During the 6:00 AM to 10:00 AM daypart, 111,800 different persons listened to WBBB-AM for a minimum of five minutes in a quarter hour.
3. During the average quarter hour in this daypart, 2.6 percent of all persons in this market were listening to this station.
4. During the average quarter hour in this daypart, 10.8 percent of all persons who were listening to any radio station in this market were listening to this station.

Source: Arbitron

FIGURE 12.8   *Example Nielsen Audio Book*

Source: Nielsen Audio 2014b

RADIO PROGRAMMING RESEARCH

Knowing how radio researches its audience can be helpful to marketers to understand how programmers define benchmarks for their decisions to add or remove music from their playlists, or increase or decrease the frequency songs are played. The key research tools used by programmers are discussed in the next chapter in the section on charts.

Many stations use panels of listeners for programming research. The panel method is a research technique in which the same people are studied at different points in time. Members of the panel are selected to reflect a representative sample of a station or format's listenership, and are periodically surveyed on their opinions of music and programming. The panel members are contacted either by telephone or email and asked to respond to song hooks played either over the phone or through the Internet. The programmer then analyzes the listener response to make decisions about the continued use of the song on the air and how frequently it will be played.

A second method for collecting listener responses to music is the auditorium test. As the name implies, listeners are brought together in a single place, most often a hotel ballroom, and asked to listen to and rate the music at the same time. The biggest advantage of the auditorium test is control, but improved sound quality is another major plus. Invited listeners record their responses either on paper or using a WiFi monitored dial and the researchers compile the data for the radio station.

Another tool being used by programmers is called MScore, which uses the audience preference tracking of the Portable People Meter to judge whether listeners switch stations while a song is playing. For programmers, it is useful feedback about the music they program; for record labels, the information is useful to continuously monitor a song in the marketplace and modify promotional strategy (Albright, 2009; mediamonitors.com).

THE CHANGING FACE OF RADIO

The first commercial radio stations, all thirty of them, were licensed by the U.S. government in 1922, twelve years before the FCC was established. The following year, in 1923, there were 556 commercial radio stations, all of them broadcasting on AM frequencies. The technology didn't change much until the 1960s when the FCC approved a new band of radio frequencies for commercial broadcasters—FM. FM promised better sound quality and fewer commercials (6 minutes per hours instead of the 18 to 20 minutes typical at the time) to gain market share.

SATELLITE RADIO

XM Satellite Radio was founded in 1992, and launched on September 25, 2001. The next year, Sirius Satellite Radio was launched. In 2008 the two companies merged. Sirius XM offers an array of music and other entertainment channels which are fed to proprietary radio receivers, meaning you must own a special receiver in order to access programming. The subscriber to Sirius XM pays a monthly fee for basic services and a higher fee for additional services, much like a satellite television service. The music channels are commercial free other than promotions for upcoming shows and other Sirius XM channels.

The opportunities for marketers of recordings with satellite radio is that the company has longer playlists within each genre of music, so there are more opportunities for new music to be played. This is, in part, because the individual channels are not ranked like a terrestrial radio station and they are not dependent upon advertising revenue. Another plus for the label is that every song that is played displays the artist and song title to the listener on the faceplate of their receiver.

As a viable service, satellite radio was slow to be adopted by consumers in part because of the cost of buying special receiving equipment for a vehicle or a home audio system, and because of the monthly fees associated with the services. The company has been profitable since 2009, and now has nearly 27 million subscribers (Sirius XM, 2014). The company appears to be well positioned to survive both HD radio and Internet radio.

HD RADIO

Opportunities for marketers of recorded music have improved with the addition of new technology for the radio broadcast industry. Commercial stations are in the process of converting their AM and FM stations to HD radio, greatly improving the quality of the signals for both. (HD does not mean "high definition." It is a term used to brand the new service.) What this means to music marketers is that songs played on the radio will deliver near-CD quality audio, and have the ability to display the artist's name and the song title on the radio receiver, just like satellite radio. Radio announcers rarely provide artist or song information to listeners, and this new technology will help consumers of recorded music to identify artists and songs. As radio stations convert to digital broadcasting, they are also given up to two additional signals that are adjacent to their primary broadcast frequency. These new "stations" give the station owners opportunities to explore experimental programming and to broaden their listener bases. An element of a lawsuit settlement with the New York attorney general in 2005 requires several major broadcast companies to give new artists not associated with major labels the opportunity to have their music heard. It often is the HD channels that are used to allow these artists be heard on commercial radio.

Digital radio availability is growing as major broadcast companies convert stations to digital broadcast signals; however, stations and listeners have been slow to adopt the new technology. Upgrading station transmitters and licensing the technology in a time of declining listenership is an expensive investment (Buc, 2014). In 2012, only about two percent of radio listeners were tuned into HD radio at any given time, according to Pew Project for Excellence in Journalism's State of the News Media 2012 (Mook, 2012). Consumer adoption was slowed by the initial high cost of after market receivers and the economic recession which slowed the sale of new cars already equipped with HD receivers, but as more HD stations came available (over 2200 as of August 2014) and new car sales rebounded, usage has reached 3.8 billion annualized listening hours (Ibiquity Digital, 2014). Now the biggest threat to the adoption of HD radio comes from Internet radio services like iHeartRadio and Internet ready in-dash receivers.

INTERNET RADIO

In an interview with Radio World, James Cridland, managing director of Media UK, said, "There's terrible confusion between a 'radio' station—something with a human connection and a shared experience—and the likes of Pandora, Slacker, iTunes Radio and music subscription services. I find it very unhelpful that the radio industry has let folks like Pandora and Apple steal the 'radio' brand, and use it to describe a poorer music juke-box experience. They are not comparable" (Careless, 2014). There are two kinds of radio on the Internet: simulcast streaming of the programming of terrestrial radio stations and non-interactive streaming services like Spotify and Apple's iTunes Radio. Services like Last.fm.com and Pandora.com offer some of the features of traditional radio while giving the listener a unique listening experience they can tailor to their specific music tastes. With the expanding availability of WiFi service, and the power of 4G networks, streaming radio-like music services over wireless devices, including in-dash receivers in cars, brings music closer to specific consumer interests and poses a competitive challenge for traditional commercial radio. An additional advantage of Internet radio is that the song information is displayed on the device's screen as the song is played, often with a link to an online retail store to facilitate the impulse purchase.

Internet radio was hailed as a boon for the independent label and the unsigned artist because of their willingness to add "anything" to their database of music, but the reality may be something less spectacular. Hypebot.com estimated in 2013 that 4 million tracks available on Spotify had never been streamed—not once (Houghton, 2013). For artists that do get played they are paid between $0.0005 and $0.0069 per stream. Microsoft's Xbox Music Service pays $0.035 per stream (Dredge, 2013).

GETTING AIRPLAY

It is the job of the record promoter to get airplay on commercial radio stations. This has become more difficult with the consolidation of radio because there are fewer music programmers, and competition for getting added to the playlist is fierce. The process of record promotion is outlined in the following chapter on promotion, airplay and charts.

APPENDIX

STATION OWNERSHIP DATA SOURCES FOR TABLE 12.2

ABC—merged with Citadel in 2007—www.thewaltdisneycompany.com
CBS/Infinity Broadcasting—126 www.cbsradio.com
iHeart Media—850 www.clearchannel.com
COX—57 www.coxmediagroup.com
Cumulus Radio—525 Cumulus bought citadel in 2011 www.bizjournal.com
Entercom—106 www.entercom.com

GLOSSARY

Add date—This is the day the label is asking that the record be added to the station's playlist.

Average quarter-hour (AQH)—The number of people listening to a radio station during a 15-minute period as measured by Nielsen Audio is called the AQH.

Cume—The total of all different listeners who tune into a particular radio station is its cume.

Format—The kind of programming used by a radio station to entertain its audience is the format.

Heavy rotation—These recordings are among the most popular songs played on a radio station.

Light rotation—These are recordings that are played fewer times on a radio station than songs in heavy rotation.

Playlist—The list of songs currently being played by a radio station makes up a playlist.

P-1—The primary core of listeners to a specific radio station are P-1s.

Program director (PD)—This is an employee of a radio station or a group of radio stations who has authority over everything that goes over the air.

Share—A share is the radio audience of a specific station measured as a percentage of the total available audience in the market.

TSL—This means "time spent listening" by radio station listeners at particular times of the day.

REFERENCES

Albright, J. "The PPM Music Test 'MScore Switch' Is On," presentation at A&O Pre-CRS Seminar, March 3, 2009, Nashville, TN.

Arbitron. "The Portable People Meter System." 2009. www.arbirton.com/portable_people_meters/home.html.

Buc, F. Personal interview, November 14, 2014.

Careless, J. "While Internet Radio Market Grows Indie Web Stations Suffer." *Radio World*, March 7, 2014. Accessed August 9, 2014. http://www.radioworld.com/article/while-internet-radio-market-grows-indie-web-stations-suffer/269220.

Denten, P. "Promotions & Marketing Manager—Job Profile." On Air with Paul Denten, n.d. Accessed November 14, 2014. http://www.pauldenton.co.uk/Promotions_and_Marketing_Manager.html.

Diven, Y. *Broadcasting & Cable Yearbook 2010*. New Providence, NJ: ProQuest. 2009. Accessed August 8, 2014. www.americanradiohistory.com.

Dredge, S. "Streaming Music Payments: How Much Do Artists Really Receive?" *The Guardian*. N.p., August 19, 2013. Accessed August 9, 2014. http://www.theguardian.com/technology/2013/aug/19/zoe-keating-spotify-streaming-royalties.

Federal Communications Commission, n.d. Accessed July 30, 2014. http://www.fcc.gov/guides/review-broadcast-ownership-rules.

Houghton, B. "4 Million Songs on Spotify Have Never Been Played." *Hypbot.com*. N.p., October 11, 2013. Accessed August 9, 2014. http://www.hypebot.com/hypebot/2013/10/4-million-songs-on-spotify-have-never-been-played-stats-infographic.html.

Ibiquity Digital. "Explosive Growth In Hd Radio™ Equipped Vehicles, Boosts Digital Radio Listening To Over 3.8 Billion Annualized Hours" [Press release]. (2014). http://www.ibiquity.com/press_room/news_releases/2014/1639.

"Market Survey Schedule & Population Rankings." Arbitron.com. Nielsen Media, Fall 2013. Accessed November 14, 2014. http://www.arbitron.com/downloads/fa13_market_survey_schedule_poprankings.pdf.

Marketing Charts. 2014. http://www.marketingcharts.com/wp/radio/radio-ad-revenue-growth-forecast-downgraded-28169/attachment/biakelsey-radio-indus-advertising-revenue-2006–2017-mar2013/.

McVay, M. McVay Media. 2014. www.mcvaymedia.com.

Mook, B. "Slow Growth for HD Radio." *Current.org For People in Public Media*. N.p., November 5, 2012. Accessed August 9, 2014. http://www.current.org/2012/11/slow-growth-for-hd-radio/.

Nielsen. "Radio Increases Year-Over-Year Reach." 2013. http://www.nielsen.com/content/corporate/us/en/press-room/2013/radio-increases-year-over-year-reach-by-more-than-700-000-according-to-december-2013-radar-report.html.

Nielsen Audio. "State of the Media: Audio Today 2014." 2014a. http://www.nielsen.com/us/en/insights/reports/2014/state-of-the-media-audio-today-2014.html.

Nielsen Audio. 2014b. http://www.nielsen.com/content/corporate/us/en/pressroom/2012/music-discovery-still-dominated-by-radio-says-nielsen-music-360.html.

Oxenford, D. "On the 15th Anniversary of the Telecommunication Act of 1996, The Effect on Broadcasters is Still Debated." *Broadcast Law Blog.* Ed. David Oxenford. N.p., February 9, 2011. Accessed July 31, 2014. www.broadcastinglawblog.com.

"Radio Today: How America Listens to Radio." Arbitron Report. 2008.

Sirius XM. *SiriusXM Reports Third Quarter 2014 Results. Investor Relations.* Sirius XM, October 28, 2014. Accessed November 14, 2014.

Tarver, C. 2005. www.udel.edu/nero/Radio/glossary.html.

Williams, G. "Review of the Radio Industry, 2007," Federal Communications Commission. 2007. http://hraunfoss.fcc.gov/edocs_public/attachmatch/DA-07-3470A11.pdf.

CHAPTER 13

Promotion, Airplay, and the Charts

From the earliest days of product marketing, sales people have constantly sought as many ways as are possible to say their wares are the best available and that theirs are the number one products in the eyes of consumers. In music especially, bragging rights of having a "number one" provide leverage for promoters at the label to ask the chart makers to "join the crowd" and move their single or album higher on their charts, ultimately impacting sales. In this chapter, we will explore how radio promotion leads to airplay, and some of the important charts that are helpful in getting exposure on radio, on television, and on the Web.

In this chapter, we will look at the importance of promotion at a record label. We have all been to the doctor's office and seen drug company marketers drag their wheeled bag of samples and a box of donuts into the office area "behind the door." They are touting benefits of medicines and building relationships with doctors so they can earn bonuses based on how many prescriptions are written in their territories. Doctors do not need medicines but their patients do. In the music business promoters seek to convince program directors to "prescribe" the label's music to the station's listeners. And like the promoter of medicine, the music promoter is paid bonuses based on how many programmers are convinced to choose the label's music to play on the air.

Airplay, in the traditional sense of music being played on terrestrial radio, is viewed by some as a dying medium and a waste of promotion money. As we saw in the last chapter, however, radio continues to draw huge audiences exposing recorded music to tens of millions every day. During one week in November 2014, the top single on *Billboard*'s top 40 airplay chart was heard

FIGURE 13.1

by a U.S. audience of over 86 million. Any medium that connects our product with its target market with such impact is far from dead and must be an important component to our marketing mix.

RADIO PROMOTION

Lobbying and lobbyists have been around as long as any one person has been responsible for a decision or a vote, and people have always wanted to influence that decision or vote in their favor. Every day, lawmakers at the national and state levels meet with representatives of special interest groups who ask them to vote on matters that are in the best interests of their groups or their clients.

The same thing happens between a radio programmer and a record promoter. Record promoters are lobbyists in the purest form. They are either on the staff of a record company or they are part of a company specifically hired by the label to promote new music.

The people lobbied are usually the program directors of radio stations that report their station's airplay list to the major trade magazines that publish the charts. Program directors, also referred to as PDs, have the ultimate responsibility for everything that a radio station broadcasts—banter by personalities, advertising, information such as news and traffic reports, and all music played by the station. Simply said, the radio programmer can decide whether a recording ever gets on the air at their station.

Decisions by radio programmers are one the keys to the life of a recording and have become the basis for savvy, smart, and creative record promotion. Programming decisions about music determine:

- Whether a new recording is added to the playlist of a radio station that reports its chart and airplay to trade magazines.
- Whether the recording receives at least light rotation on the playlist.
- Whether it eventually receives heavy rotation on the station's playlist.

PROMOTION

Traditional marketing texts teach the four P's, one of which is promotion. Those same texts tell us that the promotional mix consists of public relations, advertising, sales promotion, direct marketing, and personal selling.

The record promotion by a label comes closest to being personal selling than any other aspect of the traditional promotional mix at a label, except perhaps the label sales department. Personal selling is one-on-one, and the work of a record promoter is exactly that. The radio promotion department at a record label has the responsibility of securing radio airplay for the company's artists. Quite simply, they lobby the station decision makers—the gatekeepers—to add recordings by the label's artists to the station's playlist. With commercial radio creating hundreds of millions of music impressions every week, some label executives claim that radio is responsible for 70% of the sales of recorded music. Label record promotion people can be critical to the success of an album project. It is this close connection between airplay and record sales that creates the need for record promotion to radio as a key element of the marketing plan.

FIGURE 13.2

The effectiveness of a record promotion person hinges on the strength of the relationships he or she has with radio programmers. These relationships are built much like anyone in business that has a client base requiring regular service. The promoter makes frequent calls in person and by telephone to the programmer, arranges lunch or dinner meetings, provides the station access to the label's artists, and helps the station promote and market itself with contests, performances by artists, and giveaways. The promoter who has effectively developed a good relationship with a programmer can make the telephone call and ask the programmer to treat his current record project favorably—which is difficult for smaller labels to do because of their limited promotional resources.

If the promoter has no relationship with the radio programmer from a reporting station, it is highly unlikely that telephone calls will ever be returned. Programmers today have too many things to do and little time to listen to promotional pitches from people and companies they don't know. And because of centralized programming by some radio groups, there is more pressure on fewer programming personnel to find time to take a call from a promoter.

PROMOTING

It is important to understand that promotion by a label is focused on radio stations that program current, new music. Stations airing older music are using record company catalogues as the basis for their entertainment programming, and many fans of the artists they program already have the music in their libraries or on their MP3 players. Older albums and their related singles have been around for years, sometimes decades, and there is limited interest on the part of a label to promote them to radio. Most energy and money is invested by record companies in promoting the newest singles and album projects.

Record promotion to radio by labels typically needs to answer four questions:

1. How does it help us sell recordings?

2. What does it do for our artist?

3. What does it do to further our agenda and help us market the artist?

4. Does it make sense for the radio station?

As you create a promotional tour for an artist with a new CD, emphasis should be on the most popular radio stations in targeted cities, especially those "reporting" stations that report their airplay charts to major trade magazines such as *Billboard*. To determine which stations are rated the highest in audience shares, you will find Nielsen Audio's reported findings at www.radioandrecords.com, as well as through other online sources. Specific information about the station programmer's name, address, and telephone number is available through several media databases including Bacon's MediaSource, as well as Cision, which include radio.

After a single has been released to radio, nearly everyone at the label tracks its progress. For the promotion staff, the monitoring tools and charts discussed in this chapter help guide their work. Some labels also have staff members who monitor the "buzz" on their new music in chat rooms and in blogs. All of this information helps the label determine whether they have a hit on their hands, whether they should step up the promotional effort, or whether it's time to cut their losses and move on to the next project.

The typical business week for record promoters begins early in the week as the trade chart closings show the successes and failures of singles from the previous week and give the promotion team the information they need to allocate their time for the next seven days. With the plan in place, the promotion staff then begins its weekly cycle of contacting radio programmers to build airplay for the label's products.

One label executive says his regional label promoters are always on the job. The only time he allows a break is when his promotion team members are on vacation. Jobs as regional promoters with some experience can go to work at a starting salary in the range of $70,000 to $80,000 US, plus expenses. Vice presidents of label promotion for larger labels can easily earn well over $200,000, plus bonuses and other incentives.

HISTORY

FIGURE 13.3

Record promotion and its regulation by the Federal government began not long after the advent of commercial radio broadcasting. In 1934, Congress created and passed the Communications Act, which restricted radio licensees—the stations themselves—from taking money in exchange for airing certain content unless the broadcast was commercial in nature.

However, this early act contained nothing that prohibited disc jockeys (DJs) from taking payments in exchange for airplay. During the big band era of the 1940s, and the rock 'n' roll days of the 1950s, DJs were routinely taking money from record promoters in exchange for the promise to play a record on the air. Disc jockeys during this time often made their own decisions about which records would be included on their programs, and promoters would approach them directly to influence their record choices.

Lawmakers railed against the rampant bribes being given to DJs to play records. In 1960, Congress amended the 1934 act to include a provision that was intended to eliminate illegal bribes to play music, so called payola. Under the revised law, disc jockeys and radio stations were permitted to receive money and gifts to play certain songs, but the amendment placed a requirement that these inducements must be disclosed to the public on the air. If this disclosure was not made, it exposed the DJ and management to possible fines and imprisonment (Freed, 2005). The change in the law also created the requirement that record labels must report their cash payments and major gifts for airplay to the station. This 1960 amendment continues to guide the radio and records industries today.

Despite the stronger laws against payola, Federal investigators were called upon to investigate scandals within the record promotion business in the 1970s and 1980s. There were no major convictions despite the appearance that money, drugs, and prostitution were being used as leverage by promoters to get radio airplay for recordings (Katunich, 2002).

Entanglement by the payola laws began when radio stations asked the label promoters for something of value, whatever it was. The first question to the record company then becomes, "How does this help me sell copies of the artist's record?" One label says that everything they do today with radio stations to promote an artist "has to really pass the smell test." They require proof that promotional prizes and free concerts by artists are acknowledged on the air as being given by the label. They expect to receive affidavits from the station showing when the announcements were aired and at how much money those announcements were valued, underscoring the disclosure requirement for radio stations taking payments, as well as the record labels providing payment. In the late 1990s and early 2000, stations and labels began to break the payola law and in 2005 came under investigation by the New York Attorney General's office. The examination into these practices resulted in Sony BMG, EMI, Warner Bros., and Universal Music paying over $30 million in penalties for paying radio stations to play their music. The Federal Communications Commission followed the

From:
Sent: Tuesday, January 14. 2003 8:53 AM
To:
Subject: RE: WLIR/Long Island in on Everclear

unfortunately we had to pay his indie 1000 dollars, so there is no money for additional promo, let him know.

-----Original Message-----
From:
Sent: Tuesday, January 14. 2003 6:59 AM
To:
Subject: WLIR/Long Island in on Everclear
Importance: High

Gary Cee is asking for some promotional support for this add and moving Coldplay to 5x/day....he asked for $1,500 in tee shirts, I said there was no way, he then asked for $1,000-I said I would check, lemme know what I can do.........
please advise....

cheers,

FIGURE 13.4

New York inquiry, which resulted in $12.5 million in fines levied against Citadel Broadcasting, CBS, Entercom, and iHeart Media Communications (Barbington, 2007). Following the investigation, labels of all sizes spent considerable time and fees on lawyers to assure that promotion stayed within the guidelines of the law.

One of hundreds of emails acquired by and used by the New York Attorney General's office as evidence of pay-for-play violation of Payola Laws. (Source: NY Attorney General's website.)

GETTING A RECORDING ON THE RADIO

The consolidation of terrestrial radio has concentrated some of the music programming decisions into the hands of a few programmers who provide consulting and guidance from the corporate level to programmers at their local stations. In many cases, local programmers have the ability to add songs to their playlists based upon the preferences of the local audiences. Here is how songs are typically added to station's playlists for those stations that program new music:

1. A record label promoter or an independent promoter hired by the label calls the corporate programmer, the station music director (MD), or program director (PD) announcing an upcoming release. Radio music directors have "call times." These are designated times of the week that they will take calls from record promoters. The call times vary by station and broadcasting company and are subject to

change. For example, a MD may take calls only on Tuesdays and Thursdays, 2:00–4:00 p.m.

2. Leading up to the add date, meaning the day the label is asking that the record be added to the station's playlist, the promoter will call again touting the positives of the recording and ask that the recording be added.

3. The music director or program director will consider the selling points by the promoter, review their preferred charts for performance of the recording in other cities, consider current research on the local audience and its preferences, look at any guidance provided by their corporate programmers/consultants, and then decide whether to add the song.

4. The PD will look for reaction or response to adding the song.

PROMOTION DEPARTMENT

An important component of promoting a recording to radio is the effectiveness of the record company promotion department, or the independent promoters hired to get radio airplay. This would appear to be a simple process, but the competition for space on playlists is fierce. Thousands of recordings are sent to radio stations every year, and the rejection rate is high because of the limited number of songs a station can program for its audience. Some of the recordings are rejected from being included on playlists because they are inferior in production quality, some are inappropriate for the station's format, some may never connect with a commercial radio audience, and ultimately will lose their label support if they fail to quickly to become favorites with the radio audience.

Record promotion has been a big-dollar investment, which made it a key marketing element necessary to stimulate consumers' interest in buying new music. Costs for a label to market and promote a single easily reach $1 million or more, not including the production of the recording or any advances to the artist or producer. Even in the world of country music where annual sales are often less than 10% of all recorded music, labels invest as much as $300,000 just to get the single of a new artist into the top 20 of an airplay chart.

However, the marketer's litmus test for the viability of a recording is to honestly compare it with other songs achieving success on the various charts. If it is not at least as good as those listed on those charts, then it doesn't have any chance at all, even if it has a competitive marketing budget.

LABEL RECORD PROMOTION AND INDEPENDENT PROMOTERS

Most large labels have a promotion department whose sole purpose is to achieve the highest airplay chart position possible. While most consumers assume a number one song is the biggest seller at retail, the number one song on most Billboard charts actually is the song that has the most airplay on radio. The connection between airplay and sales is well documented, so a high chart position is critical to the success of a recording and becomes the heart of the work of a record promotion department.

Labels often have a senior vice president of promotion who usually reports directly to the label head. The senior vice president of promotion at a pop music label typically has several vice presidents of various music types based on radio formats. These vice presidents then have regional promotion people who are viewed by the label as field representatives of the promotion department. They are liaisons to key radio stations in their region. The vice president and the so-called regionals are the front line for the label attaining airplay. It is their responsibility to create and nurture relationships with programmers for the purpose of convincing them to add the company's recordings to their playlists. At a country label, the senior vice president of promotion typically has a national director of promotion and several regionals (see Appendix, p. 290–291).

Labels sometimes hire independent promoters (indies) to augment their own promotional efforts. Since promoting songs for airplay relies on well-developed relationships, indies may have developed stronger relationships with some key stations than the label has, and the record company is willing to pay indies (half of which is often recoupable from the artist) for the value of those relationships. In the 1980s, a practice of labels hiring independent promoters, or "indies" to attain airplay at radio came under very tight scrutiny by Federal law enforcement. Fred Dannen, in his book, Hit Men, found that CBS Records was paying $8–10 million per year to indies to secure airplay for their acts. By the mid-1980s, he says that amount was $60–80 million for all labels combined. The resurgence of independent promotion in the 1990s caused the New York Attorney General in 2004 to look closely at these lucrative agreements with the result that most indies shuttered their doors or went into other businesses.

Independent promoters have typically made their money this way: they would sign an agreement with a radio station to be the stations' exclusive consultant on new music for a year, and then pay the radio station for that right. The indie promoter did not require the station to play specific songs, according the typical agreement, but the station did promise to give its play-list for the following week to the indie promoter before anyone else. Then, the promoter would send an invoice back to the record company for $3,000–4,000 for each single that was charted on stations that they represented. Because of this "exclusive" arrangement with the radio station, record label promotion people had no dealings with the radio stations represented by indies. One label promotion vice president said the $4,000 can easily turn into $30,000 per single if the indie had to provide contest prizes to help promote the single at radio and to pay for other promotional expenses. Many arrangements with indie promoters included provisions for bonuses based on their success at charting records with individual stations (Phillips, 2002).

The use of independent promoters provided the record company a layer of insulation between themselves and radio. The record companies did not deal directly with certain radio stations in matters of adds and spins; rather, they dealt only with independent promoters who promote to these stations. As one executive said, "The use of independent promoters creates a clearinghouse by removing the label one step away from . . . making any compromises that some might make to get a song on the air." Another says, "This way the money doesn't go directly from the label to the radio station." (Personal interviews, Fall 2004).

Critics of the independent promotion system claimed it shut off access to radio airplay by independent labels and artists. Alfred Liggins is CEO of Radio One, a company that owns nearly 70 radio stations targeting African-American audiences. He acknowledged that their exclusive relationships with independent promoters meant that labels without an indie promoter are less likely to get a record played on his stations (ABC Television, 20/20, 2002). Nearly every record label had discontinued its relationship with independent promoters who pay radio stations for access. While the current trend is away from independent promotion, history has shown that this might just be temporary.

SATELLITE RADIO PROMOTION

Satellite radio has expanded the horizon for promoting recordings to radio. The services provided by Sirius XM have a relatively low monthly subscription fee and offer over 72 channels. Though the service requires that a vehicle or home be equipped with a special receiver or subscription with online computer service, many newer cars and trucks have satellite receivers integrated into their standard radios. Satellite radio delivers an audience over 25 million subscribers, and offers exclusive talk, live news, major sports events, and commercial free and curated music channels with sub-genres and niche formats, which gives labels great opportunity for air-play. Though record company promoters actively work with satellite music channel programmers seeking adds to their playlists of current music, the future of satellite radio no longer remains in question. As a company, Sir-ius XM turned its first profit in 2010, and with the aid of the U.S. car mar-ket turn-around, has found itself in the black with surging sales as new-car buyers sign up after their trial subscription lapse.

THE CHARTS

This section of the chapter looks at charts of all types from the viewpoint of their value to marketers of music in promoting music to the gatekeep-ers. From a historical perspective music charts have been among the most valued information provided by trade magazines and newsletters, which is why our discussion of charts begins with the trades.

The History of Trade Magazines

Billboard Magazine has, for more than a century, been the leading music business publication providing comprehensive weekly views of the record-ing industry, the music business, and commercial radio.

Billboard Magazine was first introduced in 1894 as a publication sup-plying its readership with information about advertising and, a few years later, about the carnival industry. Its reporters then began writing stories about sales of sheet music, and early in the last century it added regular features about silent films and commercial radio. In 1936, it published its first "hit parade," which was a term used at the time to rank popular songs which then became a term used by radio to denote its most popular music In 1940, *Billboard* compiled and published its first music popularity chart called "Best Selling Records Chart." The first national number one record-ing reported by *Billboard* was Tommy Dorsey's, "I'll Never Smile Again" with vocals by Frank Sinatra. It was the decade of the 1940's when *Bill-board* became a leader and ultimately the icon for music charts by publish-ing numerous charts for various genres of music. The *Billboard* Hot 100, which ranked single releases based on both sales and airplay, was launched in 1958. That chart remains as one of the continuing staples of magazine (Billboard.com).

Other national trade magazines have competed with *Billboard* over the years including the *Gavin Report,* which closed in 2002, *Cashbox,* which closed in 1995, and *Radio & Records* which closed in 2009. Gavin,

Cashbox, and R&R magazines were all key industry trade publications for many years until they were retired for economic reasons. Gavin and Cashbox relied upon "reported" airplay by radio stations which gave the publications their airplay charts for the coming week, whereas *Billboard* (currently owned by Prometheus Global Media who also owns and operates *Adweek*, *Back Stage*, *Film Journal International*, and *The Hollywood Reporter*), now relies on electronically monitored airplay by Broadcast Data Systems (BDS) to generate their charts (BDS, 2009).

Another airplay-generated chart is MediaBase, and although it does not publish a hard copy trade magazine for distribution, its data is supplied to *USA Today* every Tuesday and helps various entertainment outlets such as the American Music Awards, and the nationally syndicated show "America's Top 40 with Ryan Seacrest."

The Importance of Charts

Perhaps the most important piece of real estate a record label can own is a high position on a chart which influences gatekeepers. For example, a top-selling digital single on the SoundScan digital sales chart may influence a radio programmer's decision to add music to its airplay list or a decision by a respected blogger to mention it online. Credible, influential charts of all types are key tools used by record labels to promote music to people who can provide access to their audiences. Airplay, sales, and download charts can have the effect of demonstrating "word of mouth" interest between and among audiences, and give a label promotion department the information they need to encourage industry decision-makers to engage the music. Airplay charts are especially important for use by radio programmers to give them a basis to compare their audience offerings with those audiences of similar cities, and adjust their programming if necessary. Remember—radio is in the business to sell advertising airtime and matching listeners to the music is key.

Creating the Airplay Charts

Charts in trade magazines are defined in numerous ways, and it is important to be sure the distinction is made between sales charts and airplay charts. If a label has a number one album, it earns that designation based solely on its position on a sales chart. If a label has a number one single on the radio, that position is based upon any or all of the following criteria: the number of times a single is played on the radio during a specific week, how big the cities are in which the song is played, and how many singles it sold.

Nearly every major music genre has an airplay chart in *Billboard*. The airplay charts compile the national airplay of singles on radio stations as detected by a proprietary system, called Broadcast Data Systems (BDS). Most genres are also represented in MediaBase airplay charts, but first let's look at BDS.

Broadcast Data Systems (BDS) and *Billboard* Charts

Broadcast Data Systems (BDS) is the technology used by *Billboard* and *Canadian Music Network* magazines to detect each spin of a recording on radio in cities in which BDS has installed a computer to monitor airplay. As the spins are detected, the computers upload the number of detections to a main database that is then used to create the weekly airplay charts. Geoffrey Hull cites *Billboard* describing the system as:

A proprietary, passive, pattern-recognition technology that monitors broadcast waves and recognizes songs and/or commercials aired by radio and TV stations. Records and commercials must first be played into the system's computer, which in turn creates a digital fingerprint of that material. The fingerprint is downloaded to BDS monitors in each market. Those monitors can recognize that fingerprint or 'pattern,' when the song or commercial is broadcast on one of the monitored stations.

(Hull, G., 2004, pp. 201–202)

As the computerized airplay monitor "listens" to a song being played on the radio, it compares its digital fingerprint to that on file, and then logs it as a detected play of the song.

BDS has monitors for airplay in 140 radio markets, and claims to listen to over 1,600 stations and detect over 100 million songs each year (BDS, 2014). These detections are used to compile 25 airplay charts for *Billboard*. Additionally, the service compiles detections of airplay and streaming on satellite and cable music channels. Label marketers must be sure they register their recorded music with BDS or there will be no detections of airplay. BDS provides information on its website on how to register a song and get a digital fingerprint created for the airplay monitoring system. Without these statistics, the radio promotion department will be without some of the bragging rights they need to promote the single, and the marketing department will be without one of its key tracking tools (BDS).

Monitored: Chart / New & Active / Most Added / Most Increased / Recurrent Chart
Indicator: Chart / New & Active / Most Added / Most Increased / Recurrent Chart

billboard

Powered By nielsen BDS

MAINSTREAM TOP 40 NATIONAL AIRPLAY ©

Issue Date:
11/22/2014

TW	LW	WEEKS ON	ARTIST TITLE IMPRINT / PROMOTIONAL LABEL	PLAYS TW	+/-	AUDIENCE MILLIONS	RANK
			*** NO. 1 ***				
1	1	17	TOVE LO Habits (Stay High) ISLAND/REPUBLIC	15534	+807	86.585	1
			2 week(s) at number 1				
			◄ MOST INCREASED PLAYS ►				
2	7	7	MAROON 5 Animals 222/INTERSCOPE	14116	+2857	79.983	2
3	2	15	JESSIE J, ARIANA GRANDE & NICKI MINAJ Bang Bang LAVA/REPUBLIC	13443	-1142	64.724	5
4	3	12	TAYLOR SWIFT Shake It Off BIG MACHINE/REPUBLIC	13115	-824	73.060	3
5	8	16	ED SHEERAN Don't ELEKTRA/ATLANTIC	11668	+764	58.742	6
6	5	16	MEGHAN TRAINOR All About That Bass EPIC	11109	-851	68.151	4
7	6	15	JEREMIH FEAT. YG Don't Tell 'Em DEF JAM	10580	-1008	51.499	8
8	4	19	IGGY AZALEA FEAT. RITA ORA Black Widow TURN FIRST/HUSTLE GANG/DEF JAM	10479	-2157	56.684	7
9	9	9	CALVIN HARRIS FEAT. JOHN NEWMAN Blame DECONSTRUCTION/FLY EYE/ULTRA/ROC NATION/COLUMBIA	9274	+396	50.539	9
10	10	12	MR PROBZ Waves LEFT LANE/ULTRA/RCA	8543	+1064	49.294	10

FIGURE 13.5

When reading the singles chart, you see TW (this week's) current position, LW (last week's) position, number of weeks on the chart, artist, single title, with album title and label underneath, and important stats to follow: TW number of spins from panel of radio stations that report to *Billboard*, increase/decrease of spins from prior week, audience count (in millions), and then rank of single if based on number of listeners.

BDS gives the label's marketing department considerable information about which radio stations are spinning a single and how frequently. Combining this information with SoundScan data on sales of singles and albums, label marketers have continuing feedback on the performance of their recorded music projects. And most importantly, this feedback gives marketers the information needed to modify marketing plans in order to draw as much commercial activity out of the marketplace as possible.

Many weekly charts representing a variety of airplay, sales, and a combination of both appear in *Billboard*. Charts that rank with SoundScan sales data only are the *Billboard* Top 200 and each genre albums chart. Charts that use mixed data, meaning both sales, airplay and streaming information, are the "hot" charts that combine the data to show the action on each of the titles (billboard.com).

Some of *Billboard*'s weekly airplay charts rank singles based upon the number of times a song is played on monitored radio stations and the resulting size of the combined audience that heard the song each time it was played. For example, a song played once by Nielsen Audio's major R&B station in Los Angeles represents a larger audience than the song played once in Chillicothe, Ohio. But the combination of all plays multiplied by all audiences as measured by Nielsen Audio dayparts determines the airplay ranking for a single. Charts using an audience method are Hot Country, Hot 100 Airplay, Hot Rap Songs, and Hot Latin Songs. The remaining airplay charts base the ranking of singles only on the number of times the single is played by all monitored stations in one week, with weighting ranking by audience size. (*Billboard*, 2008).

Within the charts, *Billboard* indicates recognition for accomplishments each week with the use of special awards. Designations are as follows:

Designation	Description
Greatest Gainer	Indicates the album with the greatest increase in sales
Pace Setter	Notes the album with the largest percentage sales gain
Heatseeker Graduate	Is an album by a new artist that was removed from the Heatseekers Chart into the top half of the Billboard 200 chart
Highlighted position number	This identifies a single or an album that has shown growth in either audience or sales over the previous week

FIGURE 13.6

A heatseeker is a special designation by Billboard for developing artists and is described as:

The Top Heatseekers chart lists the bestselling titles by new and developing artists, defined as those who have never appeared in the Top 100 of The Billboard 200 chart. When an album reaches this level, the album and the artist's subsequent albums are immediately ineligible to appear on the Top Heatseekers chart.

(*Billboard* online)

The Hot 100 has been a part of *Billboard* magazine since 1956, and it has spawned numerous other genre-specific charts. The Hot 100 is a chart that is developed each week using a formula that combines the physical sales of singles, sales of digital downloaded singles, as well as the number of spins a song receives on radio along with online streaming, regardless of the genre of music or the radio format in which the song is programmed.

Other charts published weekly by the magazine include rankings by genre, created by using data gathered for airplay through its BDS reporting system. Also, each week Billboard publishes its Top 200 chart, which is a compilation of the bestselling albums ranked by the number of unit sales for the previous week. Both of these charts are especially important to the marketing effort by a label on behalf of its active new music projects. Tracking the impact the music is having at radio and at retail gives label marketers information that is helpful to control the success of singles and albums.

Among the newer charts is Billboard's reaction to the power of social media. There are three streaming charts: Streaming Songs (for non-interactive sites like Pandora), On-Demand Songs (for interactive sites like Spotify), and a YouTube chart. A few more charts have been created to reflect Twitter and the persuasion of influence: Twitter Real Time, Twitter Top Tracks, and Twitter Emerging Artist. The Real Time chart has the most "moving parts" since it is based on comparing the number of times a song is shared within the last hour to the hourly average of shares over a rolling 24-hour period. Think about this—this is a massive amount of monitoring with varying degrees of "what is considered shared" and "does the sharing have to be positive" (Billboard.com).

How are songs shared and/or mentioned on Twitter?

Song shares are tracked and incorporated into the *Billboard Twitter Real-Time Charts* by:

- the use of, or the inclusion, of a link to the song via music listening platforms, such as Spotify, Vevo and iTunes.
- the use of various track sharing notations, such as the hashtags "#nowplaying" or "#np," along with song/artist name.
- the use of various terms associated with the song and song playing, such as "music," "song," "track," "listen."

Examples

Below are some examples illustrating the types of tweets about Ariana Grande's "Problem," which could be counted towards a position on the *Billboard Twitter Real-Time Charts*:

This tweet uses the artist's official Twitter handle, the song name, and a streaming link. It has an **excellent** chance of counting towards the charts.

This tweet uses the artist's official Twitter handle and song name, but does not include a streaming link. It has a **great** chance of counting towards the charts.

This tweet includes a conversation about a song, along with the name of the artist and song title, but does not include hashtags, official Twitter handle, or a streaming link. This example has a **good** chance of counting.

FIGURE 13.7

MediaBase

MediaBase is a service owned by iHeart Media Communications that monitors the airplay of recordings on over 1,000 U.S. radio stations, and 125 in Canadian stations (Hoovers). The airplay detections by the service are then published as "America's Music Charts", as a full page advertisement in *USA Today* each Tuesday (Leeds, 2006), as well as on the MediaBase website and AllAccess.com.

While the information is very similar to that provided by BDS and is used by record labels in the same ways, there are some differences between the services. Where BDS uses computers to detect airplay, MediaBase employs people to actually listen to radio stations and log the songs played. Employees of the company who are paid to detect airplay are experts in their genres of music, they often work from their homes, and they are provided the necessary hardware and software by the company. Employees who work in airplay detection are often responsible for logging songs for the 24-hour broadcast days of eight radio stations.

Another difference in the services lies in the particular stations whose airplay is monitored. MediaBase monitors an estimated 80% of the same stations as BDS. Some record labels, artist managers, and artists see the need to subscribe to both services to be sure they are getting accurate feedback on the performance of single releases at radio (Rhodes, 2009).

Other radio services provided by the AllAccess Music Group include the "Rate The Music" chart, which acts similar to Call Out Research performed by radio. This chart tests the Top 30 songs in each radio format as determined by MediaBase and uses internet listeners to rate song familiarity, familiarity percentage, and burn (a term that is used industry wide to indicate when an audience is getting tired of a particular song.) Big Champagne is another service that provides a comprehensive chart that is based on a combination of sales, radio, streaming, and social media. AllAccess Music Group also has a point-of-sales monitoring system that counts over-the-counter sales of music known as Street Pulse that gives its charts verifiable sales data (AllAccess.com).

OTHER CHARTS THAT MEASURE POPULARITY

CMJ Charts

CMJ charts are created through the CMJ Network, which refers to itself "as connecting music fans and music industry professionals with the best in new music through interactive media, live events and print" (CMJ). Charts for CMJ are created by reports of airplay from their "panel of college, commercial and noncommercial radio stations," and tend to represent music that is not a part of the commercial mainstream. Its charts are reported in publications by the company, as well on its website and chart titles from contributing radio stations nationwide. Charts include:

CMJ Top 200—core list of top songs played that week from reporting stations to the CMJ panel

CMJ Radio Add—a list of songs "added" to song libraries at reporting stations but not necessarily added to play lists.

Hip-Hop—A top 40 chart where rap and hip hop exist. A combination of 300 stations.

AAA Top 40—Adult album alternative: 40 songs deep featuring singer-songwriters which can highlight lots a major record label artists.

Loud Rock—Broken into two charts that include Loud Rock College and Loud Rock Crucial Spins that features a most-added section that is 5 deep. Both charts are 40 songs.

Jazz—Top 40 with about 250 stations reporting. About 250 report to it. Note that it is mostly traditional jazz, and not "smooth" jazz.

New World Top 40—the chart where new-age and world music fit in. It is 40 deep and has about 200 reporting stations.

RPM Top 40—a techno/electronic chart with 350 reporting stations.

Internet Broadcast chart has recently emerged with about 10 stations reflecting their playlists.

RAM Crucial Spins Chart—RAM means "real-time airplay metrics." This is the college radio equivalent of BDS and Mediabase, where certain stations are actually monitored, and the charts thus reflect what is actually played as opposed to what is said is being played. Two RAM charts have been created: one for album and another for singles.

http://www.radio-media.com/song-album/articles/airplay6.html

iTunes Charts

iTunes issues its continuing charts of music sales which include 20 charts of the best selling albums based on genre. It also publishes its top-selling singles within the same 20 categories. With iTunes being the largest retailer of recorded music, its charts carry considerable weight with those who would use their information to create buzz among gatekeepers about the popularity of an album or single. The data has rolling 24-hour refresh so music purchased reflect most-recent activity by iTunes consumers. The iTunes "Top Songs" chart provides a listing of 100 most-purchased songs on their website and expands the vision of sales to international music downloads, underscoring the meaning of the World Wide Web as well as the global impact of music.

Amazon Charts

One of the oldest online retailers, Amazon, creates hourly charts that rank music being purchased through its website in 25 ratings categories. Their overall chart, Bestsellers in Music, is not genre-specific like their other 24 charts are. Unlike iTunes, Amazon sells digital versions of recorded music as well as physical CDs, vinyl albums, and used CDs. To the record label, the near real time charts provided by Amazon can give the promotion and publicity departments another tool to use in acquiring more exposure for the music.

YouTube

YouTube's daily charts of the most popular music videos create yet another opportunity for the record label to promote its music to gatekeepers. As most users of the website have discovered, it provides an efficient way to find the most popular music videos in a variety of genres as well as videos of those who are rising stars, whether they are on independent labels or the majors. The real-time tracking of video views creates an on-going "chart" of sorts and creates another talking position for those promoting a label's music.

OTHER SOURCES TO TRACK MUSIC POPULARITY

The number of sources available to the label to help promote music past the gatekeepers continues to grow, some of which include a fee to subscribe for their services. Social networking sites track views of videos and streams of songs with data that is available to the label at no cost. Counting "Likes" and "Followers" on Facebook, Twitter, and Instagram and streaming activity on YouTube and VEVO contributes to the overall Social Media action of an artist or their project. Several websites have created interfaces to help leverage this data in a comprehensive, easy-to-read manner that includes Big Champagne and Next Big Sound. These tools are discussed in chapter 16.

Pollstar magazine tracks the popularity of recording artists based on their ticket sales at performance venues. The magazine's online website exists to connect music fans with concert tickets; however, the magazine is a weekly subscription-based publication that lists artists, venues, ticket sales percentages, and gross revenue from single appearances. Continued sell-outs by an artist can make an important statement and talking point about an artist's growing fan base and popularity. For example, a label promotion person might use the information from the Box Office Summary chart for one of 2014's new artists like the Avett Brothers to tell a programmer that the group sold out its three day run at in October at the Louisville Palace. *Billboard*'s Box Score chart is published weekly but only reports on the top 35 grossing shows making it useful to promote major hit makers but less useful for artists that are in the developing stages of their recording careers (*Pollstar*, 2009).

APPENDIX

In Figure on the next page is an example of an organizational chart for a typical pop label record promotion department. Note that the VP's of the various formats are specific to the music. However, the individual regional promoters typically promote all current music types marketed by the label.

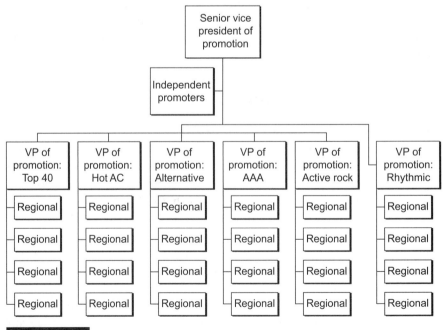

FIGURE 13.8

Organizational chart for a typical country label record promotion department:

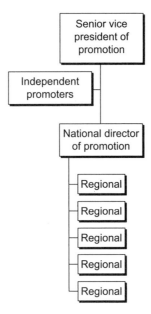

FIGURE 13.9 *Country label organization chart*

GLOSSARY

Burnout or burn—The tendency of a song to become less popular after repeated playings.

Indie—A shorthand term meaning an independent record promoter who works for radio stations and record labels under contract.

Payola—The illegal practice of giving and receiving money in exchange for the promise to play certain recordings on the radio without disclosing the arrangement on the air.

Playlist—The weekly listing of songs that are currently being played by a radio station.

Recurrents—Songs that used to be in high rotation at a station, but are now on the way down, reduced to limited spins.

Rotation—Mix or order of music played on a radio station.

Terrestrial radio—This is traditional commercial radio broadcasting using a transmitter and tower consisting of AM, FM, and HD radio

Spin—This is a reference to the airing of a recording on a radio station one time. "Spins" refers to the multiple airing of a recording.

Trades—This is a reference to the major music business trade magazines.

REFERENCES

ABC Television, 20/20. November 2002. http://www.abcnewsstore.com/store/index. cfm?fuseactioncustomer.product&product_codeT020524%2002.

Barbington, C. "Big Radio Settles Payola Charges," *Washington Post*, March 6, 2007, p. D1.

BDS. *Billboard* online. 2009. http://bdsonline.com/about.html.

BDSRadio.com. 2014.

Billboard. "Charts Legend." (October 4, 2008), p. 59.

Billboard.com. "Chart Methodologies." www.billboard.com/bbcom/about_us/ bbmethodology.jsp.

CMJ. N.d. 2014. http://www.billboard.com/articles/columns/chart-beat/6320099/ billboard-200-makeover-streams-digital-tracks.

Freed, A. 2005. http://www.historychannel.com/speeches/archive/speech_106.html.

Hull, G. *The Recording Industry*, London: Routledge. 2004.

Katunich, L. J. "Time to Quit Paying the Payola Piper." *Loyola of Los Angeles Entertainment Law Review*. April 29, 2002.

Leeds, J. "Song Tracker Finds a New Way to 'Publish' Its Charts," *The New York Times*, October 2, 2006. Section C, p. 6.

Phillips, C. "Congress Members Urge Investigation of Radio Payola." *LA Times*. May 24, 2002.

Pollstar. "Box Office Summary," January 16, 2009, p. 28.

Rhodes, R. Account Executive with MediaBase, interview February 2009 radio-media. 2009. http://www.radio-media.com/song-album/articles/airplay26.html.

CHAPTER 14

Music Videos

I want my, I want my, I want my MTV.
—Dire Straits

I'd do entire music videos in my bedroom, where I used to stand in front of my television memorizing the moves to Michael Jackson's "Beat It."

—Jimmy Fallon

VIDEO PRODUCTION AND PROMOTIONS

In today's online world, videos are a key part of any well-crafted marketing plan of a record label. Music videos create another dimension to the song and its lyrics by adding a visual element to the musical performance of the artist. The video should create strong, memorable images to pique the passion of viewers and get them more involved in the music. Ultimately, the marketer's objective is to elevate that passion into the need to possess the music in the form of a purchased CD or download.

From the point of the record label, the three functions of a music video are:

- Promotion and publicity for recordings and other aspects of the artist's career
- Licensed content for online and wireless entertainment providers, including video sites that generate revenue for the label by sharing advertising income
- A product offered for sale

CONTENTS

Music videos are just one component of the video marketing of an artist or album. This chapter is focused on gaining a better understanding of all uses of video, how they fit into a marketing plan, and how they are used strategically to stimulate retail sales of a recording, ticketed performances, merchandise sales, and the sale of ancillary products.

HISTORY OF THE MUSIC VIDEO

The earliest form of video promotion of music began in 1890. From then on and through the next 25 years, "illustrated songs" became the rage of thousands of large and small theaters across America, and were credited with selling millions of copies of sheet music. In these theaters, vocalists accompanied by bands or small orchestras would perform songs, while hand-painted glass slides were sequenced and projected on a screen depicting the story of the lyrics. Often, a vocalist would then lead the audiences in singing the song as the lyrics were projected (Norman, 2014).

When radio was introduced in the 1920s, music promotion changed. One of the earliest radio deejays was an announcer named Al Jarvis who worked in the 1930s at a radio station in Los Angeles. He created a radio show featuring recorded music mixed with chatter. It was also Al Jarvis who went on television in early 1950s with a program that featured recordings, guests, and information, which made him one of the original video deejays. His program was replicated in numerous other television markets, and became a staple of many major broadcast companies. By the mid-1950s, video deejay programs were prominent in all top 50 U.S. media markets.

Some of these early television shows featured what was known as a soundie. A soundie was a video creation by the Mills Novelty Company that featured performances by bands, and was used in coin-operated video machines. In the early 1940s, Mills Novelty created over 2,000 promotional soundies. In the early 1950s, producer Louis Snader created 750 "visual records," and Screen Gems and United Artists began creating their versions of early music videos.

The granddaddy of video deejay television shows was Bandstand, which began airing in Philadelphia in 1952 with host Bob Horn. In 1956, Dick Clark replaced Horn, and the show soon became American Bandstand on the ABC Television Network. In 1964, Clark moved the show from Philadelphia to Los Angeles "to be closer to the seat of the television industry" (Fontenot, 2014). In 1970, Soul Train became the first music television show to showcase black artists.

Television in the 1960s and early 1970s featured many shows that included musical performances, most notably, the Monkees, but it wasn't until the mid-1970s that artists like Queen and Rod Stewart created videos designed specifically to assist in the promotion of their recorded music projects. Shows like "Rock Concert" and "Midnight Special" became staples of Saturday night programming, but mostly they featured concert performance footage, not the story-based video we are used to seeing today. As disco rose to prominence, even more shows appeared. Even Casey Kasem brought a pared-down version of his top 40 radio show to television. But the link of the music video to record promotion caught fire when cable television began to flourish in the early 1980s. Among the first to begin using music videos was a program called "Night Flight" on the USA Network. Soon, scores of cable channels and hundreds of regional and local shows began to regularly feature music videos as part of their entertainment programming (McCourt and Zuberi, 2009). The most famous, of course, was MTV, which launched on August 1, 1981, with "Video Killed the Radio Star" by the British new wave band, the Buggles.

The use of music videos by cable channels has declined considerably for a number of reasons. Long before YouTube and Vevo, MTV switched their programming away from 24/7 music videos to half hour and longer programs in order to get the ratings points needed to be able to charge enough for advertising and keep the network afloat. High-speed internet has made on-demand music video capabilities provided by Vevo, MySpace, Daily Motion, YouTube, MuchMusic, Vimeo, Yahoo, and others the preferred choice for video viewing. Cable programmers realize they are unable to compete with video users who are unwilling to wait for a video to be played on television since music fans have numerous choices to find any video when they want it and where they want it. With the growth of broadband and its penetration into more than 70% of the homes in the U.S. (Broadband), the Internet has become a preferred choice for video entertainment for many music fans.

THE MUSIC VIDEO TODAY

Music videos are their own kind of entertainment and are valued far beyond their value of promoting recordings and an artist's career. Unlike in the 1980s and 1990s when video cable channels were given free and unlimited use of music videos for their programming, music videos are now licensed content to those who want to use them. Labels that create videos license their use to sites that feature music video. The license is granted in exchange for payments based on the size of the audience that views the videos, or it is based on the advertising income that the website earns from featuring the video. Depending on the online video service, advertising rates associated with music videos generate three to eight dollars for every thousand times the video's page is loaded onto someone's browser (Bruno, 2009; Miller, 2009). Forbes magazine estimated that streaming "Gangnam Style" cost YouTube $296,360 but earned them $348,285 in advertising revenue (Prabhu, 2012). By this admittedly rough estimate, YouTube made $51,925 before royalties.

According to ASCAP ". . . record companies normally charge between $15,000 and $70,000 for the use of existing master recordings in a major studio film but, depending on the stature of the artist, the length of the use, the music budget and how the recording is being used, these fees can be greater or less." (Brabec and Brabec, 2014). The television show Glee pays from $10,000 to $50,000 per song, but this is only a synchronization license. In other words, only the publisher and the writer of the song get paid. The artist and the record label make nothing because the show does not use their recording of the song (Gajewski, 2011).

In the early days of music videos the labels paid the entire cost of the video, but in 1983 that changed as labels looked to reduce costs and demand for expensive videos (Denisoff, 1991). At that point, labels paid the upfront cost of the video and then recouped a percentage, usually 50%, from album royalties and sales of the video. Unfortunately, music videos seldom sell enough copies to pay for themselves, although their availability on iTunes has probably increased sales. When the video does sell well, the label will recoup 100% of the video costs from the sales of the video before paying the artist a royalty. The royalty rate typically starts around nine percent and escalates with each new album release.

Nowadays, artists are signed under multiple rights recording contracts and the expense of a music video is treated like any other advance against royalties. The artist earns a percentage of the net profits generated from the sale and licensing of music video and, as with traditional recording contracts, that percentage usually escalates with each new album.

Nearly every social networking website gives the opportunity for members to define themselves by the kind of music they enjoy which has helped create a culture that defines individuals far beyond the answer to the question, "What's your sign?" YouTube and the links to its videos underscore the impact of music video and how it reaches potential buyers of the video, concert tickets, and related products: as of February 2014, nine of the top ten videos on youtube.com are music videos. And the growth continues— estimates are that YouTube generates 100 hours of new video content . . . every minute (Statistics, 2014)!

BUDGETING AND PRODUCING THE MUSIC VIDEO

In the past, artists and their record labels may have shot two or more videos before the first single was released. In these leaner times, labels are less willing to invest in more than one video at a time. Production budgets have also been cut. Gone are the days of million dollar videos, but since the Viacom music outlets (MTV, VH1, CMT, BET, etc.) are no longer the primary outlets for music videos, lower budget videos are getting plenty of exposure, mostly on the Internet.

Record labels recognize the importance of investing in a music video as an essential promotional piece for the artist and their recordings. Music videos are a strong branding device designed to promote the label's interests in an artist, but it also places the image of the artist in the eyes of concert ticket buyers, too. There are instances where the music and its images are inseparable, and a video is a necessity. At other times, for whatever reason (i.e. the artist may not be camera-friendly), the decision to create a video will include images that do not feature or focus on the image of the artist.

The general kinds of music videos created can be classified as shown in Table 14.1.

Table 14.1	
Kinds of Music Videos	**Table X.1**
Performance/Concert Video	Artist performs the song in a studio or live setting
Storyline Video	The video follows the storyline of the song lyrics
Lyric Video	The words of the song appear on screen in synch with the music

Source: (Friedlander, 2014)

From the record company's perspective, a number of factors are considered in the decision to make a video, but profit is not usually one of them. Creating a budget that stays within available funds is the first step in creating the music video. The budget will be determined by the stature of the artist, the genre of music and the expectations of the fans. The label will typically request proposals from video production houses for storyboards or "treatments" for a single and ask them to include their suggested concepts for the video. Video producers and production companies will propose the budget necessary to produce the video based on the suggested concept. The label either accepts the idea and its budget, or it may negotiate for a larger or smaller production and accompanying budget. Any music video that the label produces must at least be comparable to production elements used by those that make video for competing artists. You probably know of a very low-budget music videos shot on an iPhone and edited on the artist's laptop that received thousands of views, but instances like that are still rare; the label must create a music video that meets the level of creative and technical quality that matches or exceeds those currently being viewed by fans. If the goal is to get the video on a major cable channel, then the quality must meet their standards as well, and that rarely includes iPhone or consumer quality equipment. A professional quality music video can be shot and edited for a few thousand dollars, especially if the video is of the artist performing live. The more complicated the story, the more actors and set changes and edits, the more expensive the video. "Creative" can be inexpensive provided that the elements of the video production connect with the target fan of the video. For example, encouraging fans to assemble their own music video with images and video clips from their wireless devices can energize the community around an artist, but these videos will be relegated to YouTube and other Internet sites.

Budgets for music videos produced by labels rarely reach the million-dollar level these days, though that was relatively common in the past. Big budget music videos are now in the area of $200,000 to $500,000, and labels are producing fewer of them (Rosen, 2013; Anderman, 2008). And it is important to understand that professionally produced music videos represent 54% of Web streams, but they generate 94% of advertising revenue for websites—part of which is then shared with the label that produced the video. According to the Diffusion Group, user-generated music videos account for only 4% of a site's advertising (Bruno, 2009).

THE MOST EXPENSIVE MUSIC VIDEOS EVER MADE

1. Michael Jackson and Janet Jackson—"Scream" (1995) $7,000,000
2. Madonna—"Die Another Day" (2002) $6,100,000
3. Madonna—"Express Yourself" (1989) $5,000,000
4. Madonna—Bed Time Story" (1995) $5,000,000
5. Guns N Roses—"Estranged" (1993) $4,000,000
   Michael Jackson—"Black or White" (1991) $4,000,000
7. Aqua—"Cartoon Heroes" (2000) $3,500,000

8. Puff Daddy—"Victory" (1998) $2,700,000

9. MC Hammer—"Too Legit To Quit" (1992) $2,500,000

    Mariah Carey—"Heartbreaker" (1999) $2,500,000

    Janet Jackson—"Doesn't Really Matter" (2000) $2,500,000

10. Busta Rhymes and Janet Jackson—"What's It Gonna Be?" (1999) $2,400,000

    Compiled from indigoprod.com and Wikipedia November 2014

USES FOR THE MUSIC VIDEO

In a perfect world it would be a simple matter to create a music video and be able to determine the impact it has had in supporting the sale of a recorded music project. Certainly a bump in sales figures can often be seen in SoundScan data when a music video has been added to websites and into the rotation of a video channel or local cable program, but there are other opportunities for use of the video to support the project. Often, it is difficult to measure the impact of these other uses for the video, but following are some of those additional opportunities the music video offers the record label.

The publicity department and the independent publicist rely on a video to help them garner media attention for the artist. When a new artist is being pitched to a journalist or a blog for a story, the video is helpful by associating faces and performances with the music. The publicist also uses the video when vying for a performance opportunity on network television shows, including "Saturday Night Live," "Today Show," or "Ellen." This added tool gives the gatekeepers of these television shows a view of what they might expect when the artist appears.

The radio promotion staff may find the video useful, especially if the artist is new. Radio programmers are sent a link to a new artist's video as a way to initially introduce the artist, and then the station is urged to link the video to their own online website when they begin to program the recording on the radio station.

Social media departments at record labels use the video to create a presence on the Internet. Many social networking websites feature music, music samples, and links to videos posted around the Internet. The label's website includes either the video or a link to the video on the artist's site. Artist management companies that maintain an active web presence will also include links to either the label's or the artist's site to give music fans the experience of the music video.

THE IMPORTANCE OF VIDEO IN VIRAL MARKETING

You need look no further than the top 10 You Tube videos to see the importance of video to viral marketing. Remember, viral marketing is getting somebody else to promote your product for you. We do that every time we share something on Facebook or retweet a message on Twitter. Now that the Billboard Hot 100 chart includes video spins in its formula, video is even more important. Case in point, Harlem Shake by Bauer. The song had been released and uploaded to SoundCloud for eight months before a video blogger, Filthy Frank, and a few friends created a quirky, 35-second video of themselves dancing to the song. The concept was replicated on the Sunny Coast Skate YouTube page and then hundreds of times by college students, military units, talk show hosts, racecar drivers and even college administrators (Sherburne, 2013, Harlem Shake, 2014).

Inclusion of a song in a television show, a movie, or commercial can introduce an artist to millions of potential fans and boosts sales. In these cases, the advertising often pays all or part of the cost of shooting a music video at the same time the advertisement is filmed. In the case of the film or television show, everyone involved in the writing and production of the original recording my enjoy a nice boost to their income while the song receives massive exposure. Sting, Sheryl Crow, Jet, Fun, and Bob Seger have all enjoyed the benefits of having their song used by a major advertiser (Viral Video, 2014).

SALES AND LICENSING

The marketing department also has specialized uses for the music video by including it as part of a "value added" feature of the release of a physical CD. It is not uncommon for CD jewel cases to display an added bonus for the consumer in the form of video performances as a separate DVD. Packaging music video performances as part of the experience the consumer purchases with a music CD adds value to the product, and positions it to better compete with movie DVDs, which regularly include added features.

The Recording Industry Association of America (RIAA) is the industry trade association, and it annually tracks the shipment and value of various configurations of recordings. Figure 14.1 shows that since the economic downturn of 2008, the number of videos shipped and downloaded decreased despite the availability of inexpensive video downloads.

The SoundScan numbers (Figure 14.2) report actual sales of videos at retail whereas the RIAA reports only shipments. Music fans no longer need

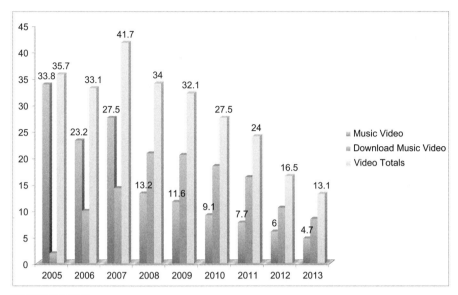

FIGURE 14.1 *Music video shipments (in millions)*

Source: RIAA

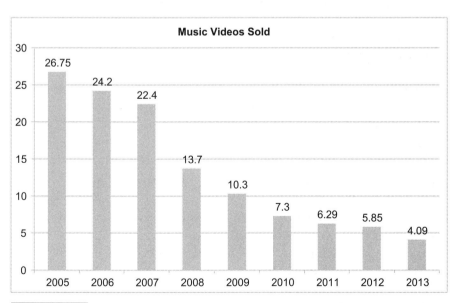

FIGURE 14.2 *Music video sales*

Source: SoundScan

to own a video if they are willing to view ads before and during the video. For those who want to own the video, iTunes is the dominant source to download video singles that can be played on wireless devices, game consoles, and computers.

We have already seen how music video licensing is quickly becoming an important source of income for the record label. While the chart above shows the steep decline in sales of music videos, income from licensing within some companies is growing at an annual rate of 80% (Chmielewski, 2008).

VIDEO STREAMING AS A SOURCE OF INCOME

Video on-demand streaming increased 35.2% from 2013 to 2014 to over 36.6 million, according to Nielsen's 2014 year-end report. There has been great improvement in the handling of user generated content (UGC) in recent years. Google's ContentID system along with other systems used by various platforms have made it much easier to identify and differentiate mashups created by users that have helped for licensing along with monetization of streaming, instead of having these valuable "tools" removed for infringing copyright. YouTube created TrueView that helps IDed user generated content, producing more revenue now than that of official videos.

But the streaming of official videos, be it concept videos or simple lyric videos—add up too. These marketing elements of creating awareness are now vital to the bottom line as revenue producers, with ad-generated monies flowing to record labels coffers through licensing agreement formed between various streaming portals and label imprints (IFPI, 2014).

PROGRAMMING

The penetration of music videos into American households through cable and satellite channels is growing each year. Cable and satellite channels that use music videos to program part of their broadcast day continue to thrive, though videos have become an increasingly smaller part of the entertainment they offer to viewers. The proliferation of online and wireless on-demand sources for music videos has reduced television viewer interest in favor of other forms of entertainment. Video channels which program music videos as part of their entertainment offerings are listed in *Billboard* each week along with their charts for videos they are programming.

Table 14.2	Cable/Satellite Music Video Channels
Cable/Satellite Channel	**Artists Programmed**
VH1	The Fray, Kanye West, Taylor Swift, Nickelback, Pink, Kelly Clarkson, Bruce Springsteen, John Legend, Ne-Yo
BET	Keyshia Cole, Kanye West, Beyoncé, Nicki Minaj, Rihanna, Soulja Boy
CMT	Blake Shelton, Dierks Bentley, Carrie Underwood, Brad Paisley, Eric Church, Keith Urban Jason Aldean, Kenny Chesney
GAC	Thomas Rhett, Billy Currington, Florida Georgia Line, Josh Turner, Eric Church, Blake Shelton, Carrie Underwood, Sara Evans, Tim McGraw
Fuse	Slipknot, Britney Spears, T.I., Kings of Leon, The Gray, Fall Out Boy, T. Pain, Beyoncé, The Killers
MTV2	Akon, Kanye West, Staind, T-Pain, N*E*R*D, Gorilla Zoe, Wale, Hoobastank, Soulja Boy
MuchMusic Canada	Ariana Grande, Taylor Swift, Death Cab for Cutie, Francesco Yates, 5 Seconds of Summer, Lights
CMT Canada	Sugarland, Higgins, Alan Jackson, The Road Hammers, Johnny Reid, Emerson Drive, One More Girl, Dean Brody

LABEL STAFFING

Labels do not do their own music video production, preferring to hire outside firms that are specialists in the field. However, all labels have someone on staff, often in the creative services department, who coordinates the creation of music videos as one of their job responsibilities. Depending on the label, video coordination may be a stand-alone job or it could be part of the creative, publicity or marketing department.

Coordination of the creation of a music video involves working with the artist, their management, A&R, and others involved in the development and branding of the artist. Once produced, publicity, marketing, radio promotion, and new media must coordinate the effort to maximize the video's exposure to the target audience. Smaller labels will hire outside companies like Aristomedia to distribute and market the video.

Contracting with an outside music video promotion company can be efficient for many record labels because the video promotion function is not one that is done on a full-time basis at most labels. Video promotion takes on many of the features of radio promotion of a single. Independent music video promotion company, AristoMedia has a 12-week program for its music video promotion clients. The company takes new music videos to cable channels and pitches them to the programming staff; puts videos on a compilation DVD or file called a video pool; services local, regional, and national video outlets; follows up with video programmers; creates tracking sheets so they know the frequency the video appears on each of the outlets to which they promote; follows the charts of each of the music video outlets; always promotes for the heaviest rotation possible.

Among the services offered by some independent video promoters is a semimonthly creation of genre-specific compilation discs, sometimes referred to as video pools, which are sent to video channels. A promotion like this is fee-based, but it is an efficient method to distribute music videos to hundreds of music video outlets. Independent video promoters also offer new media promotion services for labels that do not have staffing available to create the necessary web presence for their music videos (AristoMedia, 2009).

BEYOND THE MUSIC VIDEO

Music videos are great for branding and exposing an artist, but they are only one way to use video to market an artist. If we broaden our definition of video to include more than music videos, then we can include all television and film appearances, as well as fan-created videos. While getting your artist interviewed on a talk show may be the responsibility of the publicist, that doesn't mean you shouldn't take advantage of the appearance in every possible way. Getting permission to post or link to the interview from the artist's website is a start. Including the interview, if it is a good one, as part of the EPK may lead to even more opportunities.

Weekly or daily posting on a video blog is another great use of video that keeps the fans engaged and returning to the artist's website. According to Dr. James McQuivey of Forrester Research, one minute of video is the equivalent of 1.8 million words or 3,600 Web pages (Follett, 2012). Additionally, nearly every social media allows videos of some sort and visuals, including static pictures, attract, on average, 94% more total views than content with no image ("Its All About the Image"). This additional attention translates to more "Likes" and more comments (Corliss, 2012). Studies have shown that including a video in your Facebook posting significantly increases the chances of it being shared or reposted. According to Michael Stelzner, president of Social Media Examiner, people retain visual content 500% faster than text (Livingston, 2014)!

Videos can be a great basis for contests and to get fans involved. Asking them to upload their own covers of a song or a video entry into a contest creates an interaction between the artist and fan and may create even more exposure for the artist beyond the existing fan base.

If the artist has any acting skills, then a cameo or larger role in film or television can help create interest in their music, too. Dwight Yoakam has appeared in movies (i.e., *Slingblade*), as well as television (Under the Dome), along with other famous musicians such as Beyoncé, Harry Connick, Jr., Eminem, Dolly Parton, Ice Cube, Jennifer Hudson, and Will Smith, just to name a few.

VIDEO IS EVERYWHERE

Music videos are not limited to television and computers either. Bars and restaurants, airplanes, hotel lobbies, amusement parks, and entertainment centers like Times Square and Las Vegas are all outlets for your artist's video. Many of the outlets for music videos are regional or local programs. Although not as well known as their national competition many have a substantial number of loyal viewers. A well-established video distributor like Virool or Aristomedia will maintain and service a current list of these types of outlets as well as video pools and military service outlets. With the roll out of technology like Google Glass and GoPro cameras, the future of music and video will probably be mobile and interactive.

GLOSSARY

Broadband—This is high-speed Internet access by means other than a 56k modem.

Illustrated songs—An early form of promoting music in theaters featuring live music and projected slides depicting the lyric story line.

MPEG—A computer file format that compresses the size of the video to make it easier to view on the Internet or to send via email.

Multiple rights contract—This is a recording contract in which an artist agrees to permit the record label to share in the traditional and new revenue streams created by their career in the music business.

Soundie—Black and white short films of live performances presented in video juke-boxes in the 1940s.

Video deejay—This is someone who works on a television program that features an announcer, information, and music videos.

Video pool—A compilation of music videos that are distributed on a regular basis (semi-monthly and monthly) by independent video promoters to local, cable, and satellite video channels that use this kind of programming.

Visual records—Video film productions in the 1950s created to promote sound recordings.

A special thanks to Jeff Walker for his input and help with this chapter.

REFERENCES

Anderman, J. "The Itty Bitty Video," *The Boston Globe,* February 17, 2008, p N1.

AristoMedia and Jeff Walker. 2009. Aristomedia.com.

Brabec, J., and Brabec, T. "Music, Money, Success & the Movies: Part One." ASCAP, n.d. Web. Accessed August 15, 2014.

"Broadband Internet Penetration Deepens in U.S.; Cable Is King." Market Watch. IHS Technology, December 9, 2013. Accessed August 16, 2014. https://technology.ihs.com/468148/broadband-internet-penetration-deepens-in-us-cable-is-king.

Bruno, A. "Best Bets, Digital: The Big Payback," billboard.com, January 24, 2009.

Corliss, R. "Photos on Facebook Generate 53% More Likes Than the Average Post." HubSpot, November 15, 2012. Accessed August 21, 2014. blog.hubspot.com.

Chmielewski, D. "Labels, Websites In Video Tussle," *Los Angeles Times,* Dec 23, 2008, p. C1.

Denisoff, R. S. *Inside MTV.* New Brunswick, NJ: Transaction Publishers. 1991.

Follett, A. "18 Big Video Marketing Statistics and What They Mean for Your Business." Video Brewery, 2012. Accessed August 21, 2014. http://www.videobrewery.com/blog/18-video-marketing-statistics.

Fontenot, R. "American Bandstand Timeline: The Most Important Events in the History of Dick Clark's Legendary TV show." About.com. N.p., 2014. Accessed August 12, 2014.

Friedlander, J. P. "News and Notes on 2014 RIAA Music Industry Shipment and Revenue Statistics." *RIAA.* RIAA, 2015. Accessed June 27, 2015.

Gajewski, R. "How Much Does Glee Pay to Use a Song?" N.p., December 20, 2011. Accessed August 15, 2014. www.wetpaint.com.

Harlem Shake (meme). In *Wikipedia, The Free Encyclopedia*. August 23, 2014. Accessed August 24, 2014. http://en.wikipedia.org/w/index.php?title=Harlem_Shake_(meme)&oldid=622424566

IFPI. 2014. "Monetizing Music Videos," www.ifpi.org

"Its All About the Images." MDG Advertising. MDG, May 14, 2012. Accessed Aug. 21, 2014. http://www.mdgadvertising.com/blog/its-all-about-the-images-infographic/.

Livinston, Geoff. "Visuals Make Big Impact at Social Media Marketing World." The Vocus Blog. Mar. 28, 2014. Accessed Aug. 21, 2014. www.vocus.com.

McCourt, T., and Zuberi, N. "Music On Television." 2009 http://museum.tv/archives/etv/M/htmlM/musicontele/musicontele.htm.

Miller, M. "Some YouTube Regulars Making Income From Videos." CBS Evening News, January 8, 2009.

Norman, J. "The First Illustrated Song: Precurser of the Music Video." *Jeremy Norman's History of Inofrmation*. Ed. Jeremy Norman. N.p., n.d. Accessed August 11, 2014. http://www.historyofinformation.com/expanded.php?id=4450.

Prabhu, A. "How Much did it Cost YouTube to Stream Gangnam Style?" *Forbes,* October 29, 2012. AccessedAugust 14, 2014. http://www.forbes.com/sites/quora/2012/10/29/how-much-did-it-cost-youtube-to-stream-gangnam-style/.

Rosen, M. "How Much Does it Cost to Make a Music Video?" Indigo Productions, 23 August 23, 2013. Accessed November 10, 2014.

Sherburne, P. "How Baauer's 'Harlem Shake' Hit the Jackpot Under Billboard's New Hot 100 Rules." *SPIN*, February 21, 2013. Accessed August 24, 2014. http://www.spin.com/articles/baauer-harlem-shake-hot-100-youtube/.

"Statistics." YouTube.com. N.p., n.d. Accessed August 17, 2014. https://www.youtube.com/yt/press/en-GB/statistics.html.

Video Channels. "National Airplay Overview," R&R, (Feb 20, 2009), p. 49.

"Viral Video Sync Placements Boost Music Sales." *Storyboard Music*. N.p., Aug. 2, 2014. Accessed Aug. 24, 2014.

CHAPTER 15

Music Distribution and Music Retailing

INTRODUCTION

Historically, music distributors have been a vital conduit in getting physical and digital music product from record labels' creative hands into the brick-and-mortar and virtual retail environment. As the marketplace shifts to a digital environment, distribution companies are re-evaluating their value in the food chain and developing sales models and adding services that reflect direct sales opportunities to music consumers. By the end of 2014, more than half of all albums purchased in the U.S. were either a digital transaction or were purchased at "non-traditional" outlets, which includes digital services, Internet retailers, mail order, at "non-traditional stores" or at a concert (Nielsen SoundScan, 2008). Vinyl record sales were up 52% to 9.2 million but digital song sales fell by 250 million most likely due to the 54% increase in song streams (Smith, 2015). Whatever the sales channel, distribution companies think of themselves as extensions of the record labels that they represent.

Prior to the current business model, most individual records labels hired their own sales and distribution teams, with sales representatives calling on individual stores to sell music. It took many reps to cover retail, as Figure 15.1 shows. This model shows nine points of contact, where each label meets with each retailer. As retailers became chains, and as economics of the business evolved, record labels combined sales and distribution forces to take advantage of economies of scale, which eventually evolved into the current business model.

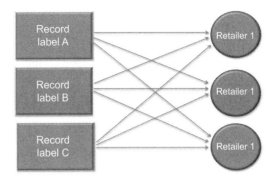

FIGURE 15.1 *Direct Contact Concept*

Figure 15.2 includes the distribution function and shows the points of contact reduced to six. In today's business model, record label sales executives communicate with distribution as their primary conduit to the marketplace, but labels also have ongoing relationships with retailers. Depending on the importance of the retailer, the label rep will often visit the retailer with their distribution partner so that significant releases and marketing plans can be communicated directly from the label to the retailer. And as deals are struck, both orders and marketing plans can then be implemented by the distributor.

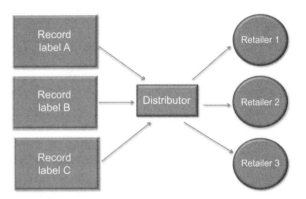

FIGURE 15.2

THE BIG 3 AND MORE

Within the last 15 years, the Big 6 have consolidated into the Big 3, reducing their number of employees to reflect the ever-decreasing size of the music sales pie. But the mergers began prior to the explosion of file sharing with the combination of Universal with Polygram in 1999, the same year as the emergence of Napster™. This merger created Universal Music & Video Distribution (UMVD), now Universal Music Group Distribution (UMGD) who has maintained their market share position at #1 since inception. In 2013, most of EMI's stable of labels and artist were added to the Universal fold, increasing the strength of this company's musical power in the marketplace.

Another recent combination of conglomerates include the blending of Sony and BMG in 2004, who consolidated their various profit centers as well as their distribution workforces but maintained much of the integrity of their imprints. They, too, gained market share, garnering the #2 position simply by merging. In 2008, Sony purchased the remaining 50% stake held by BMG in the Sony/BMG merger. The merged companies, now wholly owned by Sony, were renamed Sony Music Entertainment Inc. (SMEI), and includes all the labels that were Sony, as well as BMG.

WEA maintains a separate distribution function as part of Warner Music Group. This company is positioned as #3 of the major music conglomerates.

As noted in the market share data, independent labels continue to be a force within the music business and in 2014, independents held an approximate 17.5% market share. With the burgeoning digital storefronts, any label can now have an instant "sales" point in which to connect with customers directly, but to fulfill physical product, independent labels need to reach the brick-and-mortar outlets using traditional methods. All three of the major distributors have created an "independent" arm within their distribution division. By contracting this function of distribution to independent labels, these "independent distributors" can assist the independent label in placing their music in the mainstream marketplace. But there are many true independent distributors that are *not* tied to the Big 3 that function similarly.

UMGD—Universal Music Group Distribution

Caroline Distribution represents many independent labels
Sample labels that it distributes:

Interscope
Geffen
Island/Def Jam
Universal
UMG Nashville
Hollywood

Disney/Buena Vista
Capitol Records
Big Machine
CURB

Sony Music Distribution

Sony's RED Distribution represents many independent labels
Sample labels that Sony Music distributes:

Columbia
Epic
Arista / Arista Nashville
J Records
RCA
Jive
LaFace
Razor & Tie
WindUp
RCA Label Group Nashville

WEA

WEA's ADA Distribution represents many independent labels
Sample labels that WEA distributes:

Warner Brothers
Atlantic Records
Bad Boy
Roadrunner
WSM/Rhino
V2 Records
WEA/Fueled By Ramen

VERTICAL INTEGRATION

The Big 3 music companies share profit centers that are vertically integrated (combine two or more stages of production), creating efficiencies
in producing product for the marketplace. To take full advantage of being
vertically integrated, labels looking for songs would tap their "owned" publishing company. (Each of the major entertainment conglomerates has
a publishing company. If they only recorded songs that were published
by their sister company, more of the money would stay "in the family.")

Once recorded, the records would be manufactured at an out-sourced pressing plant (Sony is the only conglomerate that still owns CD manufacturing plants). In-turn, the pressed CDs would then be sold and distributed into retail, using the conduit of the distribution function located within the family. (Figures 15.3 and 15.4).

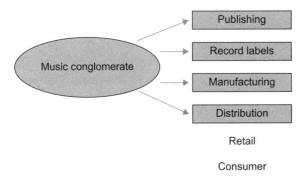

FIGURE 15.3 *Vertical Integration of Music Companies.*

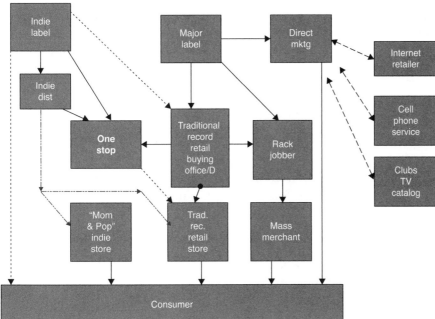

FIGURE 15.4 *Distribution Pathways*

In addition to the Big 3 companies, there are many independent music distributors that are contracted by independent labels to do the same job. Many of these independent distributors have developed a niche in marketing unique and diverse products. Ideally, the distribution function is not only to place music into retailers, but also to assist in the sell-through of the product throughout its lifecycle.

MUSIC SUPPLY TO RETAILERS

Once in the distributor's hands, music is then marketed and sold into retail. Varying retailers acquire their music from different sources. Most mass merchants like Walmart are serviced by *rack jobbers*, who maintain the store's music department including inventory management, as well as marketing of music to consumers. Retail chain stores like FYE are usually their own buying entities, with company-managed purchasing offices and distribution centers (DCs). Many independent music stores like Waterloo Records may not be large enough to open an account directly with the many distributors, but instead work from a *one-stop's* inventory as if it were their own. (One-stops are wholesalers who carry releases by a variety of labels for smaller retailers who, for one reason or another, do not deal directly with the major distributors). Retail chain stores and mass merchants will, on occasion, use one-stops to do "fill-in" business, which is when a store runs out of a specific title and the one-stop supplies that inventory. (Retail is discussed in detail later in this chapter.)

ROLE OF DISTRIBUTION

Most distribution companies have three primary roles: the sale of the music, the physical distribution of the music, and the marketing of the music, although these roles have been dramatically modified as the shift from physical to digital has advanced. The national staff of distribution companies now focus as brand extensions of the individual labels with an executive staff that acts as a wheelhouse that manages the functions of product information dissemination such as street date and sales information, manufacturing and product management, and implementation of national and regional marketing efforts to be executed at various levels. Sales information consists of setting sales goals, determining and setting deal information such as discount off of wholesale and dating of product, and soliciting and taking orders of the product from retail. Additionally, the sales administration department should provide and analyze sales data and trends, and readily share this information with the labels that they distribute.

The marketing division assists labels in the implementation of artists' marketing plans along with adding synergistic components that will enhance sales. For instance, the marketing plan for a holiday release may include a contest at the store level. Distribution marketing personnel would be charged with implementing this sort of activity. But the distribution company may be selling holiday releases from other labels that they represent. The distribution company may create a holiday product display that would feature all the records that fit the theme, adding to the exposure of the individual title.

The physical warehousing of a music product is a big job. The conglomerates have consolidated the warehousing of music with positioning of their distribution centers centrally in the United States. For instance, Sony Music's central warehouse is located in Franklin, Indiana—just outside Indianapolis. Incorporating sophisticated inventory management systems where music and its related products are stored, retailers can place an order, and it is the distributor's job to pick, pack, and ship this product to its designated location. These sophisticated systems are automated so that manual picking of product is reduced, and that accuracy of the order placed is enhanced. Shipping is usually managed through third-party transportation companies.

Because physical music sales have been reduced and many stores have either closed or minimized store floor space dedicated to music sales, music companies and their distributors have out-sourced the inventory management, fulfillment and rack-jobbing to a known-entity within the industry. Anderson Merchandisers has stepped up its role as "sales rep" for both Walmart and Best Buy, managing the inventory, as well as taking care of the in-store floor space in place of the distributors who would normally assist these accounts with sales and marketing efforts.

As retailers manage their inventories, they can return music product for a credit. This process is tedious, not only making sure that the retailer receives accurate credit for product returned, but the music itself has to be retrofitted by removing stickers and price tags of the retailer, re-shrink wrapping, and then returning into inventory.

National Structure

To optimize communication along with service, distributors need to be close to retailers. Many of the major conglomerates have structured their companies nationally to accommodate the service element of their business. Most distributors have sales representative and field marketing representatives that live and work near the major retail accounts: San Francisco and iTunes, Seattle and Amazon and Starbucks, Los Angeles and Super D, Miami and Alliance Entertainment, Minneapolis and Target and Best Buy, and Bentonville, AR—home of Walmart. Other regions of the country are covered with representatives, as well with consideration of musical transactions and population density such as Nashville, New Orleans, New York, Austin, and the northeastern territories.

Timeline

The communication regarding a new release begins months prior to the street date. Although there are varying deadlines within each distribution company, the ideal timeline is pivotal on the actual street date of a specific release. Through mid-2015, U.S. street dates occurred on Tuesdays, but in an effort to combat worldwide piracy as well as appeal to consumer's taste in receiving new music on weekends, a global street date initiative was adopted. The International Federation of the Phonographic Industries announced that in the summer of 2015, albums would be released worldwide on Fridays creating a seamless street date globally. (Flanagan)

Working backwards in time, to have product on the shelves by a specific day, product has to ship to retailer's DCs approximately 1 week prior to street date. To process the orders generated by retail, distributors need the orders 1 week prior to shipping. The sales process of specific titles occurs during a period called *solicitation* (Table 15.1). All titles streeting on a particular date are placed in a solicitation book, where details of the release are described. The solicitation page, also known as *Sales Book Copy*, usually includes the following information:

- Artist/title
- Street date
- File under category—where to place record in the store
- Information/history regarding the artist and release
- Marketing elements:
  - Publicity activities
  - Consumer advertising
  - Tour and promotional dates
  - Available POP
  - Bar Code

This information is available online on the business-to-business (B2B) sites established for the retail buyers.

Table 15.1	Sales Timetable	
Prior to SD	**Activity**	**Example dates**
8 weeks	Sales book copy due to distributor	September 27
6 weeks	Solicitation book released to retail buyers	October 11
4 weeks	Solicitation	October 25 to November 5
2 weeks plus	Orders due	November 5
1 week	Orders shipped wholesalers/retail chain	November 16 orders received
5 days	Orders shipped to one stops	November 19 orders received
Street date		November 23

One Sheet: The Solicitation Page

On the website, *MusicDish Industry e-journal*, Christopher Knab of Fore-Front Media and Music describes a one sheet as:

> *A Distributor One Sheet is a marketing document created by a record label to summarize, in marketing terms, the credentials of an artist or band. The One Sheet also summarizes the promotion and marketing plans and sales tactics that the label has developed to sell the record. It includes interesting facts about an act's fan base and target audience. The label uses it to help convince a retailer to carry and promote a new release.*

> (Knab, 2001)

The one sheet typically includes the album logo and artwork, a description of the market, street date, contact info, track listings, accomplishments, and marketing points. The one sheet is designed to pitch to buyers at retail and distribution. The product bar code is also included to assist in the buy-plugging the actual release into the inventory bar code system.

FIGURE 15.5   *One Sheet for Jason Isbell. (Source: Thirty Tigers).*

CONSOLIDATION AND MARKET SHARE

As consolidation continues, the Big 3, once the Big 6, just keep getting bigger, although the Indies have been a force to be reckoned with. Looking at market share date generated by over-the-counter sales of SoundScan, one can view how large these entities have become over time. Note that the piece of the Big 3 pie fluctuates, with Independents representing the remaining market share. Over the past 15 years, the mergers with Universal have maintained their strong market leader position, while the Sony transaction has shown signs of the turbulent times as Polygram, BMG and EMD disappearing altogether. Remember that during this time, these pieces of the pie have been shrinking, with music sales down by over 50% since its all-time high in 2000.

Included here is a breakout of 2014 mid-year sales showing the power of independent labels. All of the major's independent distribution arms,

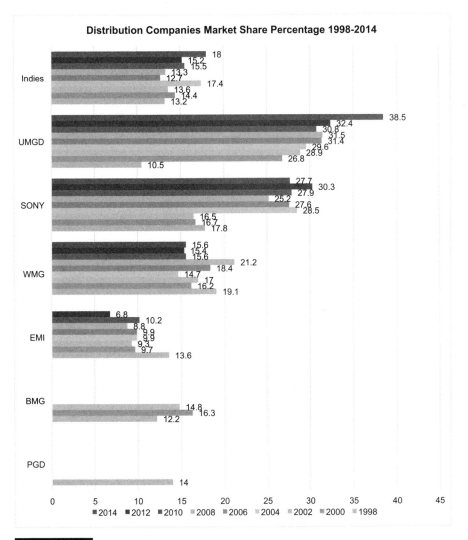

FIGURE 15.6

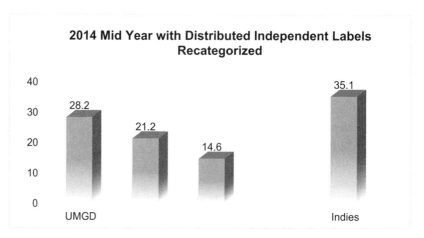

FIGURE 15.7

Warner's ADA Distribution, Sony's Red Distribution, and Universal's Caroline Distributors, distribute and market independent releases from independent labels and the conglomerates often claim that market share. If broken out independently, the independent labels would have a much larger share of the music market sales—as noted in the adjusted 2014 data below.

DIGITAL DISTRIBUTION

According to SoundScan, year-end 2014 data reveals that digital music accounts for 68% of all music consumed which includes streaming outlets. As for digital consumption alone, on-demand streaming increased 54.5% but this did not make up for loss of sales, which showed a decrease of -2% overall year-over-year. There was also a 10% increase in overall music purchases exceeding 1.5 billion units. A glimmer of good news is that Vinyl LP sales increased 52% and now comprise over 6% of physical album sales, but only hold 2% of entire sales pie.

Sales data from 2014 confirms that consumers are moving to digital consumption. As stated by the Nielsen Company 2014 Year-End Music Industry Report:

- On-Demand Streaming grew 54.5% over 2013, with Audio On-Demand (+60.5%) and Video On-Demand Streaming (+49.3%) both experiencing significant increases.
- The soundtrack to the movie *Frozen* ranked #1 for overall consumption this year (Album Sales + Track Equivalent Albums + Streaming

Equivalent Albums) with over 4.47 million album equivalent units. Taylor Swift/*1989* ranked second with 4.40 million units.

- Taylor Swift had the best-selling album of the year with 3.66 million sales for her album *1989*. The album also had the best debut week of the year and the biggest opening week for an album since 2002 with nearly 1.3 million albums sold in the first week.
- *1989* also had the second biggest digital album sales week in history. In its debut week, *1989* comprised a full 22% of all album sales for the week.
- 2014 had two albums that sold over 3.5 million units during the calendar year (Taylor Swift's *1989* and the *Frozen* soundtrack)—this is the first time since 2005 that two albums have sold over 3.5 million albums in a calendar year. While the top two albums this year performed significantly better than last year's top two albums, 2014 saw just four albums surpass 1 million units, compared to 13 last year.
- The top 10 albums in 2014 were virtually flat versus 2013, thanks to the strength of Taylor Swift and *Frozen*.
- Vinyl LPs had another record-breaking year, with 9.2 million sales, surpassing last year's record of 6.1 million units. This is the ninth consecutive year of growth for vinyl sales. Vinyl now comprises over 6% of physical album sales.
  - 27 vinyl LPs sold over 20,000 units in 2014, up from the 11 vinyl LPs to reach that level in 2013.
  - 94 vinyl LPs sold over 10,000 units in 2014, up from the 46 vinyl LPs to reach that level in 2013.
  - Rock is still the dominant genre for vinyl LPs, with 71% of vinyl LPs being classified as rock.
- The independent store strata outperformed other brick-and-mortar retailers, with album sales virtually flat against last year. The strength was led by vinyl LPs, which were up 35% at Independent stores.
  - The independent store strata had a record setting year with vinyl LP sales, with 5.2M vinyl LPs sold, making up 57 percent of all Vinyl sales.
- Genres performed differently across the different types of consumption, showing how different music fans prefer to access their favorite music.
  - Rock is the dominant genre for album sales (over 33% of albums) and of total consumption (29%). However, on a track sales basis, pop (21.1%) is nearly as big as rock (21.3%). R&B/HipHop is the dominant genre for streaming (28.5%) followed by rock (24.7%) and pop (21.1%).

- Country consumers still prefer albums (11.8%) and track downloads (12.0%) over streams (6.4%). Pop music consumers are buying individual hit songs much more than albums. While 21% of digital track sales are in the pop genre, only 10.8% of album sales are pop.
- Some genres, particularly R&B/Hip-Hop, EDM and Latin perform particularly well in streaming.
  - While R&B/Hip-Hop only comprises 13.9% of2 album sales, it makes up 28.5% of streaming. Electronic/dance (EDM) only makes up 2% of album sales, but makes up nearly 7% of streaming, making the genre a bigger share of streaming than country.
  - Latin music also performs particularly well at streaming, with 5% of streaming coming from Latin music, while just 2.4% of album sales are Latin.

DIGITAL REVENUES

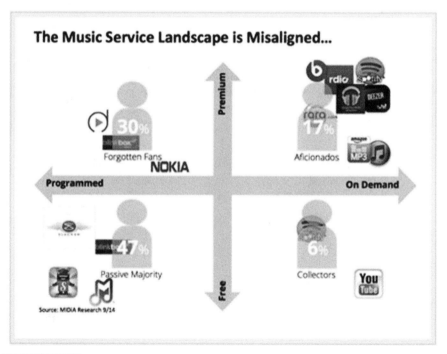

FIGURE 15.8

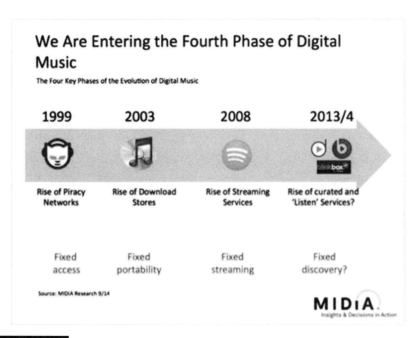

FIGURE 15.9

FIGURE 15.10

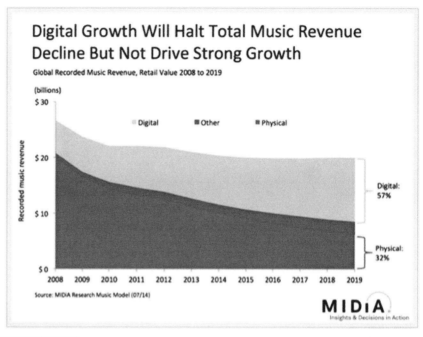

FIGURE 15.11

At the Future Music Forum in Barcelona in the summer of 2014, keynote speaker and Jupiter Researcher, Mark Mulligan presented several interesting ideas regarding the digital landscape for music consumption and how consumers may and may not be satiated in the coming marketplace. As streaming becomes the "experience" of choice and sales continue to dwindle, Mulligan recognizes that the Internet has created an environment that he calls the "Tyranny of Choice." With over 30 million tracks for listeners to choose from, the fractionalized consumer doesn't have a prayer in deciding what to listen to—so we enter into a new and fourth phase of digital music: The rise of curated and "Listen" Services.

According to MIDIA, the industry is attending to the aficionado music-phile by creating more curated outlets such as Spotify, Apple's Beats, and radio, while moving away from a traditional buyer who is not making the transition to a streaming model and is labeled the "forgotten fan." If that consumer does not make the jump, the revenue trend of negative sales will continue although the potential of conversion is there. And what will also be realized is the continued sales decline of not only CDs, but downloads as well.

MIDIA believes that the future to the music industry rests in three key factors:

1. The continued evolution of consumer behavior

2. Technology company strategy

3. Income distribution

Consumer behavior will be based on continual relationship building between Aficionados and content creators while working to convert and bridge the gap of the Forgotten Fan. But MIDIA stresses the fact that an entire generation is emerging that is known as Digital Natives who have experienced "free" all their lives. As they move forward, the more "free"—the better, which will be their comfort zone. As stressed by Mulligan, "**For a generation weaned on free, the more free you give them, the more they will crave it**. Whatever course is plotted, success will depend upon deeply understanding the needs of Digital Natives and not simply trying to shoe horn them into the products we have now that are built for the older transition generation."

So what's in it for the **tech companies**? They all have their own agendas, but it drills down to the bottom line. In 2014, Apple purchased Dr. Dre's Beats Electronics LLC for $3 billion. Beats has long been a brand leader in high-end headphones and electronics and recently launched a music streaming site. In a rare move, Apple will keep the Beats brand, coupling it with its technology to ward off the revenue encroachment that streaming threatens to the company.

Google is in the data business and uses their myriad of portals to capture information about users to sell as valuable marketing facts that will help the next brand make an impression. Anything from YouTube, Google gmail, Google Hangouts, Google Maps, Search Engines—Google is the "go to" portal for many Internet launches on computers around the world.

Amazon boasts as one of the world's largest retailers, leveraging the Internet as a "front door" to hundreds of thousands of outlets and products. Not only the largest "bookseller" in the world, the retailer sells anything from car parts, to lipstick, to computer technology, to the latest trends in fashion, and more. With the introduction of the Kindle, books, music, movies including exclusive "made for Amazon" content as well as the Internet, can be accessed.

Each of these entities uses music as bait to entice consumers to their site. Even though the sale of music through these portals generates revenue and royalties, music is not their end game. They each have another agenda that uses music as a pawn in the chess match. Their concern is not whether the label, artist or songwriter receives their portion of the $.99, but if the "experience" of the site user lingered longer on the site to perhaps purchase a phone or watch an advertisement. Music is merely a "carrot" to draw to consumer in.

The industry is witnessing download dollars being replaced by streaming pennies, and everyone is in the value-chain is suffering. The message that MIDIA recommends is that record labels become agencies or label services companies that transforms the old business model into a multi-faceted agency. This new music company would work with artists as a label and a publisher and a creative agency, as well as a product developer to dimensionalize the act in a multitude of fronts to help realize revenue from the various income opportunities that must be mined to be successful in a business of challenging opportunities.

FUTURE TRENDS

Distribution Value

Several distribution companies are exploring ways to add value to the conglomerate equation. Creating distribution-specific marketing campaigns with non-entertainment product lines helps validate distribution's existence, while hopefully enhancing the bottom line. Marketing efforts such as on-pack CDs with cereal, greeting card promotions, and ringtone services add to the branding of the participating artists, while increasing overall revenue through licensing and/or sales of primary items. The ultimate value for today's distribution companies is that of consolidator and aggregator. Distributors can consolidate labels to create leverage points within retail. To gain positioning in the retail environment, one must have marketing muscle, and by using the collective power of their various label's talent, the entire company can raise its market share by coattailing on the larger releases in the family.

THE MUSIC RETAIL ENVIRONMENT

Marketing and the Music Retail Environment

This retail environment, both physical and virtual, is designed to aid consumers in making their purchasing decision. This decision can be influenced in a number of ways, depending on the consumer. For example: "Does the store have hard-to-find releases?"; "Do they have the lowest prices?"; "Is it easy for consumers to find what they're looking for?"; "Does the store have good customer service and knowledgeable employees?" "Does the store allow the consumer to experience the music prior to purchase?" These questions should be answered, in one way or another, within the confines of the retail environment.

Music Business Association

To assist music retailers in determining business strategies, companies look for current sales trends as well as educational and support networks. Founded in 1958, NARM—The National Association of Recording Merchandisers—was conceived to be a central communicator of core business issues for the music retailing industry. This trade organization has had to evolve to represent a new day of music retailing. Rebranded as the Music Business Association, this organization looks to be a central resource for all things music commerce via the various delivery models: physical, digital, mobile, streaming, and more. Creating a digital initiative known as digitalmusic.org, this site, its resources and partners are attempting to build a future of music commerce—together. For more information regarding the Music Business Association and its activities, visit www.musicbiz.org.

FIGURE 15.12

Retail Considerations

How does a retailer learn about new releases? Distribution companies are basically extensions of the labels that they represent. To sell music well, a distribution company needs to be armed with key selling points. This critical information is usually outlined in the marketing plan that is created at the record label level. Record labels spend much time "educating" their distributor partner and retailer about their new releases.

Distribution companies set up meetings with their *accounts*, meaning the retailers. At the retailer's office, the distribution company shares with the *buyer*—that is the person in charge of purchasing product for the retail company—the new releases for a specific release date, as well as the marketing strategies and events that will enhance consumer awareness and create sales.

Purchasing Music for the Store

When making a purchase of product, the buyer will take into consideration several key marketing elements: radio airplay, media exposure, touring, cross-merchandising events, and most critical, previous sales history of an artist within the retailers' environment; or if a new artist, current trends within the genre and/or other similar artists. Depending on the importance of the release, the record label representative will accompany the distribution company sales rep with the hopes of enhancing the knowledge of the buyer of the new release. The ultimate goal is to increase the purchasing decision, while creating marketing events inside the retailer's environment. Most record labels, along with their distributors, have agreed on a forecast for a specific release. This forecast, or number of records predicted to sell, is based on similar components that retail buyers consider when purchasing product. Many labels use the following benchmarks when determining forecast:

Initial orders or IO: This number is the initial shipment of music that will be on retailers' shelves or in their inventory at release date.

90-day forecast: Most releases sell the majority of their records within the first 90 days.

Lifetime: Depending on the release, some companies look to this number as when the fiscal year of the release ends, and the release will then rollover into a catalog title. But on occasion, a hit release will predicate that forecasting for that title continues, since sales are still brisk.

Inventory Management

Larger music retailers have very sophisticated purchasing programs that profile their stores' sales strengths. Using the forecast, as well as an overall percentage of business specific to the label or genre, the retailer will determine how many units it believes it can sell. This decision is based on historical data of the artist and/or trends of the genre along with the other marketing components.

Keeping track of each release, along with the other products being sold within a store is called *inventory management*. Using point-of-sale (POS) data, the store's computer notes when a unit is sold using the Universal Product Code **(UPC) bar** code. Depending on the inventory management system, a store may have ten units on-hand, which is considered the ideal *maximum* number the store should carry. The ideal *minimum* number may be four units. If the store sells seven units, and drops below the ideal inventory number of four as set in the computer, the store's inventory management system will automatically generate a re-order for that title, up to the maximum number. This min/max inventory management system may then download the re-order through an electronic data interchange (EDI) to its supplier, and the product is then shipped to the retailer within a few days. To avoid waiting for the product to arrive, a retailer may opt to drop-ship product directly from the distribution company or a partnering one-stop supplier, avoiding the delay of processing at either the headquarters' distribution center (DC).

Turns and Returns

Know that a store's success is based on the number of units it sells within a fiscal year. Clearly, the size of the store dictates how much product or inventory it can hold. An average 2,500 square-foot store may hold 20,000 units. An average annual inventory *turn* for music may be 3.5 times. As an industry standard, this store could sell: 20,000 units × 3.5 turns = 70,000 units in a fiscal year. This does not mean that every title sells 3.5 times, but rather the store averages 3.5 sales per year for every title or unit it is holding.

To manage the real estate within the store, the best-selling product should receive the best space. To keep a store performing well, the inventory management system should notice when certain titles are not selling. Music retailers have an advantage over traditional retailers in that if a product is not selling, they can send it back for a refund, called a *return*. The refund is usually in the form of a credit, and the amount credited is based on when the product is returned along with other considerations such as if a discount had been received. It has long been an industry standard that the return average is noted at 20%. To put this statistic another way, for every 100 records in the marketplace, nearly 20 units are returned to the distributor.

Promotion

Many music consumers still find out about new music via the radio, as well as social networking as the new "word-of—mouth." On-demand streaming is having a huge impact on sales since the listening often becomes a replacement for the buy. But with premium subscription services, the hope is that these portals will fund the sales lost to streaming, although early data does not bode well for the trend.

Featured titles within many retail environments are often dictated from the central buying office of the retailer. As mentioned earlier, labels want and often do create marketing events that feature a specific title. This is coordinated via the retailer through an advertising vehicle called *coopera-tive advertising*. *Co-op advertising*, as it is known, is usually the exchange of money from the label to the retailer, so that a particular release will be featured.

Virtual end caps and digital dashboards drive online buyers to specific releases once "inside" a store's online presence. This real estate is paid for by the label and highlights new releases and focuses attention on music that the label wants the consumer to notice. Just like online, promotional efforts in-store help to highlight different releases that should aid consumers in purchasing decisions. These marketing devices often set the tone and culture of the store's environment.

Pricing and positioning (P&P)—P&P is when a title is sale-priced and placed in a prominent area within the store.

End caps—Usually themed, this area is designated at the end of a row and features titles of a similar genre or idea.

Point-of-purchase (POP) materials—Although many stores will say that they can use POP, including posters, flats, stand-ups, and so on, some retailers have advertising programs where labels can be guaranteed the use of such materials for a specific release.

Print advertising—A primary advertising vehicle, a label can secure a "mini" spot in a retailer's ad (a small picture of the CD cover art), which usually comes with sale pricing and featured positioning (P&P) in-store.

In-store event—Event marketing is a powerful tool in selling records. Creating an event where a hot artist is in-store and signing autographs of his or her newest release guarantees sales, while nurturing a strong relationship with the retailer.

Listening stations—Depending on the store, some releases are placed in an automatic digital feedback system where consumers can listen to almost any title within the store. Other listening stations may be less sophisticated, and may be as simple as using a freestanding CD player in a designated area. But all play back devices are giving consumers a chance to "test drive" the music before they buy it.

TOP 10 ACCOUNTS AND FORECASTING

The following grid is a sample forecasting and P&P planning tool used to predict initial orders and initial marketing campaign activities within a retailer's environment.

iTunes dominates the account list namely because of the volume of single downloads. As an account, they are a key player in understanding the

Table 15.2	Sales Forecasting Grid Artist Name Title Selection Number Street Date				
Account	% of Business	Target	Account Advertising P&P		Cost
iTunes	35.00%	35,000			
SoundExchange	16.00%	16,000			
Anderson (Walmart & Best Buy)	8.00%	8,000			
Amazon (Physical & Digital)	6.00%	6,000			
Spotify	5.00%	5,000			
VEVO	3.00%	3,000			
Alliance Entertainment	3.00%	3,000			
YouTube	3.00%	3,000			
Google	2.00%	2,000			
Target	2.00%	2,000			
TOP 10 Accounts	83.00%	83,000			
All Others	17.00%	17,000			
Total	100.00%	100,000			

components of the marketing plan but singles sales can be a wild card in the financial equation and are difficult to forecast.

SoundExchange is a new entry for 2014 as streaming revenue becomes more significant to the bottom line of labels and distributors. Watch this "account" continue to grow as the shift from ownership to curated "streaming" gains momentum.

Anderson now racks jobs for both Walmart and Best Buy with its "Direct Shot" entertainment merchandising arm. Anderson continues to grow is rack jobbing business beyond that of music and includes P&G, Post Foods, electronic kings Apple, Belin, Dell and Microsoft, toy icons Hasbro, Lego and Mattel, along with countless other product lines—insuring its future in the supply-chain to American's largest retailer, Walmart.

Digital and Digital Aggregators

Amazon, Spotify, VEVO, YouTube, and Google all represent the power of digital music sales and streaming and where the market is going. You should begin to wonder what the physical music landscape will look like in the near future.

FIGURE 15.13

To explain the existence of these players, you need to remember the power of the independents. Those artists and labels who look to enter the marketplace and sell through mainstream channels use digital aggregators to ease their way onto the main stage of the retail environment. Using the selling portals of a CD Baby or ReverbNation allows an artist to sell on iTunes or Amazon, which registers their sale within the Top 10 account list below.

Partial approved aggregator listing:

iTunes: Catapult, CDBaby, Ingrooves, Tunecore, The Orchard
Spotify: CDBaby, Ditto Music, Tunecore, IndigoBoom, Song Last, ReverbNation

PRICING

Although record labels set the suggested retail list price (SRLP) for a release, this is *not* what retailers are required to sell the product for. Most often, the SRLP sets the *wholesale price* or the cost to the retailer. In negotiating the order, the retailer may ask for a discount off the wholesale price. The retailer may also ask for additional *dating*, meaning that the retailer is asking for an extension on the payment due date. Each distributor has parameters in which this transaction may occur.

Generally, music product comes in box lots of 30 units. A retailer will receive a better price on product if it purchases in box lots. For example, a retailer wants to purchase 1,200 units of a new release with a 10% discount and 30 days dating. See Table 15.3.

Table 15.3	Discount Value				
SRLP	$11.98	SRLP	$11.98		
Wholesale	$7.50	Wholesale	$7.50		
x	1200 units	−10% disc.	$0.75		
	$9,000	Total/unit	$6.75	The 10% discount	
		x	1200	saved the retaller	
			$8,100	$900.00	

Money due for any purchase would normally be received at the end of the following month that the record was released. However, with an extra 30 days dating, the due date is extended, giving the retailer a longer time-frame in which to sell all the product. Adding extra dating is often a tactic of record labels that want retailers to take a chance on a new artist that may be slower to develop in the marketplace.

The price of product reflects the store's marketing strategy. The major electronic superstores look to music product as the magnet to get customers through their doors, which is why music prices in these environments are often lower than anywhere else. Often, these stores will sell music for less than they purchased it, called *loss-leader* pricing. But these stores will also raise the price after a short period of time, usually within first two weeks after the street date.

ACTUAL PRICING OF PRODUCT

Margin and Markup

Margin and markup are both calculated using the wholesale purchase price of the product. Percent margin uses the selling price as the denominator, whereas percent markup uses the purchasing (wholesale) price as the denominator for calculating.

Margin percentage on product is determined with the following calculation:

$$\text{Formula: Margin \%} = \frac{\text{Dollar Markup}}{\text{Retail}}$$

If an SRLP CD of $11.98 is purchased wholesale for $7.50 and the store wants to sell it for $9.99, the margin percentage is $9.99 −$7.50/$9.99 = 24.9%

Although there are variable ways to calculate margin, most stores use this retail markup calculation since it takes into consideration differing price lines, product extensions and customer demands in retail value.

Markup uses a similar calculation, but divides the dollar markup by the wholesale cost.

$$\text{Formula: Markup \%} = \frac{\text{Dollar Markup}}{\text{Wholesale}}$$

If an SRLP CD of $11.98 is purchased wholesale for $7.50 and the store wants to sell it for $9.99, the markup percentage is $9.99 − $7.50/$7.50 = 33.2%

There is always an arithmetical relationship between gross margin and markup.

A gross margin of 40% requires a markup of 66.67% calculated as 40 ÷ (100–40).

A gross margin of 60% requires a markup of 150% calculated as 60 ÷ (100–60).

To achieve a target gross margin of 24.9% on the previous example, based on the purchase cost, the calculations are as follows:

A gross margin of 24.9% requires a markup of 33.2% calculated as 24.9 ÷ (100–24.9) = 33.2%

Wholesale CD cost	$7.50
Margin 24.9%	× .332 = 2.49
Total	9.99

When setting prices, retailers think about markup since they start with costs and work upwards. When thinking about profitability, retailers think about margin, since these are the funds leftover to cover expenses as well as account for profit. Importantly, negotiating for the best discount off the wholesale price improves both markup and margin.

WHERE THE MONEY GOES

When the consumer plunks down their money at the cash register to purchase a CD, Figure 15.14 shows how that money is divided between all the invested parties.

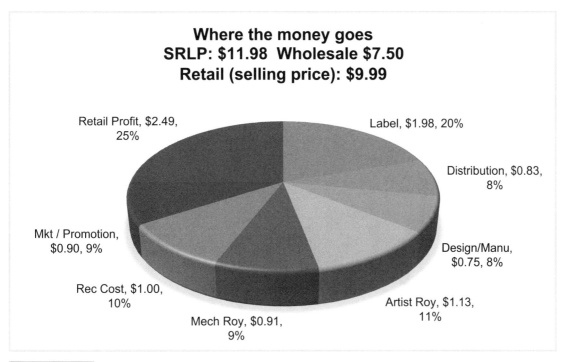

FIGURE 15.14 *Where the Money Goes Pie Chart*

Table 15.4	Where the Money Goes
SRLP	$11.98
Wholesale	$7.50
Label	$1.98
Distribution	$0.83
Design/Manu	$0.75
Artist Roy	$1.13
Mech Roy	$0.91
Rec Cost	$1.00
Mkt / Promotion	$0.90
Retail Profit	$2.49

Table 15.5 Sample Distribution Pricing Schedule

Contemporary Compact Discs

Bulk Price	List Price	Product Line
1.90	2.98	Budget
2.31	3.98	Extended Play
2.50	3.98	Budget
3.14	4.98	Budget or Frontline
3.32	5.98	Budget
4.33	6.98	Budget or Frontline
5.00	7.98	Budget or Frontline
5.61	8.98	Budget or Frontline
5.79	9.98	Nice Price
5.91	9.98	Developing Artist
6.34	9.98	Frontline
6.40	10.98	Frontline
6.84	11.98	Developing Artist
7.50	11.98	Frontline
7.81	11.98	Best Value
7.93	12.98	Developing Artist or Frontline
8.31	12.98	Frontline
7.38	13.98	Developing Artist
9.00	13.98	Frontline or Hitsavers
9.59	14.98	Frontline
10.21	15.98	Frontline
9.95	15.98	Frontline
10.62	16.98	Frontline
10.95	16.98	Frontline
11.24	17.98	Frontline
11.86	18.98	Frontline
12.05	18.98	Frontline
12.50	19.98	Frontline
12.71	19.98	Boxsets or Frontline
13.13	19.98	Frontline
13.12	20.98	Frontline
14.03	21.98	Frontline
14.51	22.98	Frontline
14.59	22.98	Frontline
15.48	23.98	Frontline

Contemporary (MEDIA DISC)

Bulk Price	List Price	Product Line
14.50	24.98	Frontline

Contemporary Vinyl LP

(All Vinyl LPs are NON-RETURNABLE unless otherwise noted.)

Bulk Price	List Price	Product Line
4.02	24.98	LP
5.51	24.98	LP
7.53	24.98	LP
8.03	24.98	LP
8.53	24.98	LP
9.03	24.98	LP
9.53	24.98	LP
10.04	24.98	LP
10.54	24.98	LP
11.04	24.98	LP
11.59	24.98	LP
12.05	24.98	LP
12.55	24.98	LP
13.06	24.98	LP
13.55	24.98	LP
14.06	24.98	LP
14.55	24.98	LP
15.06	24.98	LP
15.56	24.98	LP
16.05	24.98	LP
16.57	24.98	LP
17.06	24.98	LP
17.57	24.98	LP
18.57	24.98	LP
19.57	24.98	LP

Contemporary Compact Disc Singles

Bulk Price	List Price	Product Line
1.14	1.98	Singles
1.72	2.98	Singles
2.02	3.49	Singles
2.30	3.98	Singles
3.45	5.98	Singles
3.75	6.49	Singles
4.03	6.98	Singles
4.03	7.48	Singles

Contemporary Vinyl Singles

Bulk Price	List Price	Product Line
0.73	1.29	7" Singles
1.12	1.98	7" Singles
2.81	4.98	7" & 12" Singles
3.37	5.98	7" & 12" Singles
3.93	6.98	7" & 12" Singles
4.50	7.98	7" & 12" Singles
5.06	8.98	7" & 12" Singles
5.62	9.98	7" & 12" Singles
7.30	12.98	7" & 12" Singles
7.87	13.98	12" Singles
8.39	14.98	7" & 12" Singles
9.59	16.98	7" & 12" Singles
11.00	19.98	7" Singles
13.50	21.98	7" & 12" Singles
22.53	39.98	Singles
24.09	42.98	Singles

Bulk Price	List Price	Product Line
15.92	24.98	3Paks, Boxsets or Frontline
16.85	26.98	Frontline
18.01	27.98	Frontline
19.10	29.98	3Paks, Boxsets or Frontline
20.36	31.98	Frontline
20.99	32.98	Boxsets or Frontline
21.81	34.98	Boxsets or Frontline
23.23	35.98	Boxsets or Frontline
24.00	36.98	Boxsets
25.63	39.98	Boxsets or Frontline
25.75	39.98	Boxsets or Frontline
26.00	39.98	Boxsets
27.58	44.98	Boxsets or Frontline
29.64	47.98	Boxsets or Frontline
30.87	49.98	Boxsets or Frontline
33.96	54.98	Boxsets or Frontline
37.07	59.98	Boxsets or Frontline
38.45	63.98	Boxsets or Frontline
42.05	69.98	Boxsets or Frontline
45.04	74.98	Boxsets or Frontline
48.05	79.98	Boxsets or Frontline
50.46	83.98	Boxsets or Frontline
54.07	89.98	Boxsets or Frontline
60.07	99.98	Boxsets or Frontline
66.08	109.98	Boxsets or Frontline
72.09	119.98	Boxsets or Frontline
78.10	129.98	Boxsets or Frontline
81.10	134.98	Boxsets
84.11	139.98	Boxsets or Frontline
90.12	149.98	Boxsets or Frontline
102.13	169.98	Boxsets or Frontline
108.14	179.98	Boxsets
120.16	199.98	Boxsets or Frontline
132.16	219.98	Boxsets
150.21	249.98	Boxsets or Frontline
181.00	299.98	Boxsets
198.27	329.98	Boxsets or Frontline

Bulk Price	List Price	Product Line
20.98	24.98	LP
23.18	24.98	LP
20.57	24.98	LP
21.09	24.98	LP
21.57	24.98	LP
22.09	24.98	LP
24.09	24.98	LP
27.00	24.98	LP
30.00	24.98	LP
35.33	24.98	LP
40.00	24.98	LP
45.17	24.98	LP
49.99	24.98	LP
60.26	24.98	LP
68.08	24.98	LP
100.45	24.98	LP
134.99	24.98	LP

Contemporary Cassettes

Bulk Price	List Price	Product Line
2.25	3.98	Cassette
4.63	7.98	Cassette
5.65	9.98	Cassette
6.22	10.98	Cassette
6.79	11.98	Cassette
7.35	12.98	Cassette
11.32	19.98	Cassette

Contemporary Music DVD & Blu-ray

Bulk Price	List Price	Product Line
3.94	6.98	Frontline
4.55	7.98	Frontline
5.69	9.98	Frontline
6.85	11.98	Frontline
7.42	12.98	Frontline
8.57	14.98	Frontline or Hitsavers
9.7	16.98	Frontline
11.42	19.98	Frontline
12.56	21.98	Frontline
14.27	24.98	Frontline
15.51	29.98	Frontline
17.13	31.98	Frontline
18.27	34.98	Frontline
19.99	34.98	Frontline
22.14	38.98	Frontline
22.84	39.98	Frontline
25.63	39.98	Frontline
26.04	45.98	Frontline
28.56	49.98	Frontline
34.25	59.98	Frontline
45.59	79.98	Frontline
63	109.98	Frontline

Contemporary Music VHS

Bulk Price	List Price	Product Line
4.32	7.98	Frontline
5.41	9.98	Frontline
8.12	14.98	Frontline
10.82	19.98	Frontline
16.24	29.98	Frontline

STORE TARGET MARKET

A music store's target market or consumer generally dictates what kind of retailer it will be. To attract consumers interested in independent music, or to attract folks who are always looking for a bargain, determines the parameters in which a store operates. Music retailers have traditionally been segmented into the following profiles:

Independent music retailers cater to a consumer looking for a specific genre or lifestyle of music. Generally, these types of stores get their music from one-stops. Independent stores are locally or regionally owned and operated, with one or just a few stores under one ownership.

The **Mom & Pop** retailer is usually a one-store operation that is owned and operated by the same person. This owner is involved with every aspect of running the business and tends to be very passionate about the particular style of music that the store sells. This passion can be interpreted as being an expert in the knowledge of the genre and can be a unique resource for the consumer looking for the obscure release. Mom & pop storeowners tend to have a personal relationship with their customer base, knowing musical preferences and keeping the customers informed about upcoming releases and events.

Alternative music stores profile very similarly to mom & pop stores, but with the exception that they tend to be lifestyle-oriented. An electronic music retailer may have many hard-to-find releases along with hardware offerings such as turntable and mixing boards.

Chain stores tend to attract music purchasers who are looking for deep selection of releases along with assistance from employees who have strong product knowledge. These stores can be found in malls and cater to a broad spectrum of purchasers. These stores have been studied and replicated so that entering any store with the same name in any location feels very similar. Often, they have the new major releases upfront with many related items for sale, such as blank media and entertainment magazines. Chain stores traditionally buy their music inventory directly from music distributors, with warehousing and price stickering occurring in a central location.

Electronic superstores do not make the bulk of their profits from the sale of music. But rather, use music as an attraction to bring consumers into their store environments. By using loss leader pricing strategies, these stores often sell new releases for less than they purchased it, but for a limited time. Meanwhile, they have created traffic to the store in order to make money from the sale of all the other items offered such as electronics, computers, televisions, and so forth. Best Buy sources its music from Anderson's Direct Shot rack jobbing supplier.

Mass merchants use the sale of music as event marketing for their stores. Each week, a new release brings customers back to their aisles with the notion that they will purchase something else while there. There is little profit in the sale of music for the mass merchant, but the offering of music is looked at as a service to customers. Often, mass merchants use rack jobbers to supply and maintain music for their stores. It is the rack jobber who initially purchases the music for the mass merchant environment. Some examples of mass merchants are Wal-Mart and Meijers.

INTERNET MARKETING AND SALES

Until recently, it was the retailers who had the brand identities that were winning the Internet sales wars. Consumers went to their favorite retail store websites to browse and purchase music, meaning the actual CD that was to be delivered to the consumer's door. These well-known retailer sites left many start-up websites with unknown names with little traffic. As noted in the sales overview, downloading and file-sharing has become big "business," but not perceived as a potentially profitable business . . . until now. Downloading activities have hurt not only the labels and their distributors, but the retailers as well. With the aggressive prosecution and education campaigns to alert downloaders as to their illegal practices, consumers are beginning to use legal downloading sites to purchase music and streaming services as go-to sources for listening. Sites such as iTunes, Rhapsody, and yes, Wal-Mart are all experiencing a high volume of downloads, with more sites coming online everyday. Well-known online sites such as Amazon have also had an impact on retailing. Interesting to follow and more interesting to quantify is the impact of streaming on the bottom line of record labels.

As noted earlier, the revenue generated by SoundExchange has amounted to over 15% of the overall monies earned in 2014 and with some labels and distributors, accounts for more than that. Other streaming revenue sources such as VEVO, YouTube, and Google have caused labels to rethink their marketing efforts, with the single driving a huge portion of the bottom line. The single selection is critical and the marketing choices surrounding it and subsequent singles after it are essential to the artist's and project's success.

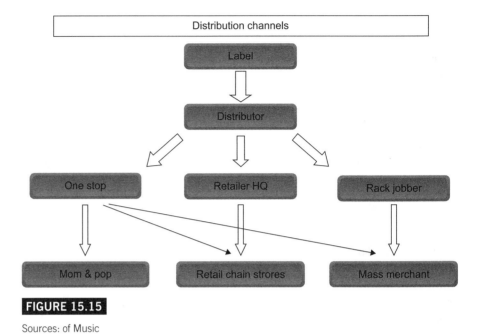

FIGURE 15.15

Sources: of Music

Figure 15.15 shows the basic flow of music as it reaches the consumer level. Recognize that one-stops' primary business is servicing independent records stores, but that they also do what is called *fill-in* business for all music retailers.

GLOSSARY

Big 3—These are the three music conglomerates that maintain a collective 85% market share of record sales: they are Universal, Sony, and Warner.

Brick and mortar—The description given to physical store locations when compared to online shopping.

Box lot—Purchases made in increments of what comes in full, sealed boxes receive a lower price. (For CDs with normal packaging, usually 30.)

Buyers—Agents of retail chains who decide what products to purchase from the suppliers.

Chain stores—A group of retail stores under one ownership and selling the same lines of merchandise. Because they purchase product in large quantities from centralized distribution centers, they can command big discounts from record manufacturers (compared to indie stores).

Computerized ordering process—An inventory management system that tracks the sale of product and automatically reorders when inventories fall below a preset level. Reordering is done through an electronic data interchange (EDI) connected to the supplier.

Co-op advertising—A co-operative advertising effort by two or more companies sharing in the costs and responsibilities. A common example is where a record label and a record retailer work together to run ads in local newspapers touting the availability of new releases at the retailer's locations.

Discount and dating—The manufacturer offers a discount on orders and allows for delayed payment. It is used as an incentive to increase orders.

Distribution—A company that distributes products to retailers. This can be an independent distributor handling products for indie labels or a major record company that distributes its own products and that of others through its branch system.

Drop ship—Shipping product quickly and directly to a retail store without going through the normal distribution system.

Economies of scale—Producing in large volume often generates economies of scale—the per-unit cost of something goes down with volume. Fixed costs are spread over more units lowering the average cost per unit and offering a competitive price and margin advantage.

Electronic data interchange (EDI)—The inter-firm computer-to-computer transfer of information, as between or among retailers, wholesalers, and manufacturers. Used for automated reordering.

Electronic superstores—Large chain stores such as Circuit City and Best Buy that sell recorded music and videos, in addition to electronic hardware.

End cap—In retail merchandising, a display rack or shelf at the end of a store aisle; a prime store location for stocking product.

Fill-in—One-stop music distributors supply product to mass merchants and retailers who have run out of a specific title by "filling in" the hole of inventory for that release.

Floor designs—A store layout designed to facilitate store traffic to increase the amount of time spent shopping (TSS).

Free goods—Saleable goods offered to retailers at no cost as an incentive to purchase additional products.

Indie stores—Business entities of a single proprietorship or partnership servicing a smaller music consumer base of usually one or two stores (sometimes known as *mom & pop* stores).

Inventory management—The process of acquiring and maintaining a proper assortment of merchandise while keeping ordering, shipping, handling, and other related costs in check.

Listening station—A device in retail stores allowing the customer to sample music for sale in the store. Usually the devices have headphones and may be free standing or grouped together in a designated section of the store.

Loose—The pricing scheme for product sold individually or in increments smaller than a sealed box.

Loss leader pricing—The featuring of items priced below cost or at relatively low prices to attract customers to the seller's place of business.

Margin—The percentage of revenues leftover to cover expenses as well as account for profitability.

Markup—The percentage of increase from wholesale price to retail price.

Mass merchants—Large discount chain stores that sell a variety of products in all categories, for example, Wal-Mart and Target.

Min/max systems—A store may have 10 units on-hand, which is considered the ideal *maximum* number the store should carry. The ideal *minimum* number may be 4 units. If the store sells 7 units and drops below the ideal inventory number of four, as set in the computer, the store's inventory management system will automatically generate a reorder for that title, up to the maximum number.

National Association of Recording Merchandiser (NARM)—The organization of record retailers, wholesalers, distributors, and labels.

One-Stop—A record wholesaler that stocks product from many different labels and distributors for resale to retailers, rack jobbers, and juke box operators. The prime source of product for small mom & pop retailers.

Point-of-purchase (POP)—A marketing technique used to stimulate impulse sales in the store. POP materials are visually positioned to attract customer attention and may include displays, posters, bin cards, banners, window displays, and so forth.

Point-of-sale (POS)—Where the sale is entered into registers. Origination of information for tracking sales, and so on.

Price and positioning (P&P)—When a title is sale priced and placed in a prominent area within the store.

Pricing strategies—A key element in marketing, whereby, the price of a product is set to generate the most sales at optimum profits.

Rack jobber—A company that supplies records, cassettes, and CDs to department stores, discount chains, and other outlets and services (racks) their record departments with the right music mix.

Returns—Products that do not sell within a reasonable amount of time and are returned to the manufacturer for a refund or credit.

Sales book—Distribution companies compile all their releases for a specific street date into a "sales book," which contains one sheet for each title that outlines the marketing efforts.

Sales forecast—An estimate of the dollar or unit sales for a specified future period under a proposed marketing plan or program.

Sell-through—Once a title has been released, labels and distributors want to minimize returns and "sell-through" as much inventory as possible.

Shrinkage—The loss of inventory through shoplifting and employee theft.

Solicitation period—The sales process of specific titles occurs during a period called *solicitation*. All titles streeting on a particular date are placed in a solicitation book, where details of the release are described.

Source tagging—The process of using electronic security tags embedded in a product's packaging.

Theft protection—Systems in place to reduce shoplifting and employee theft in retail stores. These systems may include electronic surveillance.

Time spent shopping (TSS)—A measure of how long a customer spends in the store.

Turn—The rate that inventory is sold through, usually expressed in number of units sold per year/inventory capacity on the floor.

Vertical integration—The expansion of a business by acquiring or developing businesses engaged in earlier or later stages of marketing a product.

Universal Product Code (UPC)—The bar codes that are used in inventory management and are scanned when product is sold.

REFERENCES

Bess, D. Personal Interview, September 23, 2014.

Conway, J. Personal Interview, October 15, 2015.

Knab, C. 2001. http://www.musicdish.com/mag/index.php3?id 3357. The Distributor One Sheet, March 25.

Flanagan, A. "Industry Sets Friday as Global Record Release Day." *Billboard*. N.p., Feb. 26, 2015. Accessed Mar. 1, 2015. http://www.billboard.com/articles/business/6487289/friday-global-record-release-day-ifpi.

Nielsen Company. "2008 Year-End Music Industry Report." Scoop Marketing, December 31, 2008. Nielsen SoundScan State of the Industry 2007–2012.

Smith, E. "Music Downloads Plummet in U.S., but Sales of Vinyl Records and Streaming Surge." *Wall Street Journal*. N.p., January 1, 2015. Accessed January6, 2015.

Weatherson, J. GM Nash Icons and former Executive Vice President of Universal Music Distribution, from personal interview, July 2004.

CHAPTER 16

Technology and the Music Business

INTRODUCTION

Technological advancements not only launched the music business, but have continually given it a "shot in the arm" just when it needed it most. From stereo to recordable tape to compact discs to digital files to streaming, technology continues to evolve the hardware and software of our industry. Not only has technology changed the format of the business, but technology has changed *how* the actual business gets done. Technological advancements permeate the core of the record industry, creating the scorecards for success.

BIG DATA

The term "big data" is used to describe a massive amount of both structured and unstructured data that is so "big" that it's difficult to process using traditional databases and available software. In most cases, this "big data" is so large or is moving so fast that is exceeds the current ability to process. But—this data holds the potential answers to help companies improve, from making better operational decisions to being more intelligent in the market place. And unlocking "big data" is the next big thing for many companies in the business of digital analysis and applications.

"Big data analytics" takes the process one step further by sifting through the data to uncover patterns and other applicable information. Collecting, organizing and analyzing large sets of data that are actionable to specific business decisions is big business in today's market space. Many analytic companies have emerged as business partners to help with the processing of this data by creating interactive tools, hoping to increase their partner's sales and efficiencies as well as improve operations, customer service and risk management (Beal, 2014).

In recent "data" years, Google Analytics has been a leader in this type of analysis, having created a tool that helps website managers understand the "business" of consumer traffic and how to better design a business' site to have a favorable outcome—whatever that might be. Simply put, Google Analytics tracks visitors through the website and also tracks of the performance of a marketing campaigns, following keyword searches and tracking popular pages once a searcher has landed.

But this knowledge can be basic, and to understand, as well as act on this information, it can take a big data "tool" provider such as Raven to help "fill in the blanks." Raven builds dashboards and reports that help businesses understand results of online marketing campaigns including Google Analytics, paid search advertising, search engine optimization and social media marketing. These give great at-a-glance ideas of how a campaign is performing and what needs to be fixed.

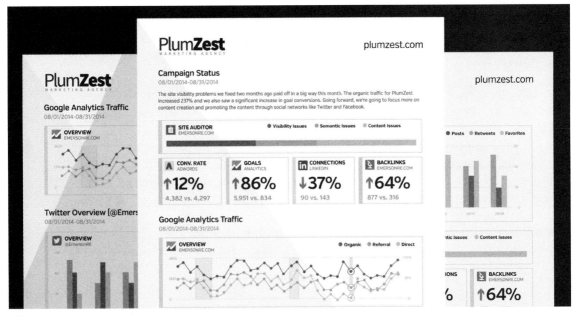

FIGURE 16.1   *Raven Tools Dashboard*

ENTERTAINMENT AND BIG DATA

The Nielsen Company has been providing important data to all kinds of products in the U.S. and abroad since the 1920s. In the mid-1930s, the company began to monitor and created an index of radio stations for the U.S. and replicated this concept for television in the 1950s as the format began to penetrate the U.S. marketplace. In doing so, the company established several key measurement devices that have aided in evaluating consumer behavior quarter-over-quarter, year after year. In doing so, Nielsen invented its own vernacular of terms that apply to the entertainment business that are important in evaluating the marketplace:

Designated Market Area (DMA)—A term used by Nielsen Media Research to identify an exclusive geographic area of counties in which the home market television stations hold a dominance of total hours viewed.

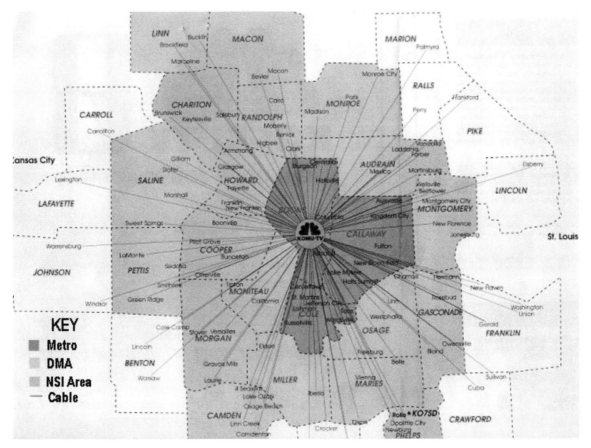

FIGURE 16.2 *DMA with County Classification*

There are 210 DMA's in the U.S. These markets are listed in order of population—meaning, stratified largest to smallest by size of city. These markets apply to the analysis of music as well.

County Size—The classification of counties according to Census household counts and metropolitan proximity. There are four county size classes "A," "B," "C," and "D." In general, "A" counties are highly urbanized, "B" counties relatively urbanized, "C" counties relatively rural, and "D" counties very rural. Why is this important? These counties establish the zones known as City, Suburb, or Rural and can help marketers determine strategic plans. As you look at this map of Columbia, Missouri, the DMA is outlined, with urbanized counties A and B in dark and mid shades, relative rural counties C in lighter shades and very rural counties D that are serviced just outlined.

Other Industry Terms

Geographic regions—states are grouped by physical location, which can lead to sales and marketing analysis based on geographic data.

Store types—As an industry standard, stores have been identified by types to aid the industry in identifying source of sales. These store types include Mass Merchants, Traditional Retailers which include chain store and electronic super store, Independent Retailers, Non-traditional Outlets that include online and venue sales, and Digital portals.

Formats—The type of music format that consumer purchase continues to evolve—be it physical or digital. Physical formats include compact discs (CD), vinyl (LP), cassette, and DVD-audio with digital formats including downloads and streaming. Streaming is not included in SoundScan except for the chart numbers.

Other technologies have aided in the data collection of the entertainment business and include the application of inventory management tools as well as digital management devices that allow for the important tracking and processing of royalty payments.

FIGURE 16.3 *Example Universal Product Code CD Bar Code*

UPC Code—The Universal Product Code contains a unique sequence of numbers that identifies a product. GS1US standards recommend a GS1 company prefix number that identifies a unique code for a specific business. The GTIN (global trade item number) initial numbers, which can vary in size depending on the variety of products the company sells, identify the business, with the last digits identifying the products. The small digit is the "check" digit which validates the barcode equation.

Many record companies designate a 5- to 6-digit number that identifies the label. The record company then assigns a 4- to 5-digit product code that identifies the release, including artist and title of the album. The eleventh digit can designate the configuration of the product. The last digit is known as the "check" digit. When scanned, a mathematical equation occurs determining if the product has been correctly scanned. The "check" digit is the "answer" to that equation, verifying an accurate scan. In the U.S., the standard UPC code contains 12 digits.

ISRC Code—For digital product, the ISRC (International Standard Recording Code) is the international identification system for sound recordings and music video recordings. Just like the UPC Code, each ISRC has a unique and permanent sequence of numbers that identifies each specific recording that can be permanently encoded into a product as its digital fingerprint. The encoded ISRC provides the means to automatically identify recordings for royalty payments, key to publishers and songwriters alike. Because these numbers are embedded into the product, this coding system is perfect for the electronic distribution of music, with the ease of adoption into the international music community that has been reliable and cost-effective.

Unlike the sophisticated algorithm of the UPC, the ISRC is an internal tracking code. The codes are simple with origin of country, label or artist identifier, year of creation, and finally song designation number.

A sample code could look like: US ASM 15 00001

To capture sales data, the company that has rights to the digital track gives tracking agencies the ISRC codes and should be imbedded during production. ISRC codes can also be provided by the digital retail portal such as CD Baby.

To learn more about UPC codes go to the GS1US website:

http://www.gs1us.org/resources/education-and-training/gs1-company-prefix-course

To learn more about ISRC codes and the application process, check out the International Standard Recording Code website:

https://www.usisrc.org/

BILLBOARD AND SOUNDSCAN

Prior to the invention of barcodes, an prior reporting system based on undocumented sales information was used by Billboard to produce the sales charts. Billboard had a panel of "reporting" retail stores that identified the bestselling record in their store, based on genre. Often times, this information was not supported through actual sales data, but was based on what store managers "thought" was their best seller. Record labels employed retail promotion teams to help influence these reports, hence the sales charts were not always valid depictions of true sales throughout the nation.

Remember, this reporting structure was what was used prior to the creation of barcode scanning systems and the use of point-of-sales data. So capturing accurate sales data was difficult, even for the retailer. The use of barcodes, or UPC (Universal Product Codes) have greatly improved product management. Not only do UPC codes assist in inventory status and reorder generation of hot selling items, the sales information captured allows the retailer to determine the best selling item by store and by chain, as confirmed through real sales data.

With the introduction of barcodes and efficient computer management of inventory, a new idea was introduced. Mike Shallet, an ex-record label promotion guy, along with Mike Fine, a statistician who had previously worked with major newspapers and magazines with a focus on surveys, conceived a revolutionary concept that would use this newfound technology to derive the top-selling records of the week. And in 1991, SoundScan was born.

SoundScan is an information system that tracks sales of music and music video products throughout the United States, Canada, and some international territories as well. Using UPC codes from point-of-sale cash registers, as well as ISRC codes from digital files, sales data is collected weekly from over 14,000 brick-and-mortar and online retail outlets. Weekly data is compiled and made available every Monday. Now owned by Nielsen, SoundScan is the sales source for the Billboard music album charts (www.soundscan.com).

Although 14,000 retail outlets sounds like a lot of stores, SoundScan does not capture all music retailers. Through analysis, SoundScan knows which retailers it does not capture and via statistical equations, is able to derive a representative total. Through nearly the end of 2014, the album that sold the most earned the #1 position for the week on the *Billboard* Top 200 chart.

But the concept of "#1" was understandably challenged since music is consumed in a multitude

---

Nielsen Entertainment

Nielsen BookScan provides weekly point-of-sale data with the highest degree of accuracy and integrity. Functioning as a central clearinghouse for book-industry data, Nielsen BookScan provides comprehensive reports from a wide variety of perspectives.

Nielsen Music Nielsen Broadcast Data Systems is the world's leading provider of airplay tracking for the entertainment industry. Employing a patented, digital-pattern-recognition technology, Nielsen BDS captures in excess of 100 million song detections annually on more than 1,400 radio stations in 130+ US markets (including Puerto Rico) and 30 Canadian markets. Nielsen SoundScan is an information system that tracks sales of music and music video products throughout the United States and Canada, and serves as the sales source for Billboard music charts. Sales data from point-of-sale cash registers is collected weekly from 14,000+ retail, mass merchant and non-traditional (on-line stores, venues, etc.) outlets.

Nielsen Games provides a full spectrum of products/services for game publishers, PC/Console manufacturers, game developers, ad agencies and advertisers to analyze the gaming habits and media consumption patterns of gamers.

Nielsen Sports provides engagement, consultation and business improvement work focused on the consumer and market intelligence about the Sports Fan to uncover the relative co-relationship between fans and media/sponsorships, while measuring marketing ROI and changes in fan behavior/brand perception.

Nielsen Home Entertainment seamlessly integrates point-of-sale (POS) data to provide a comprehensive view of the VHS and DVD sell-through business.

Nielsen Research Group (NRG) is a market research and strategic consulting company serving the entertainment industry.

FIGURE 16.4 *Nielsen Entertainment Research Products*

of ways including streaming through audio and video services and social media outlets. As of November 2014, *Billboard* changed its Top 200 chart to reflect this new consumership. Although a record that "sells" the most is a notable feat, the artist or the project that has the most "engagement," including not just sales but audio and video streaming should be considered the #1 act of the week—yes? (Project is defined as total consumption around that album physical and digital sales, TEA and SEA, and all On-Demand audio streaming activity off of that album) New challengers such as Border City Media's Buzz Angel are entering the market place utilizing additional technologies to assess these matrices. But SoundScan was the first to bring a legitimate assessment and analysis to the industry.

To register music with SoundScan, labels and artists must submit an online form to be included in sales data. http://titlereg.soundscan.com/soundscantitlereg/

A LOOK AT THE SOUNDSCAN DATA

The Charts

By compiling and organizing sales and activity data, SoundScan derives many charts that help describe the marketplace. Know that there are hundreds of charts, subdivided by specific headings. This is the actual data that drives the *Billboard* Charts.

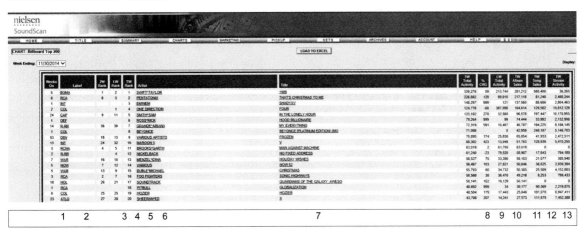

This is a sample of the Top 200 Album Chart. Each column is identified by number:

1 —Number of weeks the album has been on the Top 200 Chart
2 —Record Label
3 —Chart position 2 weeks ago
4 —Chart position last week
5 —Current chart position
6 —Artist Name
7 —Album Title
8 —This Week Total Activity
9 —% Change from Last Week to This Week
10 —Last Week Total Activity
11 —This Week's Album Sales
12 —This Week's Song's Sales (all songs from album)
13 —This Week's Stream Activity

FIGURE 16.5 *Example of a* Billboard *Top 200 (Source: Nielsen SoundScan)*

As noted in Chapter 7: The Industry Numbers, SoundScan underwent a substantial change of the Top 200 Chart by including not only album sales, but the additional data of TEA (track equivalent album) and SEA (streaming equivalent album) information.

In an attempt to more accurately compare previous years with the current sales trend, SoundScan came up with a unit of measurement called *track equivalent albums* (TEA), which means that 10 track downloads are counted as a single album. This evaluation is based on a financial equivalent, being a $.99 download x 10 would equal a $9.99 album download or CD. Thus, the total of all the downloaded singles is divided by ten and the resulting figure is added to album downloads and physical album units to give a project picture of "Total" activity reported in *Billboard*'s Top 200 chart.

Streaming songs generate licenses and royalties from the various sites and add revenue to the bottom line of copyright holders. The *streaming equivalent album* (SEA) was introduced in 2013 to measure streaming consumption. To evaluate streaming equivalents, the current industry standard is 1,500 streams of any songs from a particular album are counted as a single album. 1,500 streams x the standard royalty generated by this airplay $.005 = $7.50, being the wholesale price of an album. All of the major on-demand audio subscription services are considered, including Spotify, Beats Music, Google Play, and Xbox Music, Apple Music, Tidal, Amazon Prime, Rhapsody, Rdio and Others. Combine the streaming activity of all the singles from a particular album and divide by 1,500 and the resulting figure is added to the album downloads and physical album units to give a project picture of "total" activity known as the top 200.

Nielsen SoundScan added these two assessments of both TEA and SEA to traditional album sales creating a "total" activity evaluation known now as the Top 200. Looking at the data above, Beyoncé's new release debuted at #8 selling 42,959 units, even though the #9 position *Frozen* soundtrack sold more units of 65,054 units. When doing the "math," the single sales and online streaming of Beyoncé's record boosted her total activity to bump her release above Disney's *Frozen* soundtrack for the week.

Beyoncé *Beyoncé*:	42,959 albums
	24,618 TEA = (246,187 / 10 Songs)
	3,431 SEA = (5,146,703 / 1500 Streams)
	71,008 Total Activity
Disney's *Frozen*:	65,054 albums
	4,193 TEA = (41,933 / 10 Songs)
	1,648 SEA = (2,472,511 / 1500 Streams)
	70,895 Total Activity

Music Connect

Recognizing the value of the total picture including physical sales, down-loads, streaming, and social and web activity, Nielsen has created another look at all these metrics that allow the industry to evaluate the overall landscape. **Music Connect** blends a myriad of data points as collected by Nielsen, allowing the reader a dynamic look at an artist's performance across the various spectrums of the business—in one glance.

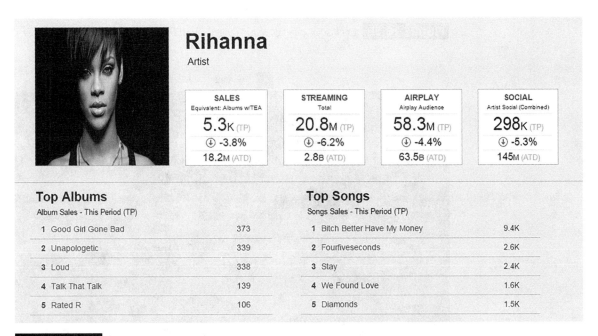

SALES Equivalent: Albums w/TEA	STREAMING Total	AIRPLAY Airplay Audience	SOCIAL Artist Social (Combined)
5.3K (TP)	**20.8M** (TP)	**58.3M** (TP)	**298K** (TP)
↓ -3.8%	↓ -6.2%	↓ -4.4%	↓ -5.3%
18.2M (ATD)	2.8B (ATD)	63.5B (ATD)	145M (ATD)

Top Albums
Album Sales - This Period (TP)

1	Good Girl Gone Bad	373
2	Unapologetic	339
3	Loud	338
4	Talk That Talk	139
5	Rated R	106

Top Songs
Songs Sales - This Period (TP)

1	Bitch Better Have My Money	9.4K
2	Fourfiveseconds	2.6K
3	Stay	2.4K
4	We Found Love	1.6K
5	Diamonds	1.5K

FIGURE 16.5A *Example of a Music Connect Artist Dashboard (Source: Nielsen)*

Artist Dashboard

Allowing the user to select a specific time period, an artist's dashboard report provides deep detail in all metrics including sales by Format: Albums with TEA/SEA and without, albums, singles, videos sales and the stream-ing activity of the tracks, as well as radio activity, streaming, and social metrics and Web activity from Facebook, Twitter, and Wikipedia. This report also shows year-to-date and activity-to-date cumulative data.

Any metric on the Artist Dashboard can be dynamically toggled on/off to be included on the integrated graph.

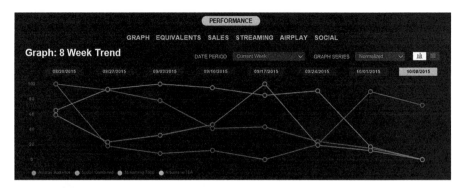

FIGURE 16.5B   *Example of graph on Music Connect Artist Dashboard (Source: Nielsen)*

Rihanna | Artist

GRAPH   EQUIVALENTS   SALES   STREAMING   AIRPLAY   SOCIAL

All Tables Contain Data for Period Ending: 10/08/2015

Equivalents

	TP	% Chg		LP	YTD	ATD
Album w/TEA w/SEA	19,199	-5.5%	↓	20,325	1,148,186	20,068,178
✓ Album w/TEA	5,324	-3.8%	↓	5,535	540,334	18,192,003

Sales

	TP	% Chg		LP	YTD	ATD
> Albums	1,357	+3.6%	↑	1,310	74,042	10,311,287
Digital Tracks	39,665	-6.1%	↓	42,253	4,662,918	78,807,155
> Singles	272	-9.6%	↓	301	2,778	81,116
> Videos	19	+113.9%	↑	9	765	91,791

Streaming

	TP	% Chg		LP	YTD	ATD
✓ ∨ Total	20,812,438	-6.2%	↓	22,184,209	911,778,688	2,814,262,997
> Provider						
∨ Play Type						
> Programmed	967,700	+82.3%	↑	530,867	27,931,070	352,475,585
> On-Demand	19,844,738	-8.4%	↓	21,653,342	883,847,618	2,461,787,412

Airplay

	TP	% Chg		LP	YTD	ATD
> Spins	10,012	-2.6%	↓	10,281	679,412	10,076,853
✓ > Audience	58,291,900	-4.4%	↓	60,962,400	3,523,216,800	63,548,332,800

Social

	TP	% Chg		LP	YTD	ATD
✓ ∨ Combined	298,428	-5.3%	↓	315,211	8,591,559	144,619,142
> Facebook Likes	-10,928	-551.4%	↓	2,421	-8,502,353	81,534,828
> Twitter Followers	215,743	-1.5%	↓	218,984	13,037,999	51,617,662
> Wikipedia Page Views	93,613	-0.2%	≡	93,806	4,055,913	11,466,652

FIGURE 16.5C   *Example of collapsed view of Music Connect Artist Dashboard (Source: Nielsen)*

Artist Chart Activity Export Save Options ▼

GROUPING: View: Table ▼

Filter Display | Show Title | Thresholds | Filter Data

Title	Artist	Chart Name	End Date	Position	+/- Chg.	Weeks On	Peak Position	Timeframe	Proc ▲
1989	Taylor Swift	Current Digital Albums	10/08/2015	7	▲ 7	49	1	Weekly	
1989	Taylor Swift	Current Physical Albums	10/08/2015	11	► 0	49	1	Weekly	
1989	Taylor Swift	Current Pop Albums	10/08/2015	1	► 0	49	1	Weekly	
1989	Taylor Swift	Digital Albums	10/08/2015	7	▲ 7	49	1	Weekly	
1989	Taylor Swift	Internet Albums	10/08/2015	26	▼ 5	49	1	Weekly	
1989	Taylor Swift	LP Vinyl Albums	10/08/2015	10	► 0	43	1	Weekly	
1989	Taylor Swift	Mass Merch & Non Trad Albums	10/08/2015	9	▲ 3	49	1	Weekly	
1989	Taylor Swift	Physical Albums	10/08/2015	11	► 0	49	1	Weekly	
1989	Taylor Swift	Pop Albums	10/08/2015	1	► 0	49	1	Weekly	
1989	Taylor Swift	Pop Albums Core Genre	10/08/2015	1	► 0	49	1	Weekly	
1989	Taylor Swift	RecordLabel UNI Albums	10/06/2015	7	▲ 1		1	Weekly	
1989	Taylor Swift	Regional Sales Albums - East North Central	10/05/2015	13	▼ 2	49	1	Weekly	
1989	Taylor Swift	Regional Sales Albums - Middle Atlantic	10/08/2015	13	▲ 1	49	1	Weekly	
1989	Taylor Swift	Regional Sales Albums - Mountain	10/08/2015	6	▲ 3	49	1	Weekly	
1989	Taylor Swift	Regional Sales Albums - Northeast	10/08/2015	8	▲ 3	49	1	Weekly	
1989	Taylor Swift	Regional Sales Albums - Pacific	10/08/2015	7	▲ 9	49	1	Weekly	
1989	Taylor Swift	Regional Sales Albums - South Atlantic	10/08/2015	11	▲ 4	49	1	Weekly	
1989	Taylor Swift	Regional Sales Albums - South Central	10/08/2015	11	▲ 1	49	1	Weekly	
1989	Taylor Swift	Regional Sales Albums - West North Central	10/08/2015	4	▲ 1	49	1	Weekly	
1989	Taylor Swift	RTD Albums	10/08/2015	143	▲ 2		143	ATD	
1989	Taylor Swift	RTD Digital Albums	10/08/2015	2	► 0		2	ATD	
1989	Taylor Swift	RTD Pop Albums	10/08/2015	28	▲ 1		28	ATD	
1989	Taylor Swift	RTD Pop Albums Core Genre	10/08/2015	21	▲ 1		21	ATD	
1989	Taylor Swift	Top Albums	10/08/2015	11	▲ 2	49	1	Weekly	
1989	Taylor Swift	Top Albums - Mkt Shr - Republic Records	10/08/2015	2	► 0		1	Weekly	
1989	Taylor Swift	Top Albums Retailers	10/08/2015	42	▼ 9	49	1	Weekly	
1989	Taylor Swift	Top Albums w/TEA	10/08/2015	6	▲ 1		1	Weekly	
22	Taylor Swift	2015 YTD Catalog Digital Tracks	10/08/2015	175	▼ 5		58	YTD	
22	Taylor Swift	2015 YTD Pop Songs Core Genre	10/08/2015	150	► 0	0	15	YTD	
22	Taylor Swift	2015 YTD Pop Songs Genre	10/08/2015	152	► 0	0	15	YTD	
22	Taylor Swift	Catalog Digital Tracks	10/08/2015	134	▲ 41	3	47	Weekly	
Back To December	Taylor Swift	RTD Country Songs Core Genre	10/08/2015	59	► 0	0	44	ATD	
Back To December	Taylor Swift	RTD Country Songs Genre	10/08/2015	61	► 0	0	46	ATD	
Bad Blood	Taylor Swift	2015 YTD Current Digital Tracks	10/08/2015	56	▲ 1		53	YTD	

FIGURE 16.5D *Example of a Music Connect Chart Activity Report (Source: Nielsen)*

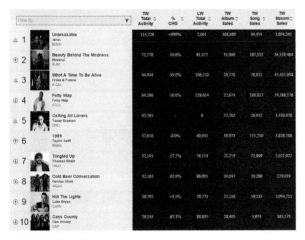

			TW Total Activity	% CHG	LW Total Activity	TW Album Sales	TW Song Sales	TW Stream Sales
1	Unbreakable	Janet / BRAU	115,728	+999%	2,061	108,689	44,434	3,894,241
2	Beauty Behind The Madness	Weeknd / R-RK	72,778	-10.6%	81,427	31,006	187,332	34,559,484
3	What A Time To Be Alive	Drake & Future / F-QM	64,944	-39.0%	106,320	26,778	78,832	45,425,098
4	Fetty Wap	Fetty Wap / ATLG	64,280	-50.0%	128,654	22,674	186,317	34,388,376
5	Calling All Lovers	Tamar Braxton / EPIC	42,761	–	0	37,782	26,932	3,430,478
6	1989	Taylor Swift / BGMA	37,650	-8.0%	40,941	19,973	151,250	3,828,766
7	Tangled Up	Thomas Rhett / VALO	32,195	-57.7%	76,154	21,259	75,899	5,021,922
8	Cold Beer Conversation	George Strait	32,161	-62.0%	86,095	30,047	19,280	279,019
9	Kill The Lights	Luke Bryan / CAPB	29,792	+3.5%	28,771	21,216	59,133	3,994,711
10	Cass County	Don Henley / CAP	29,245	-67.1%	88,835	28,405	5,974	365,573

FIGURE 16.5E *Example of a Music Connect Top 200 (Source: Nielsen)*

Chart Activity

The Nielsen Music Connect app provides a mobile access point to Music Connect, featuring the Billboard Top 200 and a myriad of other charts, Artist/Album/Song reports with sales (with and without TEA & SEA), streaming, radio airplay and social data with capability to create unlimited favorite lists and share any report, as well as industry-level stats.

Music Connect is an analytical tool utilizing the sales data from SoundScan and the radio and streaming activity from Broadcast Data Systems as well as other Nielsen data collection services which help marketers of the music industry create better strategies in reaching end consumers. Nielsen Music Connect will incorporate all of Nielsen's current capabilities in music measurement in one interface and replace their current Nielsen SoundScan and Nielsen Broadcast Data Systems tools.

But to continue on with classic SoundScan charts, here are several more samples of valuable data tools supplied by SoundScan.

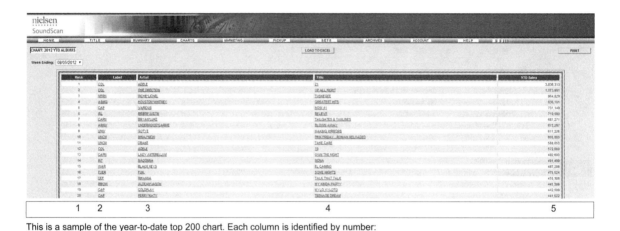

This is a sample of the year-to-date top 200 chart. Each column is identified by number:

1 —Rank in sales for the year 2012 3 —Artist 5 —Number of units sold in the year 2012
2 —Label 4 —Title of album

FIGURE 16.6 *Example of a Year-To-Date (YTD) Album Chart (Source: Nielsen SoundScan)*

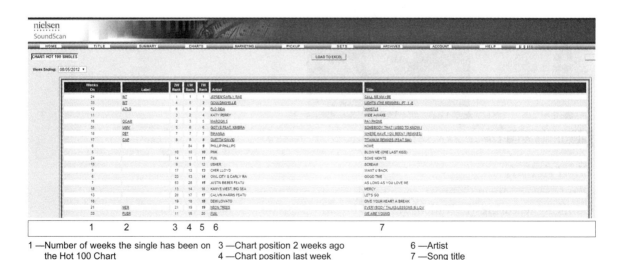

1 —Number of weeks the single has been on 3 —Chart position 2 weeks ago 6 —Artist
  the Hot 100 Chart 4 —Chart position last week 7 —Song title
2 —Record label 5 —Current chart position

FIGURE 16.7 *Example of Hot 100 Singles Chart (Source: Nielsen SoundScan)*

The Hot 100 Singles chart is created from a combination of single sales, radio airplay and video/audio streams, as calculated by Nielsen. Utilizing the sales of downloaded tracks combined with the sales of the available singles at retail, *Billboard* calculates the data with audience information captured by BDS that listens to all formats of radio—from top 40 and hip-hop to country, Latin, and rock—along with both video and audio streaming, the Hot 100 Singles chart is created.

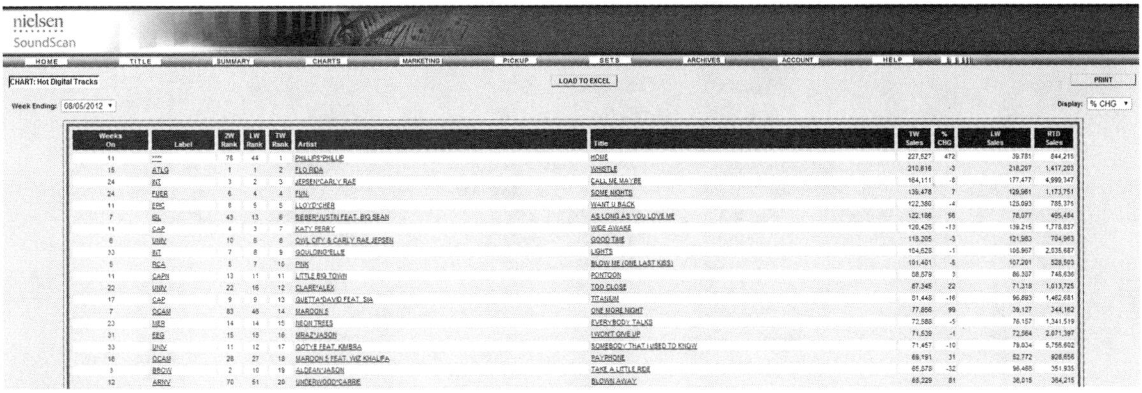

nielsen SoundScan

HOME | TITLE | SUMMARY | CHARTS | MARKETING | PICKUP | SETS | ARCHIVES | ACCOUNT | HELP

CHART: Hot Digital Tracks — LOAD TO EXCEL — PRINT

Week Ending: 08/05/2012 — Display: % CHG

Weeks On	Label	2W Rank	LW Rank	TW Rank	Artist	Title	TW Sales	% CHG	LW Sales	RTD Sales
11	[illegible]	78	44	1	PHILLIPS*PHILLIP	HOME	227,527	472	39,781	844,215
15	ATLG	1	1	2	FLO RIDA	WHISTLE	210,816	-3	218,207	1,417,203
24	INT	3	2	3	JEPSEN*CARLY RAE	CALL ME MAYBE	164,111	-8	177,477	4,999,347
24	FUER	6	4	4	FUN.	SOME NIGHTS	139,478	7	129,961	1,173,751
11	EPIC	8	5	5	LLOYD*CHER	WANT U BACK	122,380	-4	125,093	785,375
7	ISL	43	13	6	BIEBER*JUSTIN FEAT. BIG SEAN	AS LONG AS YOU LOVE ME	122,186	56	78,077	495,484
11	CAP	4	3	7	KATY PERRY	WIDE AWAKE	120,425	-13	139,215	1,778,837
8	UNIV	10	6	8	OWL CITY & CARLY RAE JEPSEN	GOOD TIME	118,205	-3	121,983	704,963
33	INT	7	8	9	GOULDING*ELLIE	LIGHTS	104,628	-1	105,967	2,035,687
5	RCA	5	7	10	PINK	BLOW ME (ONE LAST KISS)	101,401	-5	107,201	528,503
11	CAPN	13	11	11	LITTLE BIG TOWN	PONTOON	88,679	3	86,387	748,636
22	UNIV	22	16	12	CLARE*ALEX	TOO CLOSE	87,345	22	71,318	1,013,725
17	CAP	9	9	13	GUETTA*DAVID FEAT. SIA	TITANIUM	81,448	-16	96,893	1,462,681
7	OCAM	83	46	14	MAROON 5	ONE MORE NIGHT	77,856	99	39,127	344,182
23	MER	14	14	15	NEON TREES	EVERYBODY TALKS	72,588	-5	76,157	1,341,519
31	EEG	15	15	16	MRAZ*JASON	I WONT GIVE UP	71,539	-1	72,564	2,071,397
33	UNIV	11	12	17	GOTYE FEAT. KIMBRA	SOMEBODY THAT I USED TO KNOW	71,457	-10	79,034	5,756,602
16	OCAM	28	27	18	MAROON 5 FEAT. WIZ KHALIFA	PAYPHONE	69,191	31	52,772	928,656
3	BBOW	2	10	19	ALDEAN*JASON	TAKE A LITTLE RIDE	65,578	-32	96,468	351,935
12	ARNV	70	51	20	UNDERWOOD*CARRIE	BLOWN AWAY	65,229	81	36,015	364,215

FIGURE 16.8 *Example of Hot Digital Tracks Chart (Source: Nielsen)*

The Hot Digital Tracks report acts as a barometer as to the next big single or the next up-and-coming act that will break through onto the marketplace. With digital sales representing over half of all music sold, labels have retooled their business models by looking at the single (and this chart) as a money-making entity, rather than focusing on the album.

SOUNDSCAN MARKETING REPORTS

SoundScan creates various marketing charts that analyze the marketplace by segmenting sales into many categories.

When reading the YTD—Sales by Format Genre Album report, add (000) to the end of the units to accurately depict album sales by genre. Note that the CD format still dominates sales, but that digital is gaining share. Additionally, certain releases can be counted twice since they may be considered in more than one genre. Note: Core Genre reports are one and only one.

What is not calculated in this report are the genre percentages as they relate to the total. This information is in year-to-date summary reports generated by under Chart information, and can be isolated to DMA performance as well.

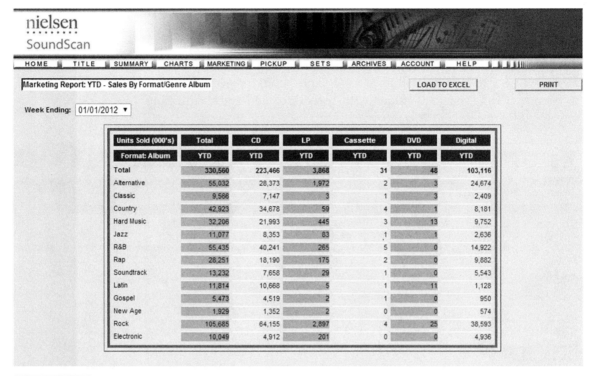

nielsen SoundScan						
HOME	TITLE	SUMMARY	CHARTS	MARKETING	PICKUP	SETS	ARCHIVES	ACCOUNT	HELP	
Marketing Report: YTD - Sales By Format/Genre Album				LOAD TO EXCEL		PRINT
Week Ending: 01/01/2012 ▼						

Units Sold (000's)	Total	CD	LP	Cassette	DVD	Digital
Format: Album	YTD	YTD	YTD	YTD	YTD	YTD
Total	330,560	223,466	3,868	31	48	103,116
Alternative	55,032	28,373	1,972	2	3	24,674
Classic	9,566	7,147	3	1	3	2,409
Country	42,923	34,678	59	4	1	8,181
Hard Music	32,206	21,993	445	3	13	9,752
Jazz	11,077	8,353	83	1	1	2,636
R&B	55,435	40,241	265	5	0	14,922
Rap	28,251	18,190	175	2	0	9,882
Soundtrack	13,232	7,658	29	1	0	5,543
Latin	11,814	10,668	5	1	11	1,128
Gospel	5,473	4,519	2	1	0	950
New Age	1,929	1,352	2	0	0	574
Rock	105,685	64,155	2,897	4	25	38,593
Electronic	10,049	4,912	201	0	0	4,936

FIGURE 16.9   *Example Marketing Report: YTD—Sales by Format/Album (Source: Nielsen)*

SoundScan Label Market Share Report

The Label Share Report shows the percentage of the market by distribution companies, as well as indies year ending on December 30, 2012. Each distributor, shown as Level 1, has "owned" labels that are part of the conglomerate, and "distributed" labels that have contracted the distributor to place their records into the marketplace, which are noted as Level 2. Label groups have sub-labels that are noted as Level 3. Each label is part of the cumulative percentage.

As an example, WEA 1) is considered the primary distributor with owned labels, 2) within the conglomerate family such as Warner Brothers and Atlantic. Through their A.D.A. Indie Distributor, Fueled By Ramen and Bad Boy Record Labels 3) are recognized, with their billing realized within the distribution family, but these labels are independents who have contracted their distribution function with WEA ADA's Independent Distributor.

From a competitive standpoint, this data allows distributors and their labels to evaluate their performance as it compares to others. This data is compiled weekly, monthly, and year-to-date.

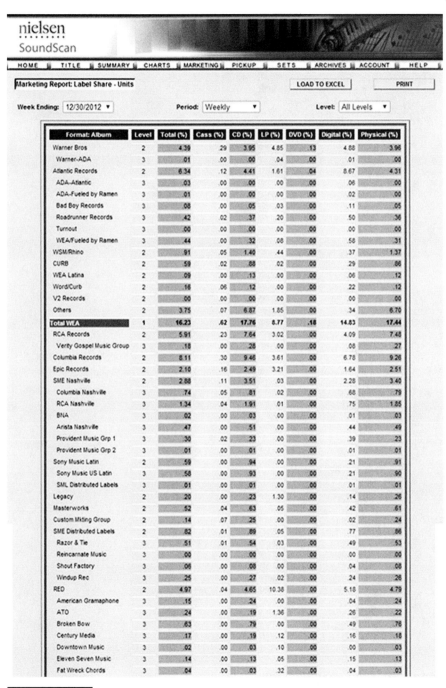

FIGURE 16.10   *Example Marketing Report: Label Shares—Units (Source: Nielsen)*

Fearless Records	3	.21	.00	.16	.06	.00	.27	.15
Glassnote	3	1.04	.00	.92	2.14	.00	1.15	.95
Lakeshore	3	.03	.00	.01	.00	.00	.06	.01
Metal Blade	3	.16	.02	.17	.22	.00	.14	.17
Mom+Pop Music	3	.07	.00	.05	.33	.00	.09	.06
MRI	3	.36	.00	.28	2.21	.00	.40	.33
New West	3	.04	.00	.05	.15	.00	.04	.05
Nuclear Blast	3	.07	.00	.12	.20	.00	.02	.12
Relapse	3	.06	.00	.06	.21	.00	.04	.07
Suburban Noize	3	.04	.00	.03	.01	.00	.05	.03
Thirty Tigers	3	.20	.00	.12	.41	.00	.28	.13
Victory Records	3	.18	.00	.11	.17	.00	.25	.11
RED Others	3	1.11	.03	.96	2.33	.00	1.25	1.00
THE ORCHARD	2	1.65	.00	.00	.06	.06	3.55	.00
Others	2	.09	.04	.07	.01	.00	.10	.07
Total SME	**1**	**27.98**	**1.01**	**30.76**	**21.72**	**.06**	**25.20**	**30.39**
Interscope/Geffen/A&M	2	8.77	.19	7.71	2.72	4.24	10.17	7.55
A&M/Octone	3	.72	.00	.68	.02	.00	.80	.66
Island Def Jam Music	2	5.15	.12	4.76	2.42	5.14	5.69	4.68
Def Jam/Def Soul	3	1.99	.06	1.82	.61	.00	2.24	1.78
Island	3	2.63	.06	2.50	1.25	.00	2.84	2.45
Lost Highway	3	.06	.00	.04	.45	.00	.06	.06
Mercury	3	.20	.00	.17	.11	5.14	.22	.18
Motown	3	.27	.00	.23	.00	.00	.32	.22
Republic Records	2	10.60	.11	8.98	3.62	.06	12.69	8.79
Republic Nashville	3	.41	.00	.49	.00	.00	.33	.47
Big Machine	3	3.33	.00	3.26	.35	.00	3.51	3.17
Valory	2	.21	.00	.24	.00	.00	.19	.23
UMG-Nashville	2	.94	.06	1.38	.11	.00	.48	1.34
MCA Nashville	3	.60	.04	.90	.00	.00	.29	.87
Mercury Nashville	3	.34	.02	.48	.10	.00	.19	.47
Show Dog Universal	2	.18	.00	.26	.00	.13	.11	.25
Universal South	3	.01	.00	.00	.00	.00	.01	.00
Show Dog Nashville	3	.18	.00	.25	.00	.13	.10	.24
S-Curve Records	2	.04	.00	.03	.00	.00	.06	.03
The Verve Group	2	1.00	.01	1.30	.45	.17	.68	1.27
Verve	3	.94	.01	1.26	.05	.17	.63	1.22
GRP	3	.05	.00	.04	.40	.00	.06	.05
Classics	2	.50	.01	.56	.05	.38	.46	.54
Univ Latin Entertainment	2	1.27	.01	2.06	.00	47.73	.35	2.08
Disa	3	.19	.00	.33	.00	.01	.05	.32
Fonovisa Records	3	.81	.01	1.33	.00	47.40	.16	1.37
Machete Music	3	.06	.00	.07	.00	.00	.05	.07
Universal Music Latino	3	.21	.00	.34	.00	.31	.09	.33
UMLE Others	3	.00	.00	.00	.00	.00	.00	.00
Hollywood Records	2	.69	.01	.62	.21	.00	.79	.61
Hollywood	3	.60	.01	.50	.21	.00	.72	.49
Mammoth	3	.00	.00	.00	.00	.00	.00	.00
Lyric Street	3	.09	.00	.12	.00	.00	.06	.12
Disney/Buena Vista	2	.90	.22	.97	.00	3.77	.84	.95
Concord Records	2	.85	.04	.94	1.33	.10	.74	.94
Peak Records	3	.01	.00	.01	.00	.00	.01	.01
Fantasy Records	3	.32	.02	.33	.97	.00	.28	.34
Telarc	3	.11	.02	.14	.03	.10	.08	.14
Hear Music	3	.15	.00	.15	.18	.00	.16	.15
Rounder Records	3	.15	.01	.18	.14	.00	.13	.17

FIGURE 16.10 *(Continued)*

Label								
CURB	2	.03	.00	.05	.00	.00	.01	.04
Curb/MCA Nashville	3	.02	.00	.04	.00	.00	.00	.04
Curb/MCA Los Angeles	3	.00	.00	.00	.00	.00	.00	.00
Curb/Universal	3	.00	.00	.00	.00	.00	.00	.00
Curb/Mercury Nashville	3	.00	.00	.00	.00	.00	.00	.00
Curb/Lost Highway	3	.00	.00	.01	.00	.00	.00	.00
Thump	2	.01	.01	.01	.00	.00	.00	.01
Universal Music Enterprises	2	.71	.20	.74	.46	.08	.69	.73
UME	3	.39	.00	.48	.08	.08	.31	.47
Now	3	.05	.00	.02	.00	.08	.08	.02
Sanctuary	3	.06	.01	.06	.19	.00	.06	.06
Special Products	3	.21	.19	.18	.19	.00	.24	.18
Varese	2	.07	.05	.06	.00	.00	.09	.05
Ark 21	2	.00	.00	.00	.00	.00	.00	.00
Bungalo Records	2	.00	.00	.00	.00	.00	.00	.00
VI Music	2	.00	.00	.00	.00	.00	.00	.00
VP Records	2	.04	.00	.05	.06	.66	.03	.06
Eagle Rock Ent.	2	.06	.00	.07	.03	.06	.05	.06
Savoy	2	.09	.00	.12	.01	.01	.05	.12
Blue Note Label Group	2	.53	.03	.50	.28	.00	.57	.50
Capitol Nashville	2	1.76	.10	2.45	.05	.00	1.05	2.37
Capitol Music Group	2	3.55	.27	3.28	10.10	.18	3.66	3.46
Capitol Records	3	2.98	.18	2.87	8.29	.07	2.95	3.01
Virgin	3	.38	.07	.31	1.06	.11	.44	.33
Astralwerks	3	.18	.02	.09	.74	.00	.22	.11
ELS & Caroline	2	.74	.06	.77	1.66	.00	.69	.79
Welk Distribution	3	.14	.01	.17	.09	.00	.11	.17
Hopeless Records	3	.11	.00	.09	.14	.00	.12	.09
Christian Music Group	2	.97	.09	.94	.04	.00	1.05	.91
EMI Latin	2	.11	.02	.15	.00	.00	.08	.14
Classics	2	.13	.00	.14	.00	.00	.12	.14
Others	2	2.54	.38	2.99	2.53	.81	2.06	2.98
Total(UMGD + EMM)	1	42.47	1.99	42.10	26.16	63.49	43.47	41.61
Madacy	2	.07	.02	.08	.00	.00	.05	.08
Tommy Boy	2	.00	.00	.01	.01	.00	.00	.01
Daptone Records	2	.01	.00	.01	.23	.00	.01	.02
Laserlight/Delta	2	.01	.04	.02	.00	.00	.01	.02
Bad Taste	2	.00	.00	.01	.00	.00	.00	.01
Roots Music	2	.00	.00	.00	.00	.00	.00	.00
TVT	2	.02	.01	.01	.02	.00	.02	.01
Rounder	2	.00	.01	.00	.00	.00	.00	.00
Rise Records	2	.23	.00	.13	.47	.00	.34	.13
Rykodisc	2	.02	.01	.01	.01	.00	.03	.01
Musart	2	.04	.00	.06	.00	.00	.01	.06
Balboa	2	.00	.00	.00	.00	.00	.01	.00
Starbucks	2	.03	.00	.07	.00	.00	.00	.06
Sub pop	2	.22	.00	.10	3.29	.00	.25	.19
Malaco/Muscle Shoals	2	.01	.01	.02	.00	.00	.01	.02
Alligator	2	.03	.00	.02	.03	.00	.04	.02
Saddle Creek Records	2	.02	.00	.01	.47	.00	.02	.02
Temporary Residence	2	.02	.00	.01	.29	.00	.02	.02
Beggars Group	2	.41	.00	.27	5.40	.00	.41	.41
Freddie Records	2	.04	.00	.08	.00	.00	.01	.08
Image Entertainment	2	.03	.00	.02	.00	.03	.06	.02
Oh boy	2	.01	.01	.01	.01	.00	.01	.01

FIGURE 16.10   *(Continued)*

Side One Dummy	2	.07	.00	.06	.34	.00	.06	.07
Kill Rock Stars	2	.02	.00	.01	.27	.00	.02	.02
Bloodshot	2	.02	.00	.01	.20	.00	.02	.02
Jagjaguwar Records	2	.08	.00	.04	1.29	.00	.09	.08
Merge	2	.17	.00	.12	1.88	.00	.18	.17
Surfdog	2	.01	.00	.02	.01	.00	.01	.02
CMH	2	.01	.01	.00	.00	.00	.01	.00
Ultra	2	.12	.00	.09	.02	.00	.15	.08
Epitaph Records	2	.20	.01	.15	.87	.01	.25	.17
Mailboat	2	.01	.00	.01	.00	.00	.01	.01
NAXOS	2	.08	.00	.10	.00	.03	.05	.10
Red House Records	2	.01	.00	.01	.00	.00	.02	.01
Sundazed Records	2	.03	.00	.01	1.46	.00	.00	.05
Collectables Records	2	.05	.00	.10	.00	.00	.00	.10
Domino	2	.07	.00	.04	.70	.00	.08	.05
Mountain Apple	2	.01	.00	.01	.00	.00	.01	.01
Relaxation Company	2	.01	.00	.02	.00	.00	.00	.02
Yep Roc	2	.02	.00	.02	.16	.00	.03	.02
Comedy Central Records	2	.12	.00	.04	.00	.00	.21	.04
Touch & Go	2	.01	.00	.00	.18	.00	.01	.01
Barsuk Records	2	.03	.00	.01	.28	.00	.03	.02
Warp Records	2	.09	.00	.06	.89	.00	.11	.08
Genius Products	2	.03	.00	.07	.00	.00	.00	.08
Secretly Canadian	2	.02	.00	.01	.26	.00	.02	.02
X5 Music	2	.12	.00	.02	.00	.00	.23	.02
E1 Entertainment	2	1.11	.03	1.20	2.37	.72	.99	1.23
E1 Music	3	.32	.01	.29	.19	.00	.36	.29
Shanachie	3	.05	.00	.06	.00	.00	.03	.06
Navarre	3	.13	.00	.19	.09	.00	.08	.18
Cleopatra	3	.03	.00	.04	.20	.00	.01	.04
E1 Other	3	.59	.02	.82	1.89	.66	.51	.65
Others	2	9.54	96.24	6.20	21.92	35.48	12.61	6.88
Total Others	1	13.32	96.39	9.37	43.34	36.27	16.50	10.56

nielsen
SoundScan

FIGURE 16.10 *(Continued)*

Marketing Report: YTD Percent Sales by DMA/Genre

The YTD Percentage of Sales by DMA/Genre report calculates the percentage of music sold in a specific DMA by genre. DMA stands for designated market area with the regions being derived by Nielsen, the company that calculates television ratings.

In the YTD Percent Sales chart shown below (Table 16.11), the New York DMA purchased 6.76% of all records sold. In other words, for every 100 records sold, nearly seven are purchased in the New York DMA.

nielsen SoundScan

HOME | TITLE | SUMMARY | CHARTS | MARKETING | PICKUP | SETS | ARCHIVES | ACCOUNT | HELP

Marketing Report: YTD - % Sales By DMA/Genre Album

LOAD TO EXCEL

PRINT

Week Ending: 12/30/2012

DMA Type: Full

Units Sold (000's)	Total	Alternative	Classical	Country	Hard Music	Jazz	R & B	Rap	Current	Top Seller	Soundtrack	Latin	Gospel	New Age	Catalog	Deep Catalog	Rock	Electronic
DMA	YTD	YTD	YTD	YTD	YTD	YTD	YTD	YTD	YTD	YTD	YTD	YTD	YTD	YTD	YTD	YTD	YTD	YTD
Total	100.00	100.00	100.00	100.00	100.00	100.00	100.00	100.00	100.00	100.00	100.00	100.00	100.00	100.00	100.00	100.00	100.00	100.00
New York, NY	6.78	7.86	9.60	2.85	6.21	10.02	7.15	6.47	6.86	6.29	6.90	4.54	6.39	6.11	6.65	6.79	7.33	9.23
Los Angeles, CA	5.24	6.17	3.58	2.67	4.64	6.35	5.46	5.51	5.31	5.01	5.55	8.75	3.52	5.09	5.17	5.23	5.49	7.33
Chicago, IL	3.06	3.55	3.88	2.02	2.67	4.10	3.15	2.86	3.18	3.11	3.14	3.79	3.69	2.78	2.92	2.92	3.21	3.82
Philadelphia, PA	2.78	2.93	3.17	1.86	2.58	3.65	3.49	2.98	2.84	2.91	2.87	1.46	4.18	2.67	2.73	2.78	2.83	2.46
SF-Oakland-San Jose	2.38	2.67	4.15	1.17	1.65	3.75	2.21	2.20	2.36	2.13	2.50	2.64	1.27	3.12	2.40	2.42	2.41	3.69
Boston, MA	2.42	2.81	3.57	1.82	2.43	3.22	1.93	2.21	2.37	2.39	2.54	.88	.69	2.78	2.47	2.51	2.78	2.69
Dallas-Ft. Worth, TX	2.29	2.07	1.81	2.49	1.92	1.78	2.57	2.71	2.29	2.27	2.41	4.72	2.85	1.97	2.29	2.28	2.01	2.18
Detroit, MI	1.39	1.37	1.63	1.14	1.41	1.72	1.69	1.49	1.45	1.52	1.47	.25	1.96	1.35	1.33	1.33	1.39	1.31
Washington, DC	2.55	2.46	3.77	1.79	1.93	3.71	3.49	2.90	2.60	2.57	2.80	1.41	4.54	2.78	2.51	2.52	2.31	2.70
Houston, TX	1.89	1.65	1.65	1.87	1.68	1.53	2.18	2.03	1.85	1.81	1.85	5.26	2.88	1.59	1.93	1.94	1.63	1.76
Cleveland, OH	1.21	1.08	1.11	1.25	1.25	1.07	1.56	1.56	1.27	1.36	1.17	.22	2.07	1.14	1.13	1.12	1.15	.93
Atlanta, GA	1.94	1.64	1.59	1.86	1.44	2.06	2.97	2.47	2.04	2.10	1.92	1.25	4.43	1.76	1.83	1.82	1.64	1.64
Minneapolis-St. Paul, MN	1.61	1.87	1.85	1.88	1.54	1.56	1.22	1.43	1.75	1.80	1.64	.53	.55	1.76	1.47	1.44	1.84	1.71
Tampa-St.Petersburg, FL	1.13	.94	1.13	1.21	1.06	1.07	1.12	1.04	1.11	1.17	1.13	1.30	1.23	1.45	1.14	1.14	1.03	1.04
Seattle-Tacoma, WA	1.89	2.32	2.40	1.46	1.91	2.29	1.46	1.67	1.92	1.85	1.94	.64	.81	2.37	1.88	1.84	2.17	2.58
Miami, FL	1.05	.79	1.50	.43	.76	1.42	1.31	1.27	1.02	.97	1.05	2.99	1.42	1.40	1.08	1.09	.89	1.72
Pittsburgh, PA	.93	.96	.91	1.09	1.08	.81	.82	.88	.98	1.04	.98	.07	.82	.81	.90	.89	1.00	.80
St. Louis, MO	1.04	1.13	.85	1.12	1.04	1.13	1.05	.90	1.03	1.05	1.05	.17	1.07	.91	1.05	1.06	1.10	.95
Denver, CO	1.55	1.89	1.68	1.48	1.59	1.44	1.09	1.31	1.55	1.52	1.60	1.44	.38	2.19	1.56	1.54	1.79	2.01
Phoenix, AZ	1.47	1.51	1.33	1.42	1.62	1.22	1.29	1.47	1.44	1.44	1.49	2.15	.49	1.89	1.49	1.49	1.50	1.53
Sacramento-Stockton, CA	1.09	1.01	1.02	.99	1.01	1.12	1.13	1.22	1.11	1.13	1.14	1.30	.70	1.30	1.06	1.06	1.03	1.09
Baltimore, MD	1.38	1.49	1.15	.82	1.22	2.05	2.12	1.72	1.38	1.28	1.23	.25	2.52	1.04	1.34	1.36	1.37	1.58
Hartford-New Haven, CT	.91	.90	1.09	.71	1.00	1.03	.82	.93	.91	.94	.95	.57	.76	1.01	.91	.91	.96	.87
San Diego, CA	.97	1.12	1.12	.58	.86	1.07	.88	.95	.97	.95	1.02	1.67	.38	1.28	.96	.97	1.01	1.34
Orlndo-Daytona Bch-Mlbrne	1.04	.90	1.01	.98	1.07	1.00	1.06	1.01	1.03	1.06	1.15	1.22	1.30	1.29	1.05	1.05	.97	1.09
Indianapolis, IN	.95	1.03	.75	1.22	.94	.79	.80	.84	1.02	1.07	.95	.37	.78	.81	.89	.87	1.02	.83
Portland, OR	1.34	1.61	2.52	1.03	1.19	2.08	.80	.86	1.31	1.11	1.16	.82	.28	1.90	1.37	1.36	1.54	1.52
Milwaukee, WI	.73	.79	.71	.68	.77	.70	.71	.70	.77	.80	.77	.48	.71	.84	.69	.68	.77	.66
Kansas City, KS-MO	.85	.84	.71	1.02	.85	.69	.86	1.08	.86	.90	.87	.39	.74	.80	.83	.81	.83	.71
Cincinnati, OH	.74	.80	.67	.92	.81	.63	.60	.62	.76	.81	.74	.16	.58	.64	.71	.70	.80	.58
Charlotte, NC	.96	.86	.60	1.07	1.00	.85	1.24	1.14	.97	.99	.87	.57	1.83	.75	.96	.96	.92	.76
Nashville, TN	.96	.92	.54	1.61	.90	.63	.81	.80	.99	1.05	.92	.31	.84	.65	.92	.92	.95	.72
Raleigh-Durham, NC	.92	.84	.82	.98	.76	.82	1.19	.99	.96	.97	.88	.83	2.38	.83	.88	.87	.83	.72
Columbus, OH	.77	.87	.61	.94	.92	.64	.66	.70	.81	.85	.77	.19	.63	.67	.73	.72	.86	.70
Greenvile-Sprtnbrg-Ashvll	.84	.55	.51	.94	.86	.47	.81	.81	.84	.86	.59	.30	.98	.62	.84	.84	.82	.46
New Orleans, LA	.51	.42	.38	.47	.45	.76	.80	.89	.49	.50	.50	.36	1.01	.34	.54	.55	.43	.46
Grnd Rpds-Klmzo-Btle Crk	.55	.57	.47	.66	.70	.42	.43	.49	.56	.61	.56	.29	.27	.58	.54	.53	.60	.47
Buffalo, NY	.48	.48	.46	.52	.57	.46	.45	.45	.50	.54	.49	.08	.58	.44	.45	.45	.51	.37
Memphis, TN	.51	.35	.26	.58	.40	.40	.66	.63	.50	.52	.48	.29	1.73	.30	.51	.51	.39	.28
Oklahoma City, OK	.72	.62	.41	1.02	.75	.44	.89	1.10	.67	.69	.70	.66	.50	.50	.77	.75	.64	.53
Salt Lake City, UT	1.07	1.28	2.01	1.16	1.27	.69	.56	.72	1.07	1.06	1.43	.87	.13	2.46	1.06	1.04	1.17	1.13
Nrflk-Prtsmth-Nwptnws-Hmp	.69	.62	.54	.62	.66	.65	1.10	.94	.71	.74	.71	.22	1.90	.61	.67	.67	.60	.58
San Antonio, TX	.81	.66	.55	.98	.80	.55	.71	.76	.77	.71	.78	3.23	.51	.68	.85	.86	.69	.82
Providence-New Bedford,RI	.51	.53	.57	.42	.54	.56	.48	.53	.50	.53	.53	.20	.16	.53	.51	.52	.55	.46
Harrsbrg-Yrk-Lncstr-Lbnon	.66	.68	.59	.80	.87	.50	.51	.57	.68	.71	.68	.28	.32	.70	.63	.62	.71	.56
Louisville, KY	.58	.53	.36	.90	.63	.37	.55	.58	.59	.64	.57	.19	.42	.44	.57	.56	.57	.38
Birmingham, AL	.55	.44	.32	.80	.50	.37	.66	.57	.55	.56	.53	.25	1.14	.35	.55	.55	.49	.32
Charleston et al, WV	.37	.31	.17	.75	.49	.22	.25	.29	.36	.40	.36	.02	.13	.19	.38	.37	.36	.22
Greensbro-WnstnSalm-HiPnt	.49	.39	.38	.67	.49	.37	.57	.50	.50	.53	.46	.45	1.03	.39	.48	.48	.43	.32
WstPlmBch-FrtPerc-VeroBch	.47	.37	.61	.39	.41	.58	.51	.50	.44	.45	.47	.67	.54	.69	.49	.49	.41	.49
Albuquerque, NM	.75	.69	.71	.81	.96	.57	.68	.93	.84	.81	.65	1.47	.15	.87	.88	.88	.75	.78
Dayton, OH	.41	.39	.34	.58	.52	.29	.35	.36	.42	.46	.41	.06	.37	.37	.41	.40	.43	.31
Albany-Schnctady-Troy, NY	.50	.49	.60	.54	.59	.55	.40	.42	.48	.52	.51	.10	.17	.56	.51	.52	.55	.42
Wilkes-Barre-Scranton, PA	.45	.48	.34	.57	.67	.31	.34	.38	.45	.49	.45	.10	.13	.37	.46	.46	.51	.35
Mobile-Pensacola, AL-FL	.43	.37	.27	.60	.44	.30	.49	.47	.42	.45	.42	.17	.79	.37	.44	.43	.39	.33
Jacksonville, FL	.51	.44	.37	.60	.47	.42	.63	.58	.51	.54	.52	.20	1.12	.48	.51	.50	.45	.43
Little Rock, AR	.48	.36	.24	.71	.49	.27	.50	.47	.41	.43	.42	.29	.82	.29	.50	.51	.40	.29
Tulsa, OK	.47	.41	.28	.78	.50	.28	.40	.48	.44	.46	.45	.24	.30	.36	.49	.49	.44	.35
Flint-Saginaw-BayCity, MI	.33	.29	.25	.46	.42	.25	.38	.32	.34	.38	.31	.05	.41	.30	.31	.31	.33	.23
Richmond, VA	.62	.45	.42	.58	.44	.90	1.19	.86	.61	.61	.53	.18	1.98	.44	.63	.64	.48	.38
Wichita-Hutchison, KS	.37	.32	.26	.60	.45	.22	.27	.32	.36	.37	.35	.60	.18	.40	.38	.38	.35	.26
Fresno-Visalia, CA	.38	.32	.27	.36	.35	.27	.39	.45	.40	.40	.38	1.35	.15	.33	.36	.36	.32	.34
Toledo, OH	.31	.29	.24	.45	.36	.23	.24	.23	.32	.35	.31	.13	.21	.31	.30	.29	.31	.21
Knoxville, TN	.44	.39	.27	.77	.49	.27	.30	.33	.44	.46	.42	.14	.26	.34	.43	.43	.44	.32
Shrvport-Txrcana,AR-LA-TX	.32	.21	.12	.51	.32	.17	.42	.35	.30	.29	.28	.38	1.19	.17	.34	.34	.24	.18
Des Moines, IA	.40	.42	.29	.55	.32	.26	.28	.38	.41	.44	.41	.24	.07	.41	.39	.38	.43	.34
Green Bay-Appleton, WI	.27	.29	.25	.36	.34	.22	.16	.21	.28	.31	.29	.20	.03	.37	.26	.26	.30	.24
Syracuse, NY	.34	.36	.36	.37	.41	.36	.28	.30	.33	.34	.33	.05	.14	.34	.35	.35	.36	.30
Roanoke-Lynchburg, VA	.36	.31	.28	.60	.40	.25	.33	.33	.36	.39	.34	.13	.43	.34	.36	.35	.35	.24
Lexington, KY	.39	.35	.22	.68	.43	.24	.28	.28	.38	.40	.36	.19	.20	.24	.38	.39	.39	.25
Austin, TX	.81	1.06	.75	.70	.67	.75	.56	.60	.79	.67	.76	1.46	.34	.77	.82	.82	.93	1.09
Rochester, NY	.34	.36	.38	.30	.35	.41	.30	.30	.34	.35	.35	.19	.29	.39	.34	.34	.37	.31
Omaha, NE	.37	.40	.27	.46	.43	.27	.30	.35	.38	.41	.39	.28	.16	.41	.35	.34	.39	.33
Portland-PolandSpring, ME	.40	.47	.42	.42	.50	.41	.26	.35	.38	.39	.37	.03	.04	.44	.41	.41	.49	.40
Champaign et al, IL	.30	.28	.27	.47	.34	.24	.21	.22	.30	.32	.31	.08	.17	.28	.29	.29	.30	.22
Pdch-CpGrdu-Hrsbg-Mrion,KY-IL	.28	.25	.13	.57	.38	.13	.22	.28	.28	.30	.27	.09	.20	.18	.31	.31	.29	.18
Spokane, WA	.40	.43	.33	.51	.55	.30	.25	.34	.37	.37	.35	.23	.06	.47	.44	.44	.48	.38
Davnprt-Rcklslnd-Moln,IL	.24	.22	.17	.38	.31	.16	.19	.23	.25	.28	.25	.12	.06	.23	.23	.23	.24	.20
Tucson, AZ	.38	.38	.45	.33	.42	.37	.33	.42	.37	.35	.37	.89	.11	.50	.38	.39	.38	.41
Hntsvlle-Decatr-Flornc,AL	.30	.25	.18	.48	.32	.18	.27	.28	.30	.33	.30	.18	.33	.23	.29	.28	.28	.21
CdarRpds-Wtrloo-Dubuqu,IA	.27	.28	.24	.40	.33	.20	.19	.25	.28	.31	.28	.10	.06	.32	.26	.26	.29	.23
Columbia, SC	.30	.22	.19	.32	.25	.23	.55	.45	.30	.32	.29	.16	1.18	.24	.30	.30	.23	.19
Springfield, MO	.30	.24	.17	.59	.33	.16	.16	.19	.28	.30	.28	.07	.09	.25	.32	.31	.28	.20
Chattanooga, TN	.28	.24	.16	.46	.32	.16	.21	.22	.27	.29	.27	.16	.22	.22	.28	.28	.28	.20
Southbend-Elkhart, IN	.23	.21	.18	.29	.27	.15	.18	.19	.23	.25	.23	.26	.16	.22	.22	.22	.22	.18
Jackson, MS	.24	.16	.12	.31	.19	.15	.39	.31	.23	.23	.22	.10	.83	.13	.25	.25	.18	.13
Tri-Cities, TN-VA	.23	.20	.13	.45	.31	.10	.13	.16	.23	.25	.22	.04	.06	.19	.23	.23	.24	.15
Johnstown-Altoona, PA	.21	.22	.18	.32	.32	.13	.12	.18	.22	.24	.21	.01	.03	.21	.21	.21	.24	.16
Youngstown, OH	.21	.20	.14	.29	.29	.16	.21	.22	.21	.23	.19	.02	.20	.15	.21	.21	.22	.14
Madison, WI	.32	.37	.40	.37	.36	.30	.21	.25	.34	.34	.33	.16	.07	.40	.31	.30	.36	.33
Las Vegas, NV	.56	.55	.49	.38	.60	.50	.67	.69	.56	.56	.57	.80	.48	.59	.56	.56	.55	.68
Brlngtn-Plattsbrgh, VT-NY	.33	.35	.43	.38	.35	.37	.19	.23	.30	.30	.31	.03	.05	.42	.36	.36	.38	.32
Evansville, IN	.23	.20	.12	.42	.31	.12	.18	.20	.23	.25	.22	.07	.08	.17	.23	.22	.23	.13
Baton Rouge, LA	.28	.22	.15	.29	.26	.22	.50	.43	.27	.27	.28	.18	.75	.18	.30	.30	.23	.21
Lincoln-Hastings-Kearney	.22	.21	.18	.34	.26	.14	.13	.18	.22	.23	.22	.19	.03	.28	.22	.21	.22	.17
Ft. Myers-Naples, FL	.29	.21	.37	.32	.25	.33	.24	.26	.27	.29	.26	.64	.13	.43	.30	.30	.24	.25
Waco-Temple-Bryan, TX	.33	.27	.20	.45	.34	.18	.38	.36	.31	.30	.32	.67	.50	.24	.35	.35	.27	.27
Springfield, MA	.23	.24	.29	.20	.26	.26	.21	.24	.23	.23	.22	.16	.12	.27	.24	.24	.25	.23
ColoradoSprngs-Pueblo, CO	.34	.37	.34	.39	.42	.26	.20	.35	.34	.34	.37	.19	.12	.46	.34	.34	.36	.38
Honolulu, HI	.45	.36	.50	.24	.37	.54	.47	.43	.40	.35	.52	.12	.24	.61	.50	.51	.37	.53
Puerto Rico & VI	.33	.12	.29	.03	.24	.24	.19	.32	.40	.23	.20	5.28	.14	.40	.27	.26	.15	.36
Others	13.78	12.59	10.10	18.76	15.68	9.42	12.08	13.20	13.20	13.33	12.89	18.66	12.82	12.70	14.38	14.33	13.38	11.45

nielsen SoundScan

FIGURE 16.11 *Sample Marketing Report: YTD—% Sales by DMA/Genre Album (Source: Nielsen)*

Additionally, the New York DMA purchased 9.6% of all classical records sold, making NY a ripe market for classical music. In this case, for every 100 records sold in the classical genre, nearly 10 records are purchased in the New York DMA and that is nearly 4% higher than the second best market, Los Angeles.

This report helps marketers to look at DMAs as a whole. Labels can compare the percentage of their artists' sales to that of the overall market, seeing if an artist's sales have over or under performed.

National Sales Summary Report

This report denotes sales by store type, geographic region, and population density. Note that Digital stores sold the most, with mass merchants being next. Chain stores include electronic superstores, as well as retail chains. Non-traditional retailers include Urban Outfitters and Hot Topic.

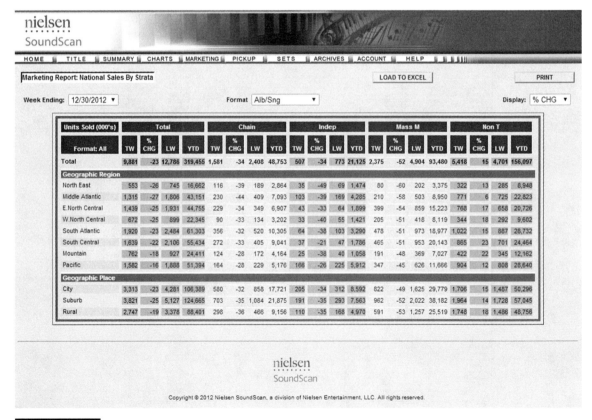

nielsen SoundScan

HOME | TITLE | SUMMARY | CHARTS | MARKETING | PICKUP | SETS | ARCHIVES | ACCOUNT | HELP

Marketing Report: National Sales By Strata   LOAD TO EXCEL   PRINT

Week Ending: 12/30/2012 ▼   Format Alb/Sng ▼   Display: % CHG ▼

Units Sold (000's)	Total				Chain				Indep				Mass M				Non T			
Format: All	TW	%CHG	LW	YTD	TW	%CHG	LW	YTD	TW	%CHG	LW	YTD	TW	%CHG	LW	YTD	TW	%CHG	LW	YTD
Total	9,881	-23	12,786	319,455	1,581	-34	2,408	48,753	507	-34	773	21,125	2,375	-52	4,904	93,480	5,418	15	4,701	156,097
Geographic Region																				
North East	553	-28	745	16,662	116	-39	189	2,864	35	-49	69	1,474	80	-60	202	3,375	322	13	285	8,948
Middle Atlantic	1,315	-27	1,806	43,151	230	-44	409	7,093	103	-39	169	4,285	210	-58	503	8,950	771	6	725	22,823
E.North Central	1,439	-25	1,931	44,755	229	-34	349	6,907	43	-33	64	1,899	399	-54	859	15,223	768	17	658	20,726
W.North Central	672	-25	899	22,345	90	-33	134	3,202	33	-40	55	1,421	205	-51	418	8,119	344	18	292	9,602
South Atlantic	1,920	-23	2,484	61,303	356	-32	520	10,305	64	-38	103	3,290	478	-51	973	18,977	1,022	15	887	28,732
South Central	1,639	-22	2,106	55,434	272	-33	405	9,041	37	-21	47	1,786	465	-51	953	20,143	865	23	701	24,464
Mountain	762	-18	927	24,411	124	-28	172	4,164	25	-38	40	1,058	191	-48	369	7,027	422	22	345	12,162
Pacific	1,582	-16	1,888	51,394	164	-28	229	5,176	166	-26	225	5,912	347	-45	626	11,666	904	12	808	28,640
Geographic Place																				
City	3,313	-23	4,281	106,389	580	-32	858	17,721	205	-34	312	8,592	822	-49	1,625	29,779	1,706	15	1,487	50,296
Suburb	3,821	-25	5,127	124,665	703	-35	1,084	21,875	191	-35	293	7,563	962	-52	2,022	38,182	1,964	14	1,728	57,045
Rural	2,747	-19	3,378	88,401	298	-36	466	9,156	110	-35	168	4,970	591	-53	1,257	25,519	1,748	18	1,486	48,756

nielsen SoundScan

FIGURE 16.12   *Example Summary Report: National Sales by Strata (Source: Nielsen)*

Although cities have higher population density per square mile, suburban areas out-purchase the city geographic marketplace, with the rural areas following close behind in purchasing power. The exception to this sales power is the independent stores, where there is less representation.

Border City Media's BuzzAngle Music

Border City Media, Inc. was formed in 2013, and is part of the "big data" movement, filling a void in the music industry by providing sophisticated tools needed to effectively analyze all the new platforms that people are using to consume music. Border City Media has developed a new, state-of-the-art service called BuzzAngle Music, which is a comprehensive tool that can analyze total music consumption, such as physical and digital album sales, song sales, streaming history, airplay history as well as social media metrics, all on a daily basis. Comparative statistics to the industry or selected artists, genres, retailers, geographies are created for benchmarking and highlight areas of opportunity.

What sets BuzzAngle Music apart is its ability to drill down on the data, answering very specific questions regarding a multitude of situations such as:

- "What are the best performing markets for on-demand audio streaming over the last six weeks?"
- "How did my label do in selling vinyl records released in the last three months at mom & pop stores in Des Moines, IA?"
- "How has the deluxe version of the album sold compared to the standard version each day after the launch of a Target advertisement featuring the deluxe version?"
- "What other songs on an album are receiving strong growth in sales and streams which could help the record label determine next single choice for radio airplay?"
- "As an artist, when should I conduct my social media promotion for a new album or song to maximize total consumption value?"

These types of questions can easily be answered utilizing BuzzAngle Music's filtering system—along with its multitude of reports that create easy-to-read graphics.

BuzzAngle Music's sources of data are comprehensive, with the majority of players from all supplier categories, including the large physical retailers, large digital retailers, large streaming services, radio airplay providers, independent music stores, artist's e-commerce websites, venues (concerts, etc.), as well as non-traditional outlets such as restaurants, clothing stores, drug stores, etc.

THE SYSTEM

BuzzAngle Music is a tool that allows the user to interact directly with the data. Rather than provide pre-processed charts and data, the system allows users to create the reports, charts, and dashboard views that are specific to their needs. Each report has options and filters that allow for modifying the data parameters along with the date of the data search. This results in an exhaustive number of possible charts and dashboards, literally in the trillions of possible combinations.

The menu system provides the ability to choose the type of report they want to see, whether a ranking of albums, songs, artists, singles, music

Border City Media BuzzAngle Music

Chart Offerings	Dashboard Summaries		
Album	**Album Chart Dashboard**		**Distribution Market Share**
Sales	Project	Genre Breakouts:	Albums & Singles Genre Breakouts:
Project	Sales	Rock	Project Rock
Songs	Digital	Pop	Sales Pop
Sales	Physical	Country	Digital Country
Streams	New Releases	Rap / Hip Hop	Physical Rap / Hip Hop
Airplay	Current	EDM	New Releases EDM
Project	Catalog		Current
Artist	**Song Chart Dashboard**		Catalog
Album Sales	Project	Genre Breakouts:	
Song Sales	Sales	Rock	
Streams	Digital	Pop	
Airplay	Physical	Country	
Project	New Releases	Rap / Hip Hop	
Single	Current	EDM	
Sales	Catalog		
Project	**Artist Chart Dashboard**		
Music Video	Album Project (All Matrices)		
Sales	Song Sales		
Project	Streams		
Distribution	Song Project (All Matrices)		
Album	Airplay		
Songs	**Industry Dashboard**		
Streams	Album History Sales by week		
Airplay	Digital / Physical Split by week		
Project	Genre Breakout by week		
	Song History by week:		
	Song Project history		
	Song Sales history		
	Song Streaming history		
	Song Spin history		
	Song Sales breakdown by Genre		

FIGURE 16.13   *Border City Media BuzzAngle Chart Offerings and Dashboard Summaries*

videos, or distributors. The sub-menu allows the user to select the criteria for the ranking, e.g., song-ranking report by streams.

Dashboards are available for artists, albums, songs, genres, geographical market areas, distributors, and industry-wide totals.

DEFINITION OF PROJECT UNITS

Border City Media has already been revolutionary in the industry by providing the first daily total consumption report using a blended, weighted-value metric they call Project Units, which more accurately measures the success of an entire musical project, including album sales, song sales, song streams, and song airplay. BuzzAngle Music uses a weighted-value formula that converts each individual disparate consumption unit into a common project unit both for albums and sales. The default weighted-value equivalents are:

1 Album Sale = 1 Album Project Unit
1 Song Sale = 1 Song Project Unit
10 Songs Project Units = 1 Album Project Unit
150 On-Demand Streams = 1 Song Project Unit
300 Programmed Streams (Internet Radio) = 1 Song Project Unit
300 Radio Spins = 1 Song Project Unit

The system also provides users the ability to change the weighted-value equivalents so that they can do their own scenario analysis.

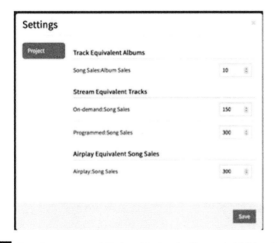

FIGURE 16.14 *BuzzAngle Settings Window Allowing for Change in Weighted-Value Analysis*

REPORTS

A sample look at the Album/Project chart reveals a comprehensive view of all matrices deriving a Top Project ranking for the week. You will see the total Project units that include album sales, song sales and song streams for each title. This graphic is unable to show all the columns that make up the total project unit number. but know this chart represents a total snapshot of an artist's album/project as it ranks against others.

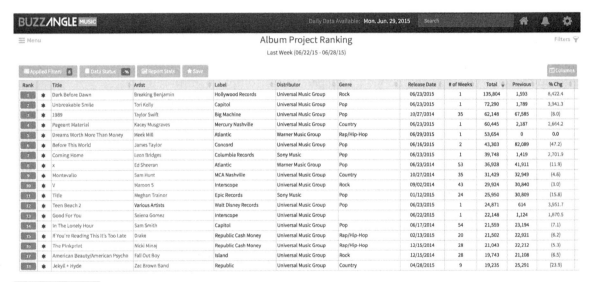

Rank		Title	Artist	Label	Distributor	Genre	Release Date	# of Weeks	Total	Previous	% Chg
1	★	Dark Before Dawn	Breaking Benjamin	Hollywood Records	Universal Music Group	Rock	06/23/2015	1	135,804	1,593	8,422.4
2	★	Unbreakable Smile	Tori Kelly	Capitol	Universal Music Group	Pop	06/23/2015	1	72,290	1,789	3,941.3
3	★	1989	Taylor Swift	Big Machine	Universal Music Group	Pop	10/27/2014	35	62,148	67,585	(8.0)
4	★	Pageant Material	Kacey Musgraves	Mercury Nashville	Universal Music Group	Country	06/23/2015	1	60,445	2,187	2,664.2
5	★	Dreams Worth More Than Money	Meek Mill	Atlantic	Warner Music Group	Rap/Hip-Hop	06/29/2015	1	53,654	0	0.0
6	★	Before This World	James Taylor	Concord	Universal Music Group	Pop	06/16/2015	2	43,303	82,089	(47.2)
7	★	Coming Home	Leon Bridges	Columbia Records	Sony Music	Pop	06/23/2015	1	39,748	1,419	2,701.9
8	★	x	Ed Sheeran	Atlantic	Warner Music Group	Pop	06/23/2014	53	36,928	41,911	(11.9)
9	★	Montevallo	Sam Hunt	MCA Nashville	Universal Music Group	Country	10/27/2014	35	31,429	32,949	(4.6)
10	★	V	Maroon 5	Interscope	Universal Music Group	Rock	09/02/2014	43	29,924	30,840	(3.0)
11	★	Title	Meghan Trainor	Epic Records	Sony Music	Pop	01/12/2015	24	25,950	30,809	(15.8)
12	★	Teen Beach 2	Various Artists	Walt Disney Records	Universal Music Group	Pop	06/23/2015	1	24,871	614	3,951.7
13	★	Good For You	Selena Gomez	Interscope	Universal Music Group		06/22/2015	1	22,148	1,124	1,870.5
14	★	In The Lonely Hour	Sam Smith	Capitol	Universal Music Group	Pop	06/17/2014	54	21,559	23,194	(7.1)
15	★	If You're Reading This It's Too Late	Drake	Republic Cash Money	Universal Music Group	Rap/Hip-Hop	02/13/2015	20	21,502	22,921	(6.2)
16	★	The Pinkprint	Nicki Minaj	Republic Cash Money	Universal Music Group	Rap/Hip-Hop	12/15/2014	28	21,043	22,212	(5.3)
17	★	American Beauty/American Psycho	Fall Out Boy	Island	Universal Music Group	Rock	12/15/2014	28	19,743	21,108	(6.5)
18	★	Jekyll + Hyde	Zac Brown Band	Republic	Universal Music Group	Country	04/28/2015	9	19,235	25,291	(23.9)

FIGURE 16.15 *BuzzAngle Album Project Ranking Chart*

FIGURE 16.16 *BuzzAngle Action Buttons for Multiview of Data*

Drill-down capability for each title is possible by clicking on an action-button menu that reveals multiple ways to view the data, such as album sales by UPC number, geography breakdown and a total performance snapshot for the songs on that album:

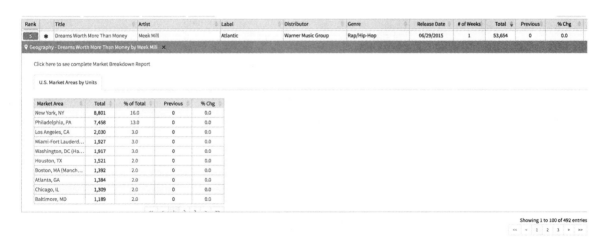

UPC/EAN	Title	Format	Configuration	Project Units	% of Total	Previous	% Chg
0602547380272	V (NEW MFIT / Deluxe)	Deluxe	Digital Album ...	1,940	26.4	1,759	10.3
0602547009739	V	Standard	CD	711	9.7	785	(9.4)
0602547382511	V [New Version][Edited]	Standard	CD	644	8.8	826	(22.0)
0602547009678	V (Jewelbox)	Deluxe	CD	614	8.4	822	(25.3)
0602547380296	V (NEW MFIT / Standard)	Standard	Digital Album ...	519	7.1	629	(17.5)
0602547382528	V [New Version][Deluxe Edition][Edited]	Deluxe	CD	513	7.0	600	(14.5)
0602547382504	V	Standard	CD	512	7.0	578	(11.4)
0602547009692	V (Jewelbox)	Standard	CD	447	6.1	594	(24.7)
0602547380265	V (NEW MFIT / Deluxe)	Deluxe	Digital Album ...	331	4.5	273	21.2
0602547382498	V [New Version][Deluxe Edition][Explicit]	Deluxe	CD	288	3.9	316	(8.9)
			Totals:	7,337		8,277	(11.4)

FIGURE 16.17   **BuzzAngle UPC/EAN Breakdown:** *V by Maroon 5*

Market Area	Total	% of Total	Previous	% Chg
New York, NY	8,801	16.0	0	0.0
Philadelphia, PA	7,458	13.0	0	0.0
Los Angeles, CA	2,030	3.0	0	0.0
Miami-Fort Lauderd...	1,927	3.0	0	0.0
Washington, DC (Ha...	1,917	3.0	0	0.0
Houston, TX	1,521	2.0	0	0.0
Boston, MA (Manch...	1,392	2.0	0	0.0
Atlanta, GA	1,384	2.0	0	0.0
Chicago, IL	1,309	2.0	0	0.0
Baltimore, MD	1,189	2.0	0	0.0

FIGURE 16.18   **BuzzAngle Geography Report:** *Dreams Worth More Than Money by Meek Mill*

REPORT FILTERS

On each BuzzAngle Music report, there is an extensive selection of filters available to further refine the data set:

FIGURE 16.19 *BuzzAngle Album Project Ranking*

Table 16.1	BuzzAngle Dashboard Setting
View Settings	# of total items to be ranked, rank criteria (units vs. growth over last period)
Timespan	Presets for Last Day, Week-to-Date, Last Week, etc. or a date range
Consumption Types	Filters for sub-elements of a consumption type; for example, album filters for CD or digital only, streaming filters for just audio or video streams, airplay filters for specific radio station formats or day parts.
Advanced Filters	Selecting releases by age of release or specific genres, artists, albums, songs, labels or distributors
Location	Selecting specific geographies or outlet types within those geographies (large physical retailers, large digital retailers, bookstores, venues, etc.)

DASHBOARDS

Dashboards are available for artists, albums, songs, genres, geographical market areas, distributors and industry-wide totals. On each dashboard, there are individual elements (widgets) that provide a specific view of the data pertinent to that artist, album or song. There are widgets for overall trends, song breakdowns, market area breakdowns, etc. Each widget has filters, which can order up a finer look at the data. The dashboard below shows an inclusive look at all of Maroon 5's releases with the sales, radio, and streaming action of all releases for the band.

One click on Maroon 5's *V* album hotlink, and the next dashboard reveals a deeper dive on just the *V* album's sales, radio, and streaming.

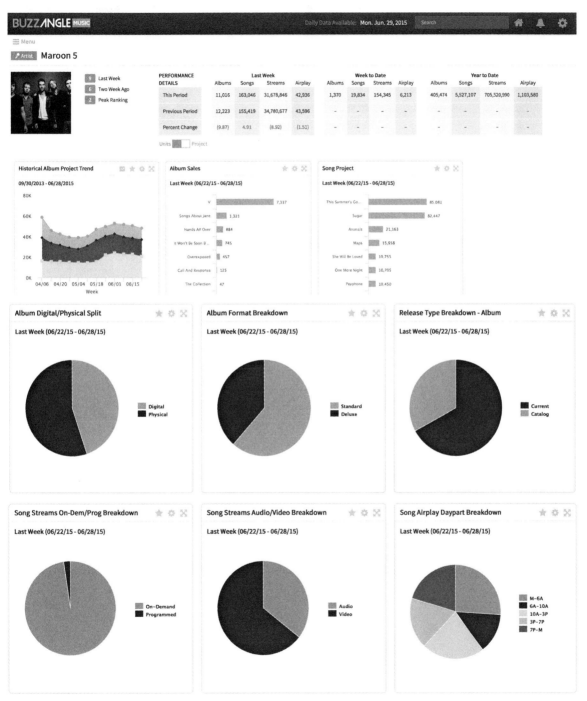

FIGURE 16.21 *BuzzAngle Artist/Album Dashboard: Maroon 5 V*

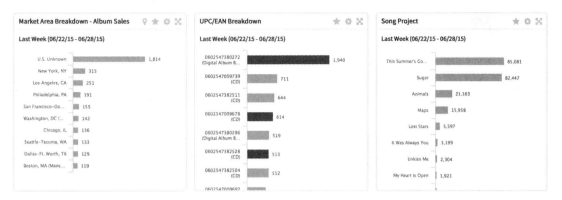

FIGURE 16.21   (Continued)

One more click on "Animals" gives the viewer a look at that song's overall performance.

Clicking on the market breakdown link will show a very detailed report for that song's individual consumption values across 200 market areas, including both a rank and an index number to compare market-by-market performance.

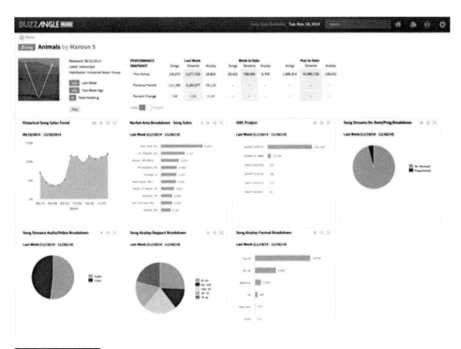

FIGURE 16.22   BuzzAngle Artist/Album/Single Dashboard: Maroon 5 V Animal

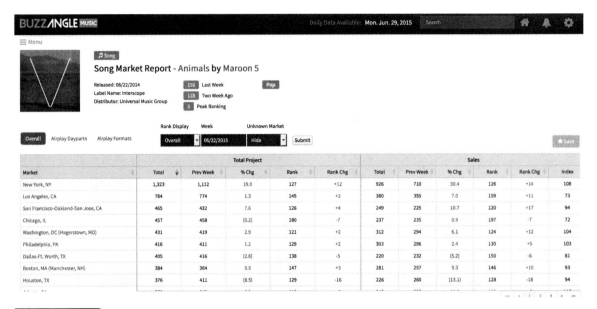

FIGURE 16.23 *BuzzAngle Song Market Report: Maroon 5 "Animal"*

The following dashboard shows the distribution market share break-down among the three major distributors for total albums sales during the previous week. Similar views are available for different consumption types, such as song sales, song streams and song airplay, as well as for varying time-frames such as year-to-date measurements.

FIGURE 16.24 *BuzzAngle Distribution Company Market Share Report*

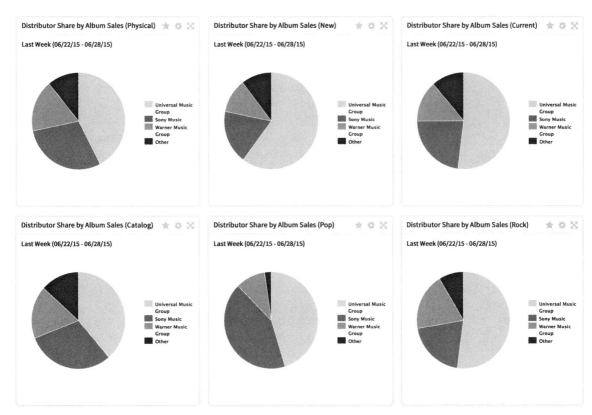

FIGURE 16.24 *(Continued)*

A DEEPER LOOK AT ALL DATA

Clearly, syndicated research summaries such as SoundScan and BuzzAngle are capable of revealing overall sales of music, a specific genre of music, a specific artist, a specific market, and many more aspects of the business as it pertains to sales of music. But by manipulating sales data beyond the scope of their pre-determined charts, marketers can better understand the marketplace and its drivers.

Seasonality and Record Sales

Like most products, there is seasonality to the sales of music. Every year, sales trends show a similar pattern with sales spikes at Valentines (second only to the end of year holidays), Easter, a lull through summer months, and then a steady rise through the fall going into the holiday selling season.

Using the weekly sales charts of combined Album and Single sales, this overlay of years of weekly sales crystallizes seasonal sales trends. Note that every year, the volume of sales slips a little bit lower from year to year.

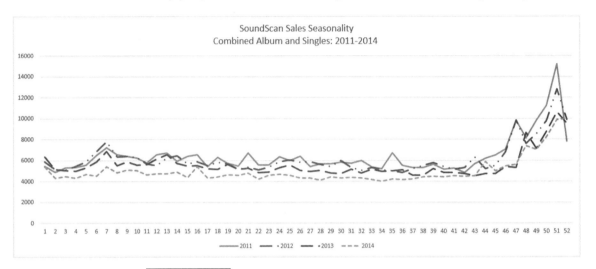

FIGURE 16.25 *SoundScan Weekly Album and Single Sales*

As students of the business, record company executives analyze when the best time of year would be to release new artist verses that of a platinum act. Strictly based on seasonality alone, most would say to release a new act shortly after the New Year to take advantage of the spring sales spikes and summer touring. And if the act has radio success or is a seasoned veteran, the fall selling season would be healthy window for a release to take advantage of the upcoming shopping season. As for the superstar act, a fourth quarter release would be perfect timing to capitalize on holiday shoppers while minimizing long-term advertising dollars.

LIFECYCLES

As discussed earlier in this text, there are four stages within the traditional product lifecycle: Introduction, Growth, Maturity, and Decline. As a product is adopted into use, and as others learn of its availability, its sales will grow. Eventually, the product will hit maturity, level off in sales, and decline. Either its maker will "reinvent" the product as "new and improved" and evolve it in some manner, or it will no longer exist.

CLASSIC PRODUCT LIFECYCLES

Product lifecycles occur in music too. Although there is an occasional exception to the rule such as a second or third single from a release being the song that drives sales, most album releases have a similar sales pattern. Once an artist is established, sales patterns rarely vary, which is why the first month of a release is so critical to the success of a record. Historically, the sales success of a release has depended on how many units were shipped initially into retail, but with the burgeoning digital market, there is a never-ending supply of a particular title-on-hand-ready for delivery. Yet the number of units initially sold is usually dependent on the pent up demand felt from the marketplace. Brick-and-mortar retail music buyers also look at the track record of the artist and their previous sales as well as considering that as consumers shift from physical to digital consumption, the current trend is approximately 55% CD album, 4% vinyl, and the remaining 41% being a digital format purchase. But physical albums are only 32% of ALL music consumed with over 68% of the remaining music sales being digital sales. This data is based on single sales converted to album equivalents, and not on transaction count.

Looking at these examples, most of the releases show a similar pattern in sales even though they are different artists, genres, sales plateaus, and time of year. Yet the similarity in sales trend is unmistakable. An established artists' sales trend shows an undeniable peak in sales early in the life cycle that tapers off within the first six weeks and three months. This is why most record labels pack their marketing strategies into this small window of time.

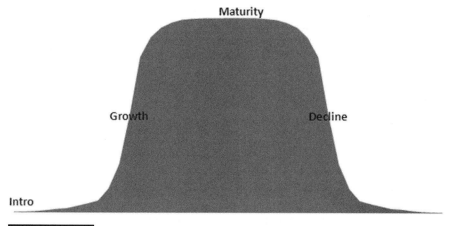

FIGURE 16.26   *Classic product life cycle*

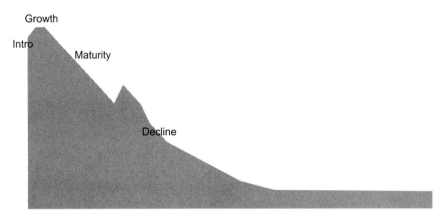

FIGURE 16.27 *Product Lifecycle for New Releases*

The exceptions to the rule are the new artists that are "breaking." Check out the first record of these now superstar acts. Most of their initial releases had a "slow boil" effect, meaning that sales did not catch on at street date but later, as consumers became more aware of the artist.

Example: Paramore started as a garage band in Franklin, TN after Hayley Williams moved into the neighborhood and met brothers Josh and Zac Farro. A big singer at the age of 13, she and the band caught industry eyes, but Williams was the only one signed to a major deal. Not getting along with the administration, Hayley was able to reformulate her agreement, and the new arrangement regrouped the neighborhood gang along with a new imprint label known as Fueled by Ramen. FBR was known for the grassroots approach to the marketplace, as well as their innovative 360 deals that enables start-up artists to hit the road with all the tools needed including booking and merchandise. Looking at their sales, the first album barely sold 2,000 units its initial week. It took two Warp Tours and the social media mojo of Fueled By Ramen to create the explosive sales that eventually happened for their second album *RIOT*, which also kicked the sales of their debut release "All We Know Is Falling" nearly two years after its street date.

Released in 2001, Room for Squares was John Mayer's "official" first album release. The hit single "No Such Thing" peaked at radio, depending on the format, in mid 2002, with the bigger hit "Your Body Is A Wonderland" driving sales even higher. Note the "slow boil" affect with the first release.

Any Given Thursday is the live album release that was to bridge sales demand while Mayer recorded the next album. Although a classic,

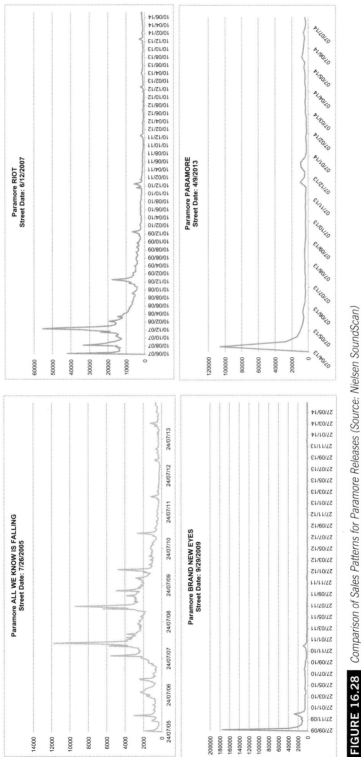

FIGURE 16.28 Comparison of Sales Patterns for Paramore Releases (Source: Nielsen SoundScan)

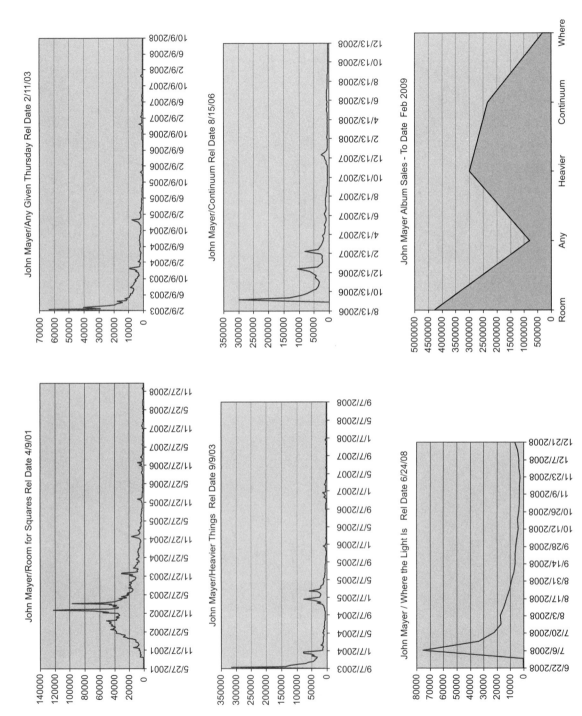

"established" artist sales profile occurs, the actual volume was much lower than that of Mayer's initial release. *Heavier Things* continued the sales trend of an established act, though showing a quick die-off in sales, mostly because of the lack of a big single at radio. *Continuum* sold well with over 2.3 million units in the market, but Mayer's summer 2008 release has not performed as well, with only 300,000+ units sold to date. To plot his cumulative solo album sales, there is a trend that vaguely shadows that of a classic product lifecycle. This is not uncommon for most artists to live a quote "product" lifecycle—it's just imperative that artists plan for such an occurrence.

DMAS AND MARKET EFFICIENCIES

The Designated Market Area (DMA) is A.C. Nielsen's geographic market design, which defines each television market. DMAs are composed of counties (and possibly also split counties) and are updated annually by the A.C. Nielsen Company based on historical television viewing patterns. Every county or split county in the United States is assigned exclusively to one DMA.

Radio audience estimates for DMAs are published in the Radio Market Reports of all Standard radio markets whose Metros are located within the DMA *and* whose names are contained in the DMA name. For example, radio audience estimates for the San Francisco-Oakland-San Jose DMA are reported in both the San Francisco and the San Jose Radio Market Reports; however, radio audience estimates for the New York DMA are reported in the New York Report, but not in the Nassau-Suffolk Report. (Katz Media Group Radio Resource Area)

The following data reflects the percentage of music sold in each of the DMAs. SoundScan lists the DMAs in order of population, with New York being the most populated area surveyed, Los Angeles being the 2nd largest populated area surveyed, and so on, with some variances based on radio or television markets. When looking at the markets, record labels should consider efficiency of the advertising dollars and marketing efforts.

Each genre of music contains a unique profile, for example: The percentage of music sold in the top 7 DMAs for Jazz equals 31.06%. Think about it—for every 100 Jazz records sold, 31 of them are sold in these top 7 DMA markets. As a marketing department, advertising in these DMAs should have a big bang for the buck, considering the efficiency of targeting to the buyers in these markets. In contrast, to sell 30 records out of 100 in the Country genre, the top 19 DMA markets are needed to achieve that percentage. Thus, to reach buyers of country music, country record labels need to spread their marketing dollars and efforts thinner, and *smarter*, to effectively reach the same percentage of buyers of the genre.

| Table 16.2 | Year-to-Date Percentage of Sales by DMA/Genre 2013: Jazz v. Country |

YTD % of Sales by DMA/Genre 2013

Mkt Rank	DMA	Jazz %	Country %
1	New York	9.44	2.8
2	Los Angeles	6.26	2.52
3	Chicago	3.42	1.96
4	Philadelphia	3.61	1.8
5	San Francisco	3.53	1.12
6	Boston	2.95	1.69
7	Dallas - Ft. Worth	1.85	2.38
		31.06	
8	Detroit	1.64	1.11
9	Washington, DC	3.25	1.77
10	Houston - Galveston	1.65	1.88
11	Cleveland	1.17	1.26
12	Atlanta	2.16	1.95
13	Minneapolis-St. Paul	1.54	1.62
14	Tampa-St. Petersburg-Clearwater	1.17	1.27
15	Seattle-Tacoma	2.29	1.43
16	Miami-Ft. Lauderdale-Hollywood	1.49	0.41
17	Pittsburg	0.88	1.12
18	St. Louis	1.1	1.1
19	Denver-Boulder	1.48	1.51
			30.7
20	Phoenix	1.29	1.4

Best Selling Markets vs. Strongest Markets

It sounds confusing—why wouldn't the best selling markets NOT be the *strongest* markets when looking at genre sales or title reports of a specific release? Let's look at how record companies can manipulate Sound-Scan data to be *smarter* marketers.

The data in table 7.4 is a new sort of the same numbers in table X.X-1. By ranking the DMAs by percentage of sales, marketing experts can now view the best selling markets in order of per capita sales. (For this book's example, only the Top 20 markets are being analyzed.) But are these the *best* markets for Jazz sales? To determine the strongest markets, using population data in the equation helps to determine where to place marketing efforts.

Here is another look at the same data. By adding 12+ population data and doing a simple ratio, the best markets emerge. The 12+ data comes from Nielsen DMA radio ratings information. The equation: sales percentage / 12+ population

Where New York and Los Angeles were ranked #1 and #2 based on population size, the strongest market for Jazz based on percentage of sales to the population of the DMA would be Philadelphia, with Boston close behind. Although Philadelphia does not sell as much Jazz as New York, the propensity of the population to buy Jazz in the Philly marketplace is over 35% greater than New York, making it a better or stronger market for Jazz music.

Table 16.3	Year-to-Date Percentage of Sale by DMA/Genre 2013 Ranked by Sale Percentage/Population

YTD % of Sales by DMA/Genre 2013

Mkt Rank	DMA	12+ Population	Jazz %	Ratio Sales %/Pop
4	Philadelphia	4558200	3.61	7.92E-07
6	Boston	4192800	2.95	7.04E-07
9	Washington, DC	4793400	3.25	6.78E-07
11	Cleveland	1774000	1.17	6.60E-07
15	Seattle-Tacoma	3638000	2.29	6.29E-07
1	New York	16157500	9.44	5.84E-07
19	Denver-Boulder	2546800	1.48	5.81E-07
2	Los Angeles	11271300	6.26	5.55E-07
5	San Francisco	6463500	3.53	5.46E-07
13	Minneapolis-St. Paul	2875600	1.54	5.36E-07
12	Atlanta	4549700	2.16	4.75E-07
18	St. Louis	2328700	1.1	4. 72E-07
14	Tampa-St. Petersburg-C	2531900	1.17	4. 62E-07
17	Pittsburg	2009200	0.88	4.38E-07
8	Detroit	3803600	1.64	4.31E-07
3	Chicago	7939500	3.42	4.31E-07
16	Miami-Ft. Lauderdale-I	3906200	1.49	3.81E-07
20	Phoenix	3419800	1.29	3.77E-07
7	Dallas -Ft. Worth	5633600	1.85	3.28E-07
10	Houston -Galveston	5362100	1.65	3.08E-07

This same type of analysis can be applied to artist DMA sales data. By looking at the sales of an artist's specific record, labels can determine where to place marketing and dollars. See chapter appendix for more analysis on market information.

OTHER TOOLS WITHIN THE INDUSTRY

Other "big data" companies have emerged as partners with labels and artists in effort to help point the way to sales and other marketing opportunities to maximize revenue potential. That potential could include corporate partnerships, touring alliances, advertising exposure, and merchandising outlets. But all digital companies are trying to promote, engage, and measure a specific target market that will aid their client: the artist and the music.

Next Big Sound

As an extension to all the drivers that move listeners to consume, Billboard has partnered with Next Big Sound (NBS) and has created the "go-to" destination among indies and majors alike to measure social interaction along with sales and touring data, creating a "big picture" for artists and their teams to act on. Online, selected artist dashboard information can be customized to not only watch, but target specific goals. Impressions and sales can be tracked with benchmarks targeted. Similar artists can be viewed in comparison to selected artist and tour marketing and support can be viewed within context of this same data. Relevant social media can be prioritized as part of a strategic plan to motivate the market, all within a click's view.

Via NBS's matrixes, *Billboard* publishes two charts that reflect social behavior of various platforms. The Social 50 is a weekly chart that ranks the most popular artists who are online on all major music sites. The Next Big Sound chart identifies the Top 15 fastest accelerating artists across the Internet that will most likely become the Next Big Sound. Using various sources, from Facebook page "likes" to actual sales, this combined data creates context to help the modern music industry make decisions.

As part of an "in touch" component, NBS partners with artists and their managers to allow free access to listening data via Spotify. Information included shows how much time Spotify clients are listening to a specific artist, the most popular tracks, and listener demographics including age, gender and location.

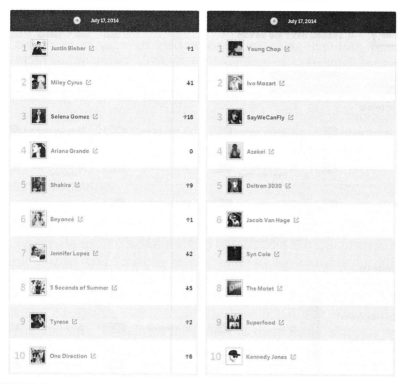

 FIGURES 16.30, 16.31, 16.32 *Next Big Sound Charts, Dashboard, and Key Metrics*

www.nextbigsound.com

http://www.spotifyartists.com/spotify-next-big-sound-artist-analytics/

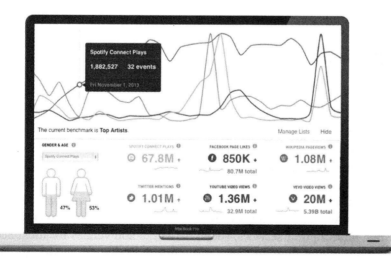

FIGURES 16.33

Bandcamp reports to SoundScan. The pricing of product is set by the artist and "free" is allowed. Playback widgets have been created so that easy imbedding can occur on social media sites such as Facebook. Capturing of emails is encouraged so that building a fan list is easy. Watching where music widgets are being imbedded and other "real time" stats is easily accessed. Bandcamp touts accurate metadata placement so that fan can find all artist info when seeking.

ReverbNation

ReverbNation works as a portal for independent musicians and labels to create traction in the digital marketplace. Co-founded by ex-record exec Lou Plaia of Atlantic and Lava Records and marketing software developer Mike Doernberg and Robert Hubbard, along with an MBA tech-head Duke grad Jed Carlson, ReverbNation offers web promotion, fan-relationship management, digital distribution, sentiment tracking, web-site hosting, and concert booking and promotion for the content creator while calculating and publishing a rolling "popularity" chart by zip code for consumers that is non-genre specific. As of August 2014, over 3.5 million musicians have created ReverbNation accounts, and the website receives over 30 million unique visitors per month.

Although ReverbNation creates its own traction, the site has learned to promote outside by creating links back to its site via widgets. Each artist can use ReverbNation as its backdoor "playback" streaming site, loading the newest music and updating the playlist as new music becomes available—all behind the scene. This "Promote It" widget allows for a seamless Facebook interface to the music, as well as other messaging through its "TuneWidget" link. The "Music For Good" allows for the sale of music via download with one half of the proceeds to be dedicated to one of 13 charities, including Fender Music Foundation, Oxfam America, World Vision, Heifer International, Zac Brown's Camp Southern Ground and CARE. When a song is purchased for $1.29, 56 cents goes to the artist, 56 cents is sent to a charity the artist selects, 12 cents goes to PayPal to process the payment, and five cents goes to ReverbNation. Other features include a "Band Equity" measurement that looks at four metrics: reach, influence, access and recency with the top 100 of each genre being recognized. Musicians can also create the own mobile app for Apple and Android devices that are customizable to include music, bio, photos, tour dates and social links. Artists can also apply for touring and festival opportunities with music being constantly evaluated for licensing opportunities for synchronization usage including ESPN, MTV, Fox, Universal, and Electronic Arts.

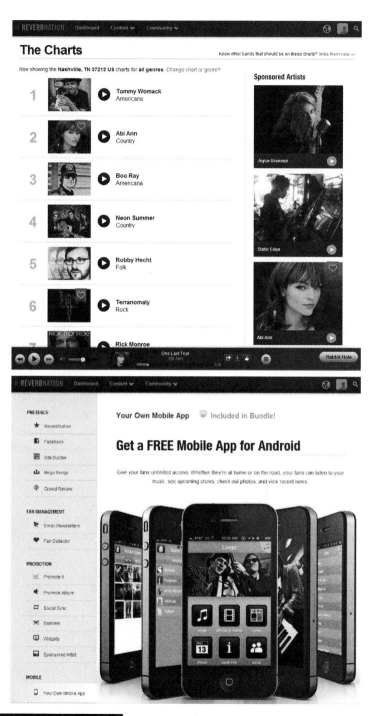

FIGURES 16.35 AND 16.36 *ReverbNation Charts and App Sampler*

Charts

There are countless charts and lists that compiled in a variety of ways, here are a few that are considered a standard because of their clout and derivation of their information.

iTunes Chart

In a rolling 24-hour "real time" chart, the iTunes chart reflects consumer behavior in a more timely fashion than that of the *Billboard* chart. Remember that SoundScan data is reported on Wednesday and then gets published the following week. So consumers receive the "latest" chart data nearly 10 days after the fact of the best-selling album of the week. iTunes data is much more current, reflecting buyer activity within a 24-hour period. What is not clear is the algorithm that drives this chart: is it strictly downloads, is it plays, do reviews and ratings affect position of single or album? iTunes does not reveal its chart positioning formula and yet ranking in a top position within the industry is highly coveted. A better position on this chart gives the artist and its music company bragging rights, helping it grease the way onto various outlets, especially radio playlists, which require consumer acceptance.

Lastfm

Last.fm is a music recommendation service that helps a listener discover new music based on the music that they are currently playing, through an algorithm known as "The Scrobbler." This proprietary method suggests new music based on the playlists of thousands of other's whose listening habits resemble your own. Charts are then derived based on frequency and popularity of new music with Top Tracks vs. Hyped Tracks identified. Indie artists are encouraged to promote their music on Last.fm with an entire tab dedicated to their cause. Video "Sessions" with live musical moments are produced, festival insights, indie charts and trends are all highlighted as part of a dedicated New Artist commitment, with free downloads of new music and promotional tracks encouraged.

APPENDIX

The analysis of artist DMA sales data can be manipulated to reveal the best performing markets based on sales / population. By looking at the sales of an artist's specific record, labels can determine where to place marketing and dollars.

This example is The Fray's, "How To Save A Life" album that was released 9/13/05. Again, ranked by DMA, then ranked by sales, then ranked by sales / population ratio, look at the variance in market strength.

The Fray
Title: HOW TO SAVE A LIFE
Rel Date: 9/13/2005

Ranked by DMA
Cummulative Sales as of 12/30/07

DMA Total	Cum Sales 2252317	12+ Pop	Ratio Sales / Pop
New York, NY	142605	15291100	9.33E-03
Los Angeles, C	120927	10826600	1.12E-02
Chicago, IL	92036	7738000	1.19E-02
Philadelphia, P	72834	4360200	1.67E-02
SF-Okland-Sar	52497	5891900	8.91E-03
Boston, MA	68987	3838300	1.80E-02
Dallas-Ft. Worl	54025	4838600	1.12E-02
Detroit, MI	36890	3888300	9.49E-03
Washington, D	51252	4176300	1.23E-02
Houston, TX	37313	4469900	8.35E-03
Cleveland, OH	29634	1794200	1.65E-02
Atlanta, GA	47912	4085000	1.17E-02
Minneapolis-St	50511	2662100	1.90E-02
Tampa-St. Pet	22381	2314300	9.67E-03
Seattle-Tacom	41236	3257200	1.27E-02
Miami, FL	21026	3533000	5.95E-03
Pittsburgh, PA	21110	1998800	1.06E-02
St. Louis, MO	25199	2282700	1.10E-02
Denver, CO	74073	2194800	3.37E-02
Phoenix, AZ	38897	3058000	1.27E-02

FIGURE 16.37   *The Fray DMA cumulative sales and population ratio*

The Fray
Title: HOW TO SAVE A LIFE
Rel Date: 9/13/2005

Ranked by Sales
Cummulative Sales as of 12/30/07

DMA Total	Cum Sales 2252317	12+ Pop	Ratio Sales / Pop
New York, NY	142605	15291100	9.33E-03
Los Angeles, C	120927	10826600	1.12E-02
Chicago, IL	92036	7738000	1.19E-02
Denver, CO	74073	2194800	3.37E-02
Philadelphia, P	72834	4360200	1.67E-02
Boston, MA	68987	3838300	1.80E-02
Dallas-Ft. Worl	54025	4838600	1.12E-02
SF-Okland-Sar	52497	5891900	8.91E-03
Washington, D	51252	4176300	1.23E-02
Minneapolis-St	50511	2662100	1.90E-02
Atlanta, GA	47912	4085000	1.17E-02
Seattle-Tacom	41236	3257200	1.27E-02
Phoenix, AZ	38897	3058000	1.27E-02
Houston, TX	37313	4469900	8.35E-03
Detroit, MI	36890	3888300	9.49E-03
Cleveland, OH	29634	1794200	1.65E-02
St. Louis, MO	25199	2282700	1.10E-02
Tampa-St. Pet	22381	2314300	9.67E-03
Pittsburgh, PA	21110	1998800	1.06E-02
Miami, FL	21026	3533000	5.95E-03

FIGURE 16.38   *The Fray DMA cumulative sales and population ratio ranked by sales*

The Fray
Title: HOW TO SAVE A LIFE
Rel Date: 9/13/2005

Ranked by Sales / Pop Ratio
Cummulative Sales as of 12/30/07

DMA	Cum Sales		Ratio
Total	2252317	12+ Pop	Sales / Pop
Denver, CO	74073	2194800	3.37E-02
Minneapolis-St	50511	2662100	1.90E-02
Boston, MA	68987	3838300	1.80E-02
Philadelphia, P	72834	4360200	1.67E-02
Cleveland, OH	29634	1794200	1.65E-02
Phoenix, AZ	38897	3058000	1.27E-02
Seattle-Tacom:	41236	3257200	1.27E-02
Washington, D	51252	4176300	1.23E-02
Chicago, IL	92036	7738000	1.19E-02
Atlanta, GA	47912	4085000	1.17E-02
Los Angeles, C	120927	10826600	1.12E-02
Dallas-Ft. Wort	54025	4838600	1.12E-02
St. Louis, MO	25199	2282700	1.10E-02
Pittsburgh, PA	21110	1998800	1.06E-02
Tampa-St. Pet	22381	2314300	9.67E-03
Detroit, MI	36890	3888300	9.49E-03
New York, NY	142605	15291100	9.33E-03
SF-Okland-Sar	52497	5891900	8.91E-03
Houston, TX	37313	4469900	8.35E-03
Miami, FL	21026	3533000	5.95E-03

FIGURE 16.39   *The Fray DMA cumulative sales and population ratio ranked by sales ratio*

Based on sales per population, Denver, the band's home town, emerges as the #1 strongest market for sales of The Fray's, "How To Save A Life." By concentrating on markets that have a stronger probability of sales, labels can better manage their marketing dollars through succinct activities which may include radio promotions, in-store events, touring, etc. The goal is maximize the market and sell records. This record also falls into the "slow boil" category, with the band being featured on the hit TV series and subsequent compilation of "Grey's Anatomy." By year end 2008, "How to Save a Life" had sold over 2 million copies.

GLOSSARY

"Big Data"—used to describe a massive amount of both structured and unstructured data that is difficult to process using traditional databases and available software. This data holds the potential answers to help companies improve, from making better operational decisions to being more intelligent in the market place.

"**Big Data Analytics**"—Collecting, organizing and analyzing large sets of data that are actionable to specific business decisions is big business in today's market space.

Designated Market Area (DMA)—A term used by Nielsen Media Research to identify an exclusive geographic area of counties in which the home market television stations hold a dominance of total hours viewed. There are 210 DMA's in the U.S. These markets are listed in order of population density—meaning, stratified largest to smallest by size of city.

County Size—The classification of counties according to Census household counts and metropolitan proximity. There are four county size classes "A", "B", "C", and "D". In general, "A" counties are highly urbanized, "B" counties relatively urbanized, "C" counties relatively rural, and "D" counties very rural.

Geographic Regions—states are grouped by physical location which can lead to sales and marketing analysis based on geographic data.

Store Types—As an industry standard, stores have been identified by types to aid the industry in identifying source of sales. These store types include Mass Merchants, Traditional Retailers which include chain store and electronic super store, Independent Retailers, Non-traditional Outlets that include online and venue sales, and Digital portals.

Formats—The type of music format that consumer purchase continues to evolve—be it physical or digital. Physical formats include Compact Discs (CD), Vinyl (LP), Cassette, and DVD with Digital formats including Downloads and Streaming.

UPC Code—The Universal Product Code contains a unique sequence of numbers that identifies a product. GS1US standards recommend a GS1 company prefix number that identifies a unique code for a specific business.

GTIN (Global Trade Item Number) initial numbers, which can vary in size depending on the variety of products the company sells, identify the business, with the last digits identifying the products. The small digit is the "check" digit which validates the barcode equation.

ISRC Code—For digital product, the ISRC (International Standard Recording Code) is the international identification system for sound recordings and music video recordings. Just like the UPC Code, each ISRC has a unique and permanent sequence of numbers that identifies each specific recording that can be permanently encoded into a product as its digital fingerprint. The encoded ISRC provides the means to automatically identify recordings for royalty payments, key to publishers and songwriters alike.

SoundScan—Now owned by Nielsen, the company that infused the use of UPC codes to count and created the current Billboard charts, utilizing technology in helping identify where sales occurred, in which outlets, which format, etc.

Track Equivalent Albums (TEA)—which means that 10 track downloads are counted as a single album. This evaluation is based on a financial equivalent, being a $.99 download x 10 would equal a $9.99 album download or CD.

Streaming Equivalent Album (SEA)— introduced in 2013 to measure streaming consumption. To evaluate streaming equivalents, the current industry standard is 1500 streams of any songs from a particular album are counted as a single album. 1500 streams x the standard royalty generated by this airplay $.005 = $7.50, being the wholesale price of an album.

Lifecycle —term used to identify the four stages within the traditional product life-cycle: Introduction, Growth, Maturity, and Decline.

Direct-to-consumer—concept of bypassing the traditional wholesale/retail outlets and selling directly to consumers from the manufacturer. In the music business, this means the artist sells his/her music directly to their fans.

REFERENCES

Bakula, D., Senior Vice President Analytics and Client Development – The Nielsen Company, Personal Interview, April 9, 2015.

"Bandcamp." *Bandcamp*. N.p., n.d. Accessed September 25, 2014. http://www.BandCamp.com/.

Beal, V. "Big Data." *Webopedia*. N.p., n.d. Accessed October 12, 2014. www.webopedia.com%2FTERM%2Fbig_data.html.

Bennett, J., Director, Music – The Nielsen Company, Personal Interview, April 8, 2015.

"BigChampagne Media Measurement." *BigChampagne Media Measurement Home*. N.p., n.d. Accessed September 19, 2014. http://www.BigChampagne.com/.

"Create Amazing Marketing Reports That Also Make You Look Amazing." *Online Marketing Tools and Reporting Software*. N.p., n.d. Accessed October 25, 2014. http://www.raventools.com/.

Dahan, C., Professor, Department of Recording Industry, Middle Tennessee State University, Personal Interview, November 4, 2014.

"Home, BuzzAngle Music." *BuzzAngle Music*. N.p., n.d. Accessed October 25, 2014. http://www.buzzanglemusic.com/.

"iTunes." *Apple*. N.p., n.d. Accessed September 25, 2014. http://www.apple.com/itunes/.

Kyle, G. A., Dr. "DMA Mapping of Columbia, MO." University of Missouri School of Journalism, n.d. Accessed January 7, 2015.

"Last.fm." *Last.fm*. N.p., n.d. Web September 25, 2014. http://www.lastfm.com/.

Lidestri, J., CEO - Border City Media, Personal Interview, October 25, 2014.

"Looking for Industry Knowledge? Our Range of Resources and Reports Cover Everything from Analysis on Digital Music, to How Labels Invest in Music." *Resources & Reports*. International Federation of the Phonographic Industry, n.d. Accessed September 20, 2014. http://www.ifpi.org/content/section_resources/iarc.html.

Muratore, C., Chief Business Development Officer – Border City Media, Personal Interview, October 25, 2014.

"Next Big Sound: Analytics and Insights for the Music Industry." *Next Big Sound - Analytics and Insights for the Music Industry*. N.p., n.d. Accessed September 20, 2014. http://www.NextBigSound.com/.

"ReverbNation: Artists First." *ReverbNation*. N.p., n.d. Accessed September 24, 2014. http://www.ReverbNation.com/.

"Solutions." *Music Sales Measurement*. Nielsen, n.d. Accessed October 12, 2014. http://www.nielsen.com/us/en/solutions/measurement/music-sales-measurement.html.

"Topspin Media." *Topspin Media*. N.p., n.d. Accessed September 25, 2014. http://www.TopSpinMedia.com/.

CHAPTER 17

Tour Support and Sponsorships

In an era when retail floor space dedicated to music is diminishing, when the single, and specifically "the file," is fast becoming the music format of choice, and consumers have so many options when it comes to spending their entertainment dollars, music companies are looking to monetize their assets in as many ways as possible. "The recording," along with the artist who produced it, has become marketable in so many more ways, via the use of Internet, licensing agreements, cross-merchandising, and using the artist's other talents to magnify and exploit all money-making opportunities. Artists need to be the "real deal" by having the talent to perform live. Some artists in the past have been "created" in the studio, therefore handicapping themselves from being able to replicate their music live. By being a strong performer, artists and their business partners have the chance to realize a broader spectrum of financial gain but getting started can be a challenge.

TOUR SUPPORT

The risk of artist development falls on the financial shoulders of the record label. The label acts as a bank and "advances" the artist monies to jump start the artist's career with the idea that the label/bank will get paid back once revenue begins to be generated. As awareness of the artist begins to increase through the various marketing activities of the label such as internet marketing, social media, and radio airplay, the

CONTENTS

artist may have the opportunity to become the opening act for a larger tour that would give the developing act great exposure to a big audience. Often times, the developing artist can "piggy back" on a tour of a much larger act that is on the same label, keeping it "in the family" and growing relationships internally, as well as the potential revenue. Payment for the opening act slot may not cover the expenses of the developing artist so the label often covers these costs associated with going on the road as "tour support" which is recoupable in most record deals. Once revenue begins to be flow, the label will pay itself back prior to paying royalties to the artist.

As noted here by recent research of the International Federation of the Phonographic Industry, investment in an artist's career can be sizable with no guarantee of success. And if an artist does not have a hit, no tour, no merchandise—and the relationship and contract dissolves, the artist owes no money to the label and walks away from the debt (IFPI.org). The research and development (R&D) is extensive in the music industry. Think about how many artists do not become household names. This is why many music companies are struggling to maintain the current business model.

Breakdown of Record Label Investment

Advance	US$200,000
Recording	US$200,000–300,000
Two or Three Videos	US$50,000–300,000
Tour Support	US$100,000
Marketing and Promotion	US$200,000–500,000
Total	US$750,000–1.4 Million

(Source IFPI)

FIGURE 17.1

TOUR SPONSORSHIP

To aid in offsetting the expenses of touring, many artists look for help from tour sponsors. This relationship tends to be a win-win between artist and sponsor, who is looking for a business relationship to help create a branding opportunity between a company's product and a target market to which the artist performs. In exchange for access to would-be consumers, the artist receives funds to help pay for touring expenses—everyone wins.

As this band landed on U.S. territory in 2013, Nabisco inked a tour sponsorship deal with One Direction, helping the band brand their image on key products that targeted their consumer—tween and teen consumers . . . and their moms. The deal included 20,000 in-store displays and 8,000 endcaps at Walmart, Targets, and Walgreens. Over 20 million packages of Oreos, Ritz Bits, Cheese Nips, Honey Maids, and Chips Ahoy snacks were imaged with One Direction music, tour information, and ticket give-away sweepstakes. A VIP artist app was created, branding both band and products that cost nearly "seven figures in investment in social media" spending. The band was also promoted to the 33 million Facebook fans of Oreo cookies—a branding match made in heaven. Uniquely, bloggers were chosen through an online contest to go on the road with the band to report on the various activities of the 35-city tour and were given exclusive tour access, $40,000 and other perks.

One Direction Activation Components	
1DVIP.com	Dedicated website offers ticket giveaways, band interviews and other exclusive content.
Consumer sweepstakes	Consumers that visit IDVIP.com can enter a sweeps dangling VIP tickets and private access to one of three concerts.
1D VIP tour correspondent	Blogger is creating and posting content from the road.
On-pack promotion	Nabisco is running an on-pack promo dangling voicemails, virtual photo booths and other band content. The snack foods giant is touting the promo on 21 million packages of Oreo, Chips Ahoyl, Cheese Nips, Honey Maid and Ritz products.
ID VIP App	Fans download a free cell phone app to access the on-package content. Fans can download the app from the Apple or Google Play stores.
On-site activation	Tour activation includes never-seen-before videos, photo booths and product sampling.

Sources
Mondelez International, Inc., Tel: 847/943-4000

FIGURE 17.2

One Direction continued promotional efforts with Pepsi and the NFL's Drew Brees, Colgate toothpaste and their singing toothbrush line, Target Stores with an exclusive deluxe release of their album "Take Me Home," and a 3-D documentary of their tour and their U.S. musical "invasion." Their U.S. sales have been a phenomena with the lion's share being mostly albums with over 6 million sold between the albums released: *Up All Night, Take Me Home,* and *Midnight Memories. Four,* their fourth album is set to release November 2014 (Hampp, 2013).

This partnership has paid off with this Brit Pop Quintet positioning themselves as the highest grossing tour of all time, earning over $230 million since hitting the road in South America in the spring 2014. This feat tops Beyoncé and Jay-Z's 2014 "On The Run" tour and exceeds Bruce Springsteen's "Magic Tour" of 2007–2008.

Targeting a completely different demographic, Citibank sponsored the Rolling Stones' "50 & Counting" tour of 2013. Citi sponsored the official Rolling Stones app, which offered breaking news, back-stage footage, and the ability for concert-goers to push song requests to the band in real time during concerts, among other features. Instagram photo booths designed to look like the Stones green room from 1971 allowed for uploading of images during concert events as well as custom Instagram prints of their photos.

Special opportunities for Citi card members included pre-sale ticket purchases, backstage tours and other hospitality experiences like access to the vintage-style custom lounges reserved for VIPs and other guests. The sponsorship built on a music platform Citi has worked to build over the past few years, according to Jennifer Breithaupt, Senior Vice President of Entertainment Marketing at Citi. What is unclear is the value of the relationship. How much did Citi pay to become a sponsor of the Rolling Stones? Did card membership go up in order for members to gain access to exclusive packages? No doubt, There is a "cool" factor that is "priceless" when it comes to access to an artist (Kirkpatrick, 2013).

A different kind of sponsorship is the tour or festival scene where multiple sponsors are involved with a multi-artist/multi-day event. The Vans Warped Tour has been running for over 20 years. Starting as a punk rock and ska music fest, this daylong, multi-stage jam now attracts very young rockers to the twenty-something that can bring their folks who can hang in the "parent day-care," if necessary. The reasonable $60 average ticket affords the ticket buyer music from 9:00 a.m. to 9:00 p.m. with various stages being sponsored by over 80 national brands including Vans, Kia, Journeys, Monster, and Beatport. Most of the artists performing are on the brink a national prominence, but the alignment to the age demographic is the payoff to the sponsor with an average of 16,000 to 20,000 in attendance at every concert stop (Vans Warped Tour).

The trend in brand sponsorship continues to grow, according to IEG Sponsorship Consultants. Sponsorship in music venues, festivals, and tours for 2014 will top $1.34 billion, an increase of 4.4% from 2013, according to IEG, LLC (Sponsorship.com). Many top corporations have a keen interest in EDM: Anheuser-Busch InBev in 2013 landed a multi-million dollar deal with SFX Entertainment, Motorola partnered with Live Nation EDM festivals, and Heineken aligned with trance DJ Armin van Buuren to support its "Dance Fast, Drink Slow" campaign.

The Coca-Cola Co. is the most active sponsor of music with 27% of properties reporting a partnership with the beverage giant.

Beer companies are 10 times more likely to sponsor music than the average of all sponsors with spirits and banks in close pursuit.

Anheuser-Busch is by far the most active sponsor of music festivals with 31% of properties reporting a partnership with the company (Sponsorship.com).

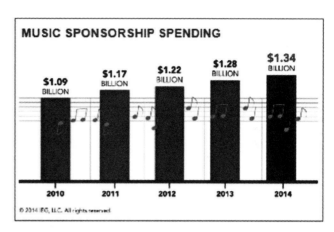

FIGURE 17.3

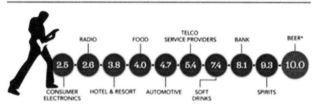

The Most Active Categories Sponsoring Music (North America)

FIGURE 17.4

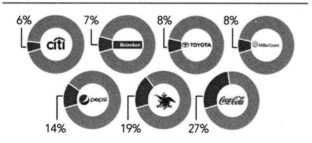

The Most Active Companies Sponsoring Music (North America)

FIGURE 17.5

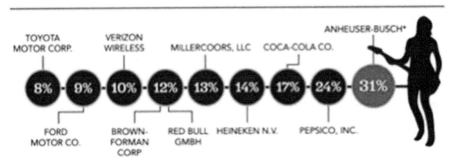

The Most Active Companies Sponsoring Music Festivals (North America)

FIGURE 17.6

PRODUCTS IN THE MUSIC

So which comes first: the chicken or the egg? The success of Fireball Whiskey is quite suspect, considering its origins in marketing. According to the drink's "plan," the marketing strategy was to focus on college markets, specifically Nashville and Austin—only the two biggest songwriting towns in America, and give "free" rounds of the liquor to entire local bars. Next was to enlist celebrities to use their social media and post about the drink. And the third initiative was to create drinking contest, encouraging bars to post pictures to Twitter and Facebook accounts. No wonder Fireball has made it into countless country songs, specifically one by the leading 2013 country duo Florida Georgia Line. The single "Round Here" has been viewed over 19 million times on YouTube with their album containing the same single "Here's To The Good Times" selling way over 2 million units. The drink's revenues increased from $1.9 million in 2011 to $61 million in 2013, about to become the number one whiskey in the U.S., overtaking Jaegermeister. So is it the drink, the social media strategy, or the song that made the whiskey a success? The "stars" aligned and put this liquor on the map (Lutz, 2014).

As part of the launch of this song, Jennifer Lopez's "On The Floor" video featuring Pitbull highlighted several products including BMW, Swarovski Jewelry, and Crown Royal liquor. These were not afterthoughts, but deliberate product placements with the hope that targeted demographics would see these items. Over 798 million viewers have streamed the

VEVO video of "On The Floor," being one of the fastest viral videos of 2011 to be watched. The cache of these product "endorsements" for the brands is invaluable but the cash exchange for these placements is not readily known.

BRANDED MUSIC

There was a time in the not-so-distant past when music companies had departments known as "Special Products and Special Markets." The efforts of these staffs were to look for opportunities and projects that were not attached to traditional sales and whose profits were considered "gravy" to the label and not projected funds to the bottom line. But with record sales still on the decline, labels look for these products and markets to make up for revenue that has since been lost in the traditional market place.

LIFESTYLE

With the purchase of $60.00 or more, Gap *consumers would* receive a gift compilation *CD containing eight songs along with a director's cut of the television commercials and exclusive behind-the-scenes footage of the making of the* advertisement (Jeans and Music, 2005).

FIGURE 17.7

With the emergence of lifestyle and specialty product retailers, a new outlet for music has evolved. When grabbing a cup of coffee from Starbucks, it's hard to ignore the "cool" music playing and CD display right under your nose. A visit to Pottery Barn or Restoration Hardware elicits more than just home décor, but a ambience that is enhanced by the music playing—which you can replicate in your own home via the purchase of the "branded" CD.

Retailers are creating "branded" CDs as product extensions of their retail offerings. Pottery Barn has offered over 70 CD compilations which have sold millions of units ("Case Studies: Pottery Barn," 2014), all of which are an "easy way to put lifestyle in a home," according to Patricia Sellman, marketing vice-president at Williams-Sonoma (Strassburg, 2003). The Gap fashion retailer wanted to drive sales of their merchandise through the use of music. In-store play, as well as exclusive music and video CDs and DVD offers were enhanced by using the artists as part of the overall advertising campaigns. The initial Madonna and Missy Elliott combo spawned several other creations that included Jason Mraz, John Legend, Joss Stone, and Keith Urban—all that included high visibility promotions with TV and print ad campaigns that Gap attributes to a 30% rise in sales for their fashion line alone ("Case Studies: The Gap").

Rock River Music, the compilation company that created the Gap CD, has evolved with the times and now delivers pre-loaded USBs, mp3 players, as well as digital programming where consumers are consuming. Ford brought the challenge to connect with younger females as they introduced the new Ford Focus. Rock River targeted consumers via musical lifestyles through a streaming webradio player and a music to-go section based on lifestyle activities. This widget was anchored on the Focus website but also syndicated on all Ford websites including MySpace and Facebook, logging thousands of hours of streaming and tens of thousands of downloaded playlists, making the connection to the targeted demographic.

CustomChannels.net helps various outlets "brand" their environments with a musical ambiance while managing the musical licensing for the business. This resource is yet another example of connecting musical "style" with the consumer. CustomChannels created the Bonnaroo Festival Channel that is a downloadable app, streaming content from festival artists who participated in most recent event while providing information on ticket sourcing, accommodations, shuttle services, social media links, and most importantly artist line-up for concert series. This "channel" exists long after the Bonnoaroo Festival is over so that festival-goers can continue to re-enjoy their experience. CustomChannels marries the sound recording and artist image with the right outlet to enhance the "experience" with consumers (Custom Channels, 2014).

THE BENEFITS

Why, then, are artists and music companies aligning themselves with retailers? What is the benefit to them and their existing record label partners? Initial response would be money. The licensing of existing recordings from labels can be very lucrative. Because labels never know the value of their recordings long term, record companies have found it hard to calculate the value of licensed product. Historically, the licensing of music has been considered "gravy"—monies generated beyond the "meat and pota-toes" of the business. But to address the burgeoning market, as well as sagging record sales, segments of marketing departments are devoted to mining the catalog of their holdings and finding new, innovative homes beyond traditional music channels.

SOUNDTRACKS AND COMPILATIONS

Since moving pictures became "talkies," music has been an integral element to the success of the movie business. Early in the evolution of movies, the industry would hire actual singers to play a part, integrat-ing music and story on-screen. Often, the music from these movies would become nationwide hits and the popularity of the singing stars would explode, making many of them classic voices of the genre. Movie soundtracks have continued to evolve, with the 1960s delivering the Beatles *Hard Day's Night,* and *The Graduate* with its hits "Mrs. Robin-son" and "Scarborough Fair." by Simon and Garfunkel. The 70s included *Urban Cowboy, Grease*, and *Saturday Night Fever*, all containing a com-mon thread of one very hot actor, John Travolta. The 1980s brought about *Footloose, Dirty Dancing*, and Roy Orbison famed *Pretty Woman*. The 1990s elevated Whitney Houston to a new level of superstardom with her film debut in *The Bodyguard*. And Celine Dion reached new heights with her mega-hit "My Heart Will Go On" from the epic *Titanic.* All of these soundtracks contained songs that would eventually become huge pop hits, selling millions of units and igniting a new kind of con-sumer to purchase music. And in some cases, a single song has helped to launch a career and create a mega-star.

Soundtracks have learned the lessons of the past while keeping up with the purchasing practices of today's music consumers. Many music purchasers have become their own disc jockeys, curating their "personal soundtracks" containing the individual songs that they want to hear. By compiling songs that capture this spirit while representing the theme of a movie, soundtrack managers have fashioned a "genre" of music that has produced sales.

Overall sales for the soundtrack genre have maintained in recent years, holding about a 4%–5% market share of all album sold between 2010 and 2013, according to SoundScan. But the holiday season of 2013 launched a game-changer for Disney and what started as a flurry has become an avalanche. *Frozen* was introduced to young audiences but has been embraced by movie goers of all ages. And as only Disney can generate, are all the accouterments that adorn their movie flicks that include memorable characters singing songs that have movie goers humming the tunes as they exit the theater. The *Frozen* soundtrack has topped the top 200 *Billboard* chart for most of 2014, generating sales of over 3.4 million, which is just the tip of the iceberg. *Frozen* has spawned several spin-offs for the entertainment giant including a Broadway version of the show, a television series, oodles of character-themed products from stuffed animals to pajamas, and entire sections of the Disney theme parks are being planned and retrofitted to resemble Arendelle—the land of *Frozen.*

But Disney didn't stop there, and their soundtrack domination continued with *Guardians of the Galaxy*, driving the soundtrack market share to 5.4% and comprising 39% of all soundtrack sales in 2014. The *Guardians* soundtrack is unique in that all the tracks have been previously released. Although the movie is targeted for young viewers, the music resonates with the parents who brought the younger audience to the theater, thus driving a double-fisted purchase. *Pitch Perfect* is another recent movie hit with a great original soundtrack documenting the a capella lives of university choirs. This movie helped to highlight the musical prowess of actress Anna Kenrick while creating a national phenomena among school kids of stacking cups. The *Pitch Perfect* soundtrack has sold nearly 1.2 million units.

Soundtracks are not restricted to the movie theaters. The "Glee" television series has been singing all the way to the bank since 2009. As the story line has evolved and as characters have come and gone, the "soundtrack" of their lives have been documented and released for consumers to purchase to the tune of 7.8 million albums sold. And of course, Disney doesn't seem to sign a child actor who doesn't sing. Several of the television shows have spun off record releases including Ross Lynch of "Austin and Ally" who has released several records under the television title as well as his family band name R5. Together, his releases have sold over 500,000 units for Hollywood/Disney Buena Vista Records.

ENHANCING CAREERS

"Matching music to an automobile's image is essential to developing a consistent and credible message," says Rich Stoddart, Ford Division's marketing communications manager. In recent years, car advertisers have paired

Led Zepplin with the new generation of Cadillac, Aerosmith's "Dream On" with Buick LaCrosse, and Jimi Hendrix's rendition of the "Star-Spangled Banner" with the 2005 Ford Mustang. While music "won't sell the car," says Gordan Wangers, president of AMCI Inc., an automotive consulting firm in Marina Del Rey, California, "it is capable of grabbing the attention of the right audience to say, 'Here's my new car and here's my message'" (LaReau, 2004).

Alicia Keys used the deep pockets of Citibank to help launch her November 2012 *Girl on Fire* release. With a tour and single of the same name, Citibank aired a national commercial featuring Keys as the anchor artist with her music as the bed prior to album release date, helping consumers connect the dots that a new single and album were available from this quintessential artist. The album is approaching platinum in over-the-counter sales, according to SoundScan.

Placement of music in national ad campaigns can put an artist on the map, but may not always convert to mega-sales. The band American Authors, landed a robust ad campaign with Lowes Home Improvement Stores where their neat "Best Day Of My Life" single scored the soundtrack to a 30-second commercial. Consumers resonated with the song, seeking repeat hits to the band's Sound Cloud with 10 million streams and 36 million views on their VEVO video, but sadly, the album only sold 85,000 units. While the band did not realize big sales, this connection with consumers could set them up for a major release with their next studio album.

PRODUCT EXTENSIONS AND RETAIL EXCLUSIVES

Many labels are attempting to capture more of the same consumer's dollar by creating product extensions of a specific artist's release. The concept is to cover the losses being generated by file-swapping or streaming by creating "super-size" versions of the bigger releases and targeting hard-core fans that are willing to dig deeper into their pockets in order to have the "special" release. According to research done by the Handleman Company, a former rack jobber to mass merchants Walmart and Kmart, "23 percent of music buyers account for an estimated 62 percent of album sales, buying an average single CD every month." (Leeds, 2004). More recent research conducted for South By Southwest 2014 Nielsen presentation stated similar findings. The "aficionado fan," identified as the top-tier music consumer, is willing to spend their money on all formats of music including artist merchandise, concerts, and online streaming services. The "aficionado fan" is noted as being 14% of the total music population and yet they consume 34% of the total music sold. Also revealed was that 68% of aficionado fans are very likely or likely to contribute to a campaign for exclusive content (Nielsen, 2013).

In this era, it seems that every artist has some sort of exclusive "give-away with pre-sale" purchase through the artist's website, but a big amount of buzz has been granted to indie artist DeadMau5 with his AllAccess website. An "aficionado fan" dream, access to DeadMau5 has taken on three tiers: free, $4.99/month, and $44.99/year. With these various levels of subscription, fans have been given access to rare cuts of music, photos, videos postings, message "boreds," live streams of the artist creating, and even chat time with the creator (www.deadmau5.com).

As an example of upsizing, who can forget the "special" release of Interscopes landmark 2004 release of U2's *How To Dismantle An Atomic Bomb*? This record had three versions: the 11-track, $10 basic version, or for $32, a fan could purchase the "collector's edition" including a DVD and 50-page hardcover book. The mid-priced version had the DVD but no book. But if a consumer bought a new "U2" Apple iPod, they got the pre-loaded version of "How To Dismantle An Atomic Bomb" for free.

Ten years later, U2 and Apple have continued their relationship. On the street date of September 9, 2014, iTunes downloaded U2's 13th studio album, *Songs of Innocence* to half a billion iTunes subscribers—for free. U2 fans were happy but a huge push back from non-U2 fans, as well as privacy

... WHILE CONCERT REVENUE IS RISING

RANK	GROSS	BAND	TOUR	YEAR	SHOWS	ATTENDANCE/SHOW	
1	$772 million	U2	U2 360° Tour	2009–11	110	66,110	
2	$635 million	The Roling Stones	A Bigger Bang Tour	2005–07	144	32,500	The Five top-grossing tours of all time
3	$495 million	The Roling Stones	Voodoo Lounge Tour	1994–95	124	51,103	
4	$477 million	AC/DC	Black Ice World Tour	2008–10	167	29,023	
5	$484 million	Roger Waters	The Wall Live	2010–13	219	18,858	

FIGURE 17.8

protectionists began, since the download was not initially delete-able. As predicted, older U2 albums began to sell, as noted on the iTunes album charts, but Apple had to issue an app to disgruntled iTunes customers that allowed for the removal of the newly acquired tracks that had not asked for permission to "come aboard."

So what's in exchange? U2 realized a need to be relevant in an age that rewards the young and relentless. The band and Apple are betting on the future, with a transference in interest for their music to an experience. They're no dummies—the band sees that sale of traditional music sales continue to slide, but U2's 360 Tour is one of the all-time top-grossing tours worldwide. So to attract attention to their latest musical effort, as well as bring new followers into the fold, U2 partnered with Apple who agreed to market their newest effort for $100 million (the band won't say how much Apple paid for *Songs of Innocence* itself), which included a retail release later scheduled for October 13, 2014 containing an acoustic version of the September download with bonus tracks. Its first week sales debuted the album in the Top 10 at #9 with over 28,000 units sold, even after the album was automatically loaded to 500 million iTunes user accounts. Although 2 million downloaded the entire album, others were not pleased and asked for iTunes to either remove the album or give them a way to remove the album from their devices. Although somewhat controversial, these efforts cued up for the holiday selling season along with the 2015 U2 World Tour. Additionally, *Songs of Innocence* is a companion album to its secondary release called *Songs of Experience*, to come out in 2016. This release is a partnership with Apple, showcasing a new format of audiovisual interaction that can't be pirated. (Catherine Mayer, 2014)

BRAND PARTNERSHIPS

Taking a page out of the soundtrack playbook, non-movie compilations have emerged as powerful sales items as well as marketing tools. The Now! Series is considered the most successful compilation collection in the history of music sales. Initially, the U.S. participating collaborators included EMI Recorded Music, Universal Music Group, Sony Music Entertainment, and the Zomba Group, with the albums rotating among Sony, Universal, and EMI for marketing and distribution. The U.S. series is now released under its own imprint and distributed by Universal and has distributed 51 compilations selling an estimated 50 million U.S. units since its inception in 1998. The key to its success has been the hits—since the brand relies on the top singles to drive sales.

Intangible is the impact on sales of participating artists. Someone purchasing a Now! CD might know three or four of the artists on the package, but the consumer most likely is introduced to a new act that they might also enjoy. The residual purchase of the full-length product of that new act is the result of this cross-promotional item.

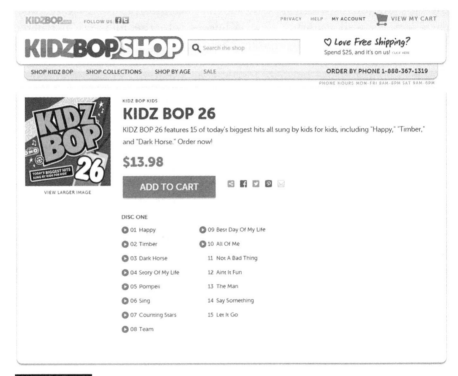

FIGURE 17.9

The KidzBop series has picked up on the Now! phenomena, but realized that some "prime time" singles contained language that many parents would not find suitable for children. So creating a "clean" version with kids at the microphone, these "sound alikes" of current-day hits have marketed their way home nearly 15 million times. (SoundScan, 2014)

GLOSSARY

Branded CDs—a CD sponsored by and sporting the brand logo or mention of a company not normally associated with the release of recorded music. Examples include product designed for and sold at Pottery Barn, Pier 1 Imports, and Victoria's Secret.

Brick-and-mortar stores—businesses that have physical (rather than virtual or online) presences, in other words, stores (built of physical material such as bricks and mortar) that you can drive to and enter physically to see, touch, and purchase merchandise.

CIMS—the Coalition of Independent Music Stores

Compilations—a collection of previously released songs sold as a one album unit, or a collection of new material, either by single or multiple performers, sold as a collaborative effort on one musical recording.

Cross-promotion—using one product to sell another product, or to reach the market of the other product.

Exclusives—retail exclusive marketing programs that are not offered to other retailers, but arranged specifically through one retail chain.

Private label—a label unique to a specific retailer.

Tour sponsorship—a brand or company "sponsors" a concert tour by providing some of the tour expenses in exchange for product exposure at the events.

Superior products—a value-added version of a product that is sold in the "regular" version elsewhere.

Value-adds—to lure consumers into a specific store, music retailers offer exclusive product with added or "bonus" content or material for their customers.

REFERENCES

"Case Studies: Pottery Barn." *Rock River Music*. Rock River, n.d. Accessed November 16, 2014.

"Case Studies: The Gap." *Rock River Music*. Rock River, n.d. Accessed November 16, 2014.

"Custom Channels Beyond Background Music." Custom Channels, n.d. Accessed November 16, 2014. www.customchannels.net.

Hampp, A. "Exclusive: One Direction Teams with Nabisco for Multi-Brand Tour." May 21, 2013. Accessed November 16, 2014. http://www.billboard.com/biz/articles/news/branding/1563052/exclusive-one-direction-teams-with-nabisco-for-multi-brand-tour.

"Jeans and Music—The Perfect Fit at Gap this Fall." *PR News Wire*. PR Newswire, August 22, 2005. Web. Accessed November 16, 2014.

Kirkpatrick, R. "Citi's Rolling Stones Tour Sponsorship Touches All Fans." *Event Marketer*. Event Marketer, May 30, 2013. Accessed November 16, 2014.

LaReau, J. "Music is Key to Carmakers Marketing." *Automotive News*. December 20, 2004. Accessed November 16, 2014.

Leeds, J. "$10 for a Plain CD or $32 with the Extras." *New York Times*, December 27, 2004. Accessed November 16, 2014.

Lutz, A. "How Fireball Whiskey Became the Most Successful Liquor in Decades." *Business Insider Strategy*. Business Insider, April 28, 2014. November 16, 2014.

Mayer, C. "U2 and Apple Are Working On A New Digital Music Format." *Time*. September 18, 2014. Accessed November 16, 2014. http://time.com/3393297/u2-apple-new-digital-format/.

Nielsen, U.S. Music Industry, Year-end Review, 2013.

"Our Industry." *IFPI.org*. IFPI, n.d. Accessed November 16, 2014.

SoundScan. Computer software. Nielsen Entertainment, n.d. Accessed November 16, 2014. http://nielsen.soundscan.com/.

Sponsorship.com. IEG, n.d. Accessed November 16, 2014. www.sponsorship.com.

Strasburg, J. "Bands to fit the brand / SF company's CDs sell for likes of Pottery Barn / SF company leading trend of 'lifestyle CDs in nonmusic Retailer." *SFGate. San Francisco Chronicle*, July 25, 2003. Accessed November 16, 2014.

Vans Warped Tour. N.p., n.d. Accessed November 16, 2014. www.deadmau5.com.

CHAPTER 18

Grassroots Marketing

INTRODUCTION

Every day, we're bombarded with more commercial messages from more sources than ever before. The competition for our attention ranges from a subtle product placement in our favorite television show or online video to top down, in-your-face ads from car salesmen on the radio. Advertisements pops-up at us while we are online, they precede theater movies, they appear on our wireless devices, they're on seat backs at stadiums, and they're even in public restrooms where graffiti once covered the doors and walls. It is this competition for our attention that has driven marketers to find alternative methods to connect products and services with billfolds and credit cards.

Grassroots marketing creates a bottom-up awareness of product, or in our case, an artist, single or album. The term refers to campaigns that start with a few, passionate people, usually at a local level, and eventually spread through the masses. In grassroots marketing we energize and motivate our most loyal fan base to do much of the promotional work for us. Grassroots marketing generally refers to a street level, door-to-door campaign of peer-to-peer selling. And the most basic element of grassroots marketing is "word of mouth."

The terms grassroots marketing and guerrilla marketing are often used interchangeably, but there are subtle differences. Entrepreneur Media describes **guerrilla marketing** as "going after conventional goals of profits, sales and growth but doing it by using unconventional means." Instead of

CONTENTS

investing money, "guerrilla marketing suggests you invest time, energy, imagination and knowledge instead" (Guerilla Marketing). Grassroots marketing is focused on energizing your artist's fan base to take action on their behalf while guerilla marketing, which may include grassroots marketing efforts, is about using resources other than money in unconventional ways to market the artist.

THE POWER OF WORD OF MOUTH

Very few concepts in business are as universal as is the power of word of mouth (WOM) in marketing. Everett Rogers, in the book, *Diffusion of Innovations*, talks about the role of "opinion leaders" in the diffusion process (see chapter 2) and how these trendsetters can be used in facilitating word-of-mouth communication messages about a new product or other innovation (Rogers, 1995). His ideas are as relevant today as they were when he wrote them in the 1990s. Word of mouth is more effective at closing a deal with a consumer than any pitch from any paid medium. Word of mouth is simply someone you know whose opinion you trust saying to you, "you gotta try this!"

A recent study by Nielsen found that 92% of people trust recommendations from friends and family and 77% are more likely to buy a new product when learning about it from friends and family. As marketers, we want to encourage our loyal fans to help us spread the word about the music we are promoting (Nielsen).

From the company's standpoint, there are several basics to developing a plan to reach new consumers through word of mouth. They are:

Educating people about your product or services
Identifying people most likely to share their opinions
Providing tools that make it easier to share information
Studying how, where, and when those opinions are being shared
Looking and listening for those who are detractors, and being
prepared to respond

(Intelliseek, 2005)

In word of mouth marketing for music, we want to identify the fans most likely to share their opinions with others, then be sure those people are properly educated about the product so that he or she is spreading the correct information. We also need to provide the tools that will make it easier for them to promote our product and look for the people who don't like the product and be prepared to respond to them as well.

Among the most popular ways to effectively use word-of-mouth marketing and promotion in the recording industry is through the artist fan club. The club is a network of people who have a passion for the music of the artist and who are willing to be actively involved in promoting the career of the artist. Members of the club are regular chat room visitors, bloggers, video bloggers, message board respondents, YouTube posters, online social networkers, and hosts of discussion groups. Labels often have someone who is responsible for coordinating an artist's promotion with fan clubs, which in many cases, has taken the place of the traditional street team. The label coordinator uses their database to give club members updates on tour dates and to initiate contests for the community built around the artist. (Harbin, 2009)

This discussion of word-of-mouth marketing is not intended to be a short course on fan club development, but the elements of it as a catalyst for "spreading the word" are among the best central strategies a label can use to draw from the strength of the idea. Word of mouth is often compared to the work of the evangelist, and fan club members easily fit the definition. Energizing them and arming them with available tools has the potential to spread positive information about the artist in geometric proportions.

Word-of-mouth marketing takes several forms. Among those are:

Word-of-Mouth Style	Elements of the Style
Buzz Marketing	This concept uses high-profile entertainment (celebrities, events, etc.) or news to get people to talk about your brand.
Cause Marketing	Cause or 'charity' marketing is when you support social causes to earn respect and support from people who feel strongly about the cause.
Conversation Marketing	Interesting or fun advertising, catch phrases, emails, entertainment, or promotions that are designed to start word of mouth activity fall under this category.
Community Marketing	This form of marketing supports niche communities with similar interests about the brand and providing them with tools, content, and information, e.g., fan clubs, groups and forums. Take a look at the new Google+ communities.
Evangelist Marketing	Evangelism marketing is when customers believe in your brand so much that they will happily promote your products to others to buy and use it. There's no contest here, hands down, Apple wins. Even after Steve Jobs' passing, the giant company continues to make products that stir the passions of devoted customers.
Grassroots Marketing	By definition, it means organizing and motivating volunteers to engage in personal or local outreach.
Influencer Marketing	Identify key communities and opinion leaders who are likely to talk and have the ability to influence the opinions of others.
Viral Marketing	Viral marketing is the word of mouth or word of mouse spreading of a message directly from consumer to consumer. It is usually aided by the marketer with easily shared photos, videos, and sound files. "Gangham Style" by Psy certainly took advantage of the ease of sharing YouTube videos to become a worldwide sensation.

Source: (Kulkarni, 2012)

Word-of-mouth promotion as defined by Malcolm Gladwell in his book, *The Tipping Point*, shows the marketer that this seemingly simple strategy can actually have a deep-rooted sophistication. He suggests there are three rules to its effectiveness. First, there is the law of the few people who are connected to many others, and who can spread the word about a product in ways that have epidemic proportions.

MALCOLM GLADWELL'S THREE RULES

For Effective Word of Mouth Marketing

1. Get your message to the social Connectors, Mavens, and "Salesmen"
2. Make your message memorable or "sticky"
3. Be sensitive to the environment, to the conditions in which you are introducing the message

—from *The Tipping Point*

Second, the law of stickiness suggests that the idea shared by the connector has to be memorable and must be able to move people to action. And third, the power of context means that people "are a lot more sensitive to their environment than they may seem" (Gladwell, 2000, p. 29) and the effectiveness of word of mouth has a lot to do with the "conditions of, and circumstances of, the times and places" (Gladwell, 2000, p. 139) where it happens.

A great deal of care must be taken to set up a word-of-mouth promotion to make it effective. The sources of word-of-mouth campaigns will ultimately determine whether the effort was effective, so it is important to keep participants reminded about the ethics of this type of promotion.

For the record label, coordinating the energy and passion for artists through their fans can generate sales. Dave Balter, owner of BzzAgent, a word-of-mouth marketing firm, says, "The key is about harnessing something that's already occurring. We tap into people's passions and help them become product evangelists" (Strahinich, 2005). And that is the essence of the word-of-mouth marketing strategy as it is applied to the recording industry.

STREET TEAMS

The idea of street teams is an adaptation of the strategy of politicians everywhere: Energize groups of volunteers to promote you by building crowds, creating local buzz, posting signs, using the strength of the

Internet, and rallying voters. In very practical ways, there isn't a lot of difference between political volunteer groups and artist street teams.

For the record label, the ultimate rallying point is getting consumers to purchase the artist's music. While many artists use street teams today, they were originally formed in order to promote music that was not radio-friendly. Without radio airplay, creators of alternative music sought other ways to connect consumers with their music, and employing street teams became an effective way to do that. And the commissioning of street teams today continues in keeping with their origins—where the Internet has its e-teams, non-mainstream labels rely on street teams to promote its roots and alternative artists. And there are occasions when the major labels will hire third party street team companies as mentioned later in this chapter. (Harbin)

For the artist, team members are "marketing representatives" who promote music at events and locations where the target market or consumer can be found. Often times, team members make up the core of the fan base of the artist and have the deepest passion for the music and message of the artist. Beyond being a fan, they can be friends with the artist too, who promote music at events and locations where the target market can be found. These places are tied to the lifestyle of the target market such as at specialty clothing stores, coffee houses, and at concerts of similar acts. A key to effective street teams is for members to understand where to find the target market of the artist and to provide tools and guidance on how to communicate to the target (Tiwary, 2002).

Street teams originally were formed as a way to reach consumers by companies who did not have the resources for mass-media marketing and to reach segments of the market which were not as responsive to mass mediated messages as they are to peer influence (Holzman, 2005). Now, major marketing and advertising agencies acknowledge street teams as an effective form of youth marketing.

Case study: Columbia Records and Switchfoot

"UBI was enlisted to help introduce the band Switchfoot, and their latest album, "The Beautiful Letdown", to the masses. Switchfoot had found success in the Christian Rock format, but their label was interested in creating crossover appeal by targeting college campuses, lifestyle/community locations, clubs/bars, and record stores in 10 major cities across the United States. The goal was to increase awareness of the band, to bring attention to their live performances and to drive sales of their new release.

"Universal Buzz Intelligence mobilized their corps of highly trained operatives for an intense 9 week promotional campaign in 10 major markets. The campaign focused on the distribution of promotional materials and other collateral marketing materials throughout college campuses and the communities that surround them. The guerilla marketing effort focused heavily on driving retail store sales for "The Beautiful Letdown" album, as Well as promoting the band's live tour dates.

"By all accounts the Switchfoot release "The Beautiful letdown" far exceeded label expectations with over one million copies sold to date (the band sold an aggregate of 100,000 albums for their previous three releases). Of the 10 markets covered by UBI, the band realized 7 sold out performances. In addition, six of their top ten selling markets were markets covered by UBI (which had not previously been strong markets for the band), with the remaining 4 cities in the UBI campaign falling within the bands top 25 selling markets. The bands crossover popularity has brought them video exposure on MTV an The Fuse Network."

Source:
http://www.universalbuzzintelligence.com/case/switch.htm

FIGURE 18.1   *Universal Buzz case study*

Source: Universal Buzz Intelligence

Among the tools that may be provided to the street team members are postcards and flyers, email lists to contact local fans, small prizes for local contests, links for music samples, advance information about tour appearances, and release dates for new music. Street team members may engage in sniping—the posting of handbills in areas where the target market is known to congregate. Some labels economize on printing by emailing flyers and posters to local street team coordinators and ask them to arrange for printing.

Street teams require servicing. This means having continuing communication with key members and finding ways to say "thanks" to the street members. Volunteers for causes require a measure of recognition for their effort in order to keep them energized, and it is no different with those working for free on behalf of an artist. The first thing the label coordinator must do is to regularly communicate with core members of the team. Keep them current on the planned activities of the artist, and make them feel they are important. Provide incentives to the extent the budget will allow. Incentives can be free music, tee shirts, meet-and-greets with the artist, and free tickets.

The hierarchy of street team management begins at the label with a coordinator, who recruits regional street team captains who then coordinate at the local level. Data captured by artists through their pages on social networking sites as well as on their primary website become an important building block for street teams. Viral marketing is driven by data captured about the target market and then effectively using it to meet the needs of the artist's fans for information and entertainment.

THE BLURRED LINE FROM GRASSROOTS TO GUERILLA MARKETING

Guerilla marketing is another term closely associated with grassroots marketing. Whereas grassroots marketing refers to a street-level, door-to-door campaign of peer-to-peer (P2P) selling, guerilla marketing refers to low-budget, under-the radar niche marketing, using both P2P and any inexpensive top-down methods.

Grassroots marketing for large companies may involve a large budget to pay street teams. On the Internet, grassroots marketing includes placing conversational presentations of the label's product or including the artist in discussion forums and message boards, placing self-produced videos and music videos online, and encouraging bloggers to write about your artist. This is done by recruiting fans to be on the artist's street team or e-team.

Large companies hire members of the target market to spread the word to peers about the artist or product. On the Internet, these online street teams seek out Web locations where the target market tends to congregate and infiltrates these areas to introduce the marketing message.

Online street team members often use the following activities to promote recording artists:

1. Posting on Socials. Visiting social media sites and posting materials including music, artwork, and videos is mandatory for street teams. Major record labels use interns and young entry-level employees to maintain their artists' presence on social networking sites. These young marketers are responsible for setting up the artist's page on each of these sites, fielding requests from fans to be added to the social group or "friends network," providing updated materials (music, news, photos, videos, etc.), and visiting related pages to engage in street team promotions. Some labels have teamed up with these sites to conduct promotions in the form of contests and giveaways such as encouraging fans to create their own YouTube music video for an artist.

2. Visiting and participating in chat rooms. Chat rooms allow for real-time interaction between members, whereas bulletin boards allow individuals to post messages for others to read and respond to. Because these members or users have a mutual interest in the site topic, user groups offer an excellent way for labels to locate members of the artist's target market. Marketing professionals, as well as web surfers, often engage in "lurking" behavior when first introduced to a new users group. Lurking involves observing quietly—invisibly watching and reading before actually participating and making yourself known. Often, user groups have their own style and "netiquette" (Internet etiquette), and the label needs to learn these rules before jumping in.

3. Blogging or successfully submitting information to bloggers. Successful bloggers (those with a substantial audience) are opinion leaders, and their message can influence the target market. Online street team members will seek out blogs that discuss music and reach out to the bloggers to check out their artist and write about the music, much the same way a publicist will seek out album reviews in traditional media. Blogging, however, has the potential for street cred(ibility) that is sometimes absent in mainstream media.

4. Pitching/promoting to online media. This is much like pitching to bloggers only more formal. Under these conditions, professional

media materials must be supplied to the publication, whether it's an online-only publication or one that also has a traditional media presence. The label may supply completed articles and press releases to busy journalists who don't have time to conduct their own research and whose decision to run an article may be influenced by convenience. See the publicity chapter in this book for additional strategies to reach bloggers and other online media.

5. Finding fan-based web sites and asking the site owners to promote the artist. We address this strategy in the section on fan-based sites in the New Media chapter. It is often up to online street team members to seek out and identify these fan-based sites. Sometimes they engage in dialogue with the site owner, but often the list of fan-based sites is compiled by the street team member and then passed along to a more senior member of the marketing team. Then these sites are managed by more experienced web marketers who can supply RSS feeds and other technical web assets to the site owner.

6. Researching sites that attract the target market and then working with those sites (see reciprocal links information in Research chapter). The following section on turning an artist's competition into partners outlines many of the street team tasks performed in this area. The foremost task is finding and identifying where the target market can be found engaging others on the Web—what web sites do they frequently visit? Then the label can make marketing decisions about how to work with those sites, whether through advertising on those sites or setting up reciprocal links to generate cross traffic.

7. Writing reviews of the artist or album on sites that post fan reviews. Many retail sites such as Amazon allow customers to post product reviews, including music reviews. It is not uncommon for artists to ask their fans to post reviews on these sites or for a label to have street team members post favorable reviews.

Any discussion of guerilla marketing must acknowledge the contribution of Conrad Levinson. He is the author of the best-selling book, *Guerilla Marketing*, first published in 1984. Levinson is credited with coining the term, which generally means using nontraditional marketing tools and ideas on a limited budget to reach a target market. In Levinson's words, guerilla marketing is "achieving conventional goals, such as profits and joy, with unconventional methods, such as investing energy instead of money" (www.gmarketing.com, 2014).

BASIC TOOLS FOR STREET TEAMS INCLUDE

Post cards, flyers, concert posters, and stickers
Small prizes for local contests
CD samplers and music links to give away at appropriate events
Advance information about tour appearances
Behind-the-scenes photos to distribute
Mix tape giveaways
Release dates for new music

GUERILLA TACTICS

Guerrilla Marketing was a book and term used in 1984 by Jay Conrad Levinson. The term initially referred to the idea of implementing an unconventional system of promotions that rely on time, energy and imagination rather than a big marketing budget. Guerrilla Marketing is under-the-radar and often implemented in a localized fashion and the objective is to create a unique experience that generates buzz. Guerrilla marketing targets consumers in unexpected places to elicit the best reactions.

The lines between the terms Grassroots and Guerrilla Marketing have definitely blurred in recent years. And in the entertainment business, the line is more than fuzzy.

One of the main differences that people site between the two is that guerrilla marketing sometimes has a negative connotation in the promotions industry because the tactics are usually carried out without approval of the appropriate people. The new term coined for this type of tactic is ambush marketing. For instance, setting up a booth to promote an artist at a local festival would be considered grassroots marketing. If you set up a stage just outside of the footprint of the festival without the permission of the people running the event and without paying any fees to the organizing group, that would be considered ambush marketing. Unofficial sponsors of a major televised event, such as the Super Bowl or the Olympic games, will buy local or cable advertising during the broadcast in order to appear as official sponsor.

So today, guerrilla marketing is loosely defined to refer to the use of non-mainstream tactics and locations to gain attention and generate the maximum amount of buzz. We want to do something that's going to break through the clutter and make someone pay attention to us. An important note—today, guerrilla marketing stunts can be quite elaborate and expensive.

FIGURE 18.2

Guerrilla marketing tactics include sticker bombing, flash mobs, graffiti or reverse graffiti as pictured here, wild posting campaigns and wait marketing like this luggage carousel at an airport obviously promoting a casino in the area. This Colgate effort and the 3M security glass display would definitely make pedestrians stop in their tracks.

Christmas Food Court Flash Mob, Hallelujah Chorus - Must See!

Alphabet Photography

Subscribe 9,894

44,689,421

+ Add to < Share ••• More 113,940 3,134

Uploaded on Nov 17, 2010
http://www.AlphabetPhotography.com - On Nov. 13 2010 unsuspecting shoppers got a big surprise while enjoying their lunch. Over 100 participants in this awesome Christmas Flash Mob. This is a must see!

FIGURE 18.3

Many of us are aware of flash mobs, where groups of people "look" to spontaneously create an unlikely event, like sing the "Hallelujah" chorus in the food court in the mall. This photo shows a flash mob that was orchestrated by Alphabet Photography, a company out of Ontario, Canada that creates personalized alphabet letter art. The video was produced in 2010 and has been viewed more than 44 million times, generating a lot of awareness and buzz for the company and its products.

A good example of what not to do can be found with the Cartoon Network's 2007 campaign to promote the cartoon Aqua Teen Hunger Force. The cable network planted dozens of blinking electronic devices in 10 cities. Unfortunately, they forgot to inform local authorities of their efforts. A worried resident in Boston called police thinking the devices were explosives.

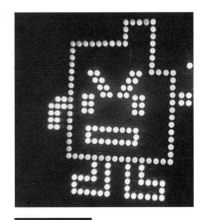

FIGURE 18.4

The incident turned into a full-blown terrorism scare complete with bomb squads being sent in to investigate and the shutting down of Boston-area bridges. In the end, the stunt cost the head of the network his job and Turner Broadcasting $2 million in compensation (*USA Today*, 2007).

Basic grassroots marketing, whether you call it word of mouth, street teams, e-teams, online teams, peer-to-peer, guerilla marketing, viral marketing, or the latest term du jour, can create an environment that can set a record label's marketing plan and results apart from those of competing artists. One of the biggest challenges of a label-marketing department is to find new and unique ways to present its artist and music to consumers. Including an element of grassroots in the marketing plan can create a plan that steps away from overused templates to add a unique element.

GLOSSARY

Bloggers—Short for weblogs, and refers to those who write digital diaries on the Internet.

Grassroots marketing—A marketing approach using nontraditional methods to reach target consumers.

Guerilla marketing—Using nontraditional marketing tools and ideas on a limited budget to reach a target market.

Marketing representatives—Another term sometimes used for members of a street team.

Sniping—The posting of handbills in areas where the target market is known to congregate.

Street teams—Local groups of people who use networking on behalf of the artist in order to reach their target market.

Video bloggers—Also known as vloggers, they are the video counterparts to bloggers except the content contains audio and video.

Viral Marketing—the use of social networking services to increase brand awareness.

REFERENCES

Gladwell, M. *Tipping Point*, New York: Little, Brown, and Company. 2000.

"Guerilla Marketing." Entrepreneur. Entrepreneur Media, n.d. Web. November 15, 2014. http://www.entrepreneur.com/encyclopedia/guerrilla-marketing.

Harbin, J. Personal interview. 2009.

"Head of Cartoon Network Resigns Over Marketing Stunt." *USA*. USA Today, February 9, 2007. Accessed November 15, 2014.

Holzman, K. "Effective Use of Street Teams." Music Dish Network. February 20, 2005. Accessed November 15, 2014. http://www.indiemusician.com/2005/02/effective_use_o.html.

Intelliseek. "Word-of Mouth in the Age of the Web-Fortified Consumer." 2005. Accessed November 15, 2014. www.intelli-seek.com.

Kulkami, K. "8 Types of Word of Mouth Marketing." *Weblaa*. N.p., December 12, 2012. Accessed November 15, 2014.

Rogers, E. M. *Diffusion of Innovations* (4th edition), New York: Free Press. 1995.

Strahinich, J. "What's All the Buzz; Word-of-Mouth Advertising Goes Mainstream." *The Boston Herald*. January 23, 2005. (LexisNexis Academic).

Tiwary, V. J. "Starting and Running A Marketing/Street Team." 2002. Accessed July 7, 2015. http://www.thomashutchison.com/IRI/Starpolish/Starting%20and%20 running%20street%20teams.pdf.

Word of Mouth Marketing Association, "Word of Mouth 101." 2005. www.womma. org. www.gmarketing.com. www.gmarketing.com/what_is_gm.html.

CHAPTER 19

Advertising in the Recording Industry

BASICS OF ADVERTISING

Advertising is a form of marketing communication. As described by University of North Carolina at Pembroke, "**Advertising** is mass media content intended to persuade audiences of readers, viewers or listeners to take action on products, services, and ideas. The idea is to drive consumer behavior in a particular way in regard to a product, service or concept" (Canno, 2014).

The fact that advertising is paid for and persuasive (i.e., the seller controls the content and message) separates it from other forms of mass mediated communication such as publicity, news and features. Advertisements are usually directed toward a particular market segment, and that dictates which media and which vehicles are chosen to present the message. A *medium* refers to a class of communication carriers such as television, newspapers, magazines, outdoor, and so forth. A *vehicle* is a particular carrier within the group, such as *Rolling Stone* magazine or MTV network. Advertisers determine where to place their advertising budget based on the likelihood that the advertisements will create enough of an increase in sales to justify their expense, in other words, their **return on investment** (ROI). Advertisers must be familiar with their market and consumers' media consumption habits in order to be successful in reaching their customers as effectively as possible.

The most basic market segments for advertising are: 1) consumers, and 2) trade, which are people within the industry that make decisions affecting the success of your marketing efforts. The first market (consumers) targeted through radio, television, out-of-home which includes billboards,

CONTENTS

FIGURE 19.1

street furniture such a bus benches, transit media including bus and train signage, along with alternative placement such as on-building wraps, direct mail, magazines, newspapers and the Internet. Consumer advertising is directed toward potential buyers to create a "pull" marketing effect (see Chapter 2).

The second target market is the *trade*, and consists of wholesalers, retailers, and other people known as gatekeepers who may be influenced by the advertisements and respond is a way that is favorable for the marketing goals. This creates a "push" marketing effect (again, see Chapter 2). In the recording industry, this would include radio program directors that may be influenced by an advertisement in *Billboard*, to more favorably consider a particular advertised song for inclusion in the station's weekly playlist. *Trade advertising* is usually done through direct mail and through *trade publications* (usually magazines or email) aimed at people who work in the industry. Trade advertising is not common through television, radio and newspapers because they are too general in nature—not targeted enough to effectively reach the industry.

CONSUMER ADVERTISING: THE MEDIA BUY

When considering a media buy, labels consider many factors:

- Media
- Budget
- Target market
- Timing
- Partners
- Artist Relations

There are advantages and disadvantages to the various mediums and why each might be chosen for a particular message.

COMPARISON OF MEDIA OPTIONS FOR ADVERTISING

The most complex issue facing advertisers involves decisions of where to place advertising. The expansion of media has increased the options and complicated the decision. The following chart in Table 19.1 represents a basic understanding of the advantages and disadvantages of the various media options.

Table 19.1	A Comparison of Media	
Media	**Advantages**	**Disadvantages**
Television	■ Reaches a wide audience, but can also target audiences through use of cable channels ■ Benefit of sight and sound ■ Captures viewers' attention ■ Can create emotional response ■ High information content	■ Short life-span (30–60 seconds) ■ High cost ■ Clutter of too many other ads; consumers may avoid exposure ■ Can be expensive
Magazines	■ High quality ads (compared to newspapers) ■ High information content ■ Long life-span ■ Can target audience through specialty magazines	■ Long lead time ■ Position in magazine uncertain ■ No audio for product sampling (unless a CD is included at considerable expense)
Newspapers	■ Good local coverage ■ Can place quickly (short lead) ■ Can group ads by product class (music in entertainment section) ■ Cost effective ■ Effective for dissemination of information, such as pricing	■ Poor quality presentation ■ Short life-span ■ Poor attention-getting ■ No product sampling
Radio	■ Is already music-oriented ■ Can sample product ■ Short lead, can place quickly ■ High frequency (repetition) ■ High quality audio presentation ■ Can segment geographically, demographically and by music tastes	■ Audio only, no visuals ■ Short attention span ■ Avoidance of ads by listeners ■ Consumer may not remember product details
Out-of-Home: Billboards, Street Furniture, Transit, Alternative	■ High exposure frequency ■ Lower cost ■ Can segment geographically	■ Message may be ignored ■ Brevity of message ■ Not targeted except geographically ■ Environmental blight

(Continued)

Table 19.1	(Continued)	
Media	**Advantages**	**Disadvantages**
Direct Mail	■ Best targeting ■ Large info content ■ Not competing with other advertising	■ High cost per contact ■ must maintain accurate mailing lists. ■ Associated with junk mail
Mobile	■ Good for "just-in-time" promos ■ Good for proximity marketing ■ Can reach targeted demos	■ Limited message capacity ■ Consumers must opt-in

Online	■ Highly targeted
	  ■ Psychographics
	  ■ Demographics
	  ■ Geographics
	■ Potential for audio and video sampling; graphics and photos
	■ Can be considered point-of-purchase if product available online
	■ Can use cost-per-click (CPC) instead of impressions for setting rates
		■ Cluttered environment
		■ Internet is vase and adequate coverage is elusive

Television Advertising

Television reaches a broad audience and now with cable, it's easier than ever before to target by interest: home owners, gardeners and DIYers might be reached through HGTV, those who love and/or own animals can be reached through the Animal Channel and music lovers might be able to be reached through MTV, CMT or GAC. "Might" because these channels are reducing their music programming hours and increasing reality TV and other programs that appeal to a wider audience. TV also has the benefit of sight and sound, which can be attention grabbing and can create an emotional connection to the product. Disadvantages are the high cost for placement and production of the spots and the short life span—it's hard to say everything you need to say in 30 seconds. Also, DVR penetration continues to grow with nearly 50% of all television households now having at least one DVR. And what's their favorite way to use those DVRs? You know it—to skip that great promo spot you just developed.

Terminology: **ROS or Run-of-schedule**. ROS just refers to the scheduling of a promo spot at any time the station wants to run it within a specified timeframe. Typically, this is much less expensive than if you specifically request a daypart or placement within a particular program. For example, a ROS spot may run at 10:45 a.m. on the station one day then 9:45 p.m. the next day.

Road Blocking is another term that is referenced mostly in television but can take place in other mediums as well. Have you ever switched a channel to avoid an annoying commercial only to see it running on the other major networks at the same time so that you really can't escape it? That's called road blocking.

FIGURE 19.2

PAGE 1

RATES AND DISCOUNTS
National Editions

	PEOPLE REGULAR	PEOPLE FEATURE ISSUES*†	PEOPLE SEXIEST MAN ALIVE‡	PEOPLE BEST OF 2014‡	NEWSSTAND SPECIALS
Rate Base	3,475,000	3,600,000	3,700,000	3,900,000	350,000
4-Color Rates:					
Full Page	$337,400	$352,000	$361,800	$381,300	$46,700
⅔ Page	$288,000	$300,400	$308,800	$325,400	$39,900
½ Page	$227,800	$237,600	$244,300	$257,400	$31,600
⅓ Page	$155,300	$162,000	$166,500	$175,400	$21,500
⅙ Page	$94,500	$98,600	$101,400	$106,800	$13,100
Cover 2	$421,800	$440,000	$452,300	$476,700	$58,400
Cover 3	$371,200	$387,200	$398,000	$419,500	$51,400
Cover 4	$455,500	$475,200	$488,500	$514,800	$63,100
Black & White Rates:					
Full Page	$236,200	$246,400	$253,300	$267,000	$46,700
⅔ Page	$201,600	$210,300	$216,200	$227,900	N/A
½ Page	$159,500	$166,400	$171,000	$180,300	N/A
⅓ Page	$108,700	$113,400	$116,600	$122,900	N/A
BRC Insert Cards:					
Reg - Supplied	$270,000	$281,600	$289,500	$305,100	$37,400
Reg - We-Print	$337,400	$352,000	$361,800	$381,300	$46,700
Oversize - We-Print	$388,100	$404,800	$416,100	$438,500	$53,800
Oversize - Supplied	$303,700	$316,800	$325,700	$343,200	$42,100

Circulation includes the print and digital editions of the Magazine. Qualified full-run advertisements will run in both editions. See MAGAZINE ADVERTISING TERMS AND CONDITIONS for additional information including opt-out and upgrade options.

*Feature Issues: Half Their Size, Oscar, 50 Most Beautiful, Hollywood's Hottest Bodies, The Style Issue
†On newsstand for two weeks

UPDATED: 5.28.14

FIGURE 19.3

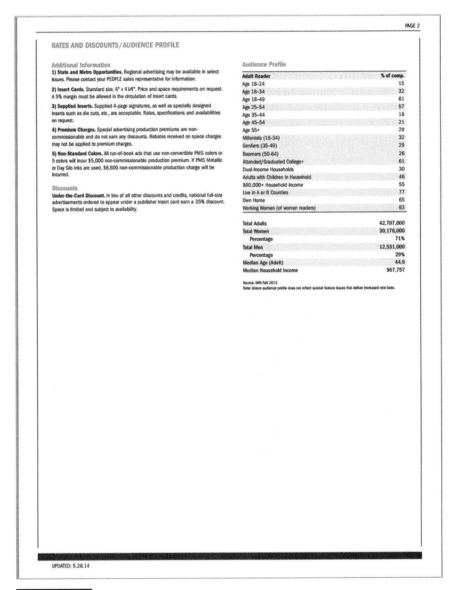

RATES AND DISCOUNTS/AUDIENCE PROFILE

Additional Information

1) State and Metro Opportunities. Regional advertising may be available in select issues. Please contact your PEOPLE sales representative for information.

2) Insert Cards. Standard size, 6" x 4 1/4". Price and space requirements on request. A 5% margin must be allowed in the circulation of insert cards.

3) Supplied Inserts. Supplied 4-page signatures, as well as specially designed inserts such as die cuts, etc., are acceptable. Rates, specifications and availabilities on request.

4) Premium Charges. Special advertising production premiums are non-commissionable and do not earn any discounts. Rebates received on space charges may not be applied to premium charges.

5) Non-Standard Colors. All run-of-book ads that use non-convertible PMS colors or 5 colors will incur $5,000 non-commissionable production premium. If PMS Metallic or Day Glo inks are used, $6,800 non-commissionable production charge will be incurred.

Discounts

Under-the-Card Discount. In lieu of all other discounts and credits, national full-size advertisements ordered to appear under a publisher insert card earn a 35% discount. Space is limited and subject to availability.

Audience Profile

Adult Reader	% of comp.
Age 18–24	15
Age 18–34	32
Age 18–49	61
Age 25–54	57
Age 35–44	18
Age 45–54	21
Age 55+	29
Millenials (18-34)	32
GenXers (35-49)	29
Boomers (50-64)	26
Attended/Graduated College+	61
Dual-Income Households	30
Adults with Children in Household	46
$60,000+ Household Income	55
Live in A or B Counties	77
Own Home	65
Working Women (of women readers)	63

Total Adults	42,707,000
Total Women	30,176,000
Percentage	71%
Total Men	12,531,000
Percentage	29%
Median Age (Adult)	44.9
Median Household Income	$67,757

Source: MRI Fall 2013
Note: Above audience profile does not reflect special feature issues that deliver increased rate base.

UPDATED: 5.28.14

FIGURE 19.4

Print Advertising

Magazines fall under the print medium and have the advantage of being able to communicate a lot of information. They also have a long life span—some magazines are passed along to others and some are kept for years. You can also target particular audiences through specialty publications like *Car and Driver* or *Seventeen* or *Country Weekly*. Disadvantages are the long lead time—you have to plan for an ad in a magazine and turn in ad copy and creative art work weeks prior to the printing. And unless you pay for premium positioning, your ad might be placed in a less than desirable location—like next to another ad or next to unrelated editorial.

This is an example of a print rate card for *People* magazine. You can see that is provides information like the rate base—which means the minimum number of subscribers and over the counter purchasers projected for those specific issues, then the rate for a four-color ad and black and white.

Typically on any type of rate card they will list the open rate, which is the highest rate you would be charged. The rates are almost always negotiable, especially in the print world. The rate card also provides an audience profile that details who reads the magazine. You can see at the bottom that they pulled this information from MRI, one of the syndicated research services discussed earlier.

Newspapers are another form of print advertising. Newspaper ads are great if you have short lead time and you can run an ad in the entertainment section, for instance, to reach people interested in those types of products. Disadvantages are the poor print quality, short life span, and the inability to really do any type of product sampling except if you were to print a digital code for a download or special content that they have to go online to "redeem." Also, the printed version of newspapers skews to an older buyer so knowing your target demographic is really important.

Although print advertising has decreased substantially in recent years and it is not a big expense, typically, for Record Labels, there are occasionally partnership opportunities that might provide exposure in print. Knowing print terminology will aid you when the occasion arises.

The center spread is the ad unit in the center of a publication. If you dropped a magazine on a table, it might naturally open to this spread because of the way the magazine is bound. **Circulation**, or circ, is the average number of copies per issue. **Closing date** is the final deadline that you may reserve space in a publication. A **double page** spread refers to an

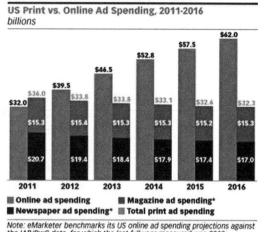

US Print vs. Online Ad Spending, 2011-2016
billions

	2011	2012	2013	2014	2015	2016
Total print ad spending	$32.0	$39.5	$46.5	$52.8	$57.5	$62.0
Magazine ad spending*	$36.0	$33.8	$33.8	$33.1	$32.6	$32.3
Newspaper ad spending*	$15.3	$15.4	$15.3	$15.3	$15.2	$15.3
Online ad spending	$20.7	$19.4	$18.4	$17.9	$17.4	$17.0

■ Online ad spending ■ Magazine ad spending*
■ Newspaper ad spending* ■ Total print ad spending

Note: eMarketer benchmarks its US online ad spending projections against the IAB/PwC data, for which the last full year measured was 2010; eMarketer benchmarks its US newspaper ad spending projections against the NAA data, for which the last full year measured was 2010; *print only
Source: eMarketer, Jan 2012

136019   www.eMarketer.com

FIGURE 19.5

ad that spreads across two facing pages. **FP4C** is shorthand for full page four color. **Guaranteed position** ensures a particular placement in a publication (as opposed to **run of print**, which means your ad could be placed anywhere).

The cover of a mag is referred to as cover 1. When you open up the cover and are looking at the page on the left, that is referred to as the **inside front cover, or cover 2. Cover 3** is the inside of the back cover, and **cover 4** is the back of the magazine. These are all referred to as "**premium positioning**" which means that they generally cost a lot more than just a run of print ad. Why? Because more people actually look at those than other pages in the publication.

If you want to place an ad in a magazine, or even if it's an ad that's being offered as a promotional trade, you will be asked to fill out an insertion order specifying the date of the issue you want. **The specs** list the mechanicals that you will want to provide to the designer. **The primary** reader is the one who receives the subscription or pays for the issue at newsstand.

The readers per copy take into account the initial subscriber or purchaser, as well as any other person who may have read the issue. These people are called **secondary readers**. For instance, the primary reader might leave an old magazine at her doctor's office where 10 additional secondary readers may thumb through it as well.

Radio Advertising

Radio Advertising and its reach has already been covered extensively in Chapter 12, but advantages are that radio is obviously music oriented and that makes product sampling really easy. It also boasts a short lead time, high quality and the ability to segment by musical taste or genre and by geographic area. But, radio spots are generally only 30 or 60 seconds long, they don't offer visuals and there's a lot of clutter with other ads so it's easy for a message to get lost.

Out-of-Home Advertising

Out-of-home advertising (OOH) refers to advertising that attempts to target the consumer outside of his or her home. There are four main categories: billboards; street furniture like bus shelters or kiosks; transit advertising like buses, ads on subways, ads in taxis, and airport advertising; and alternatives like bike racks, building wraps, stair wraps and gas pumps.

IMAGE 19.6

IMAGE 19.7

IMAGE 19.8

IMAGE 19.9

Billboard advantages are that there are a large number of people reached but a low cost-per-thousand people reached. It is also easy to target geographically using billboards. Disadvantages are, hopefully anyway, drivers are more concerned with watching the road than reading your billboard. And there really isn't a way to target a particular demographic group so there is the potential to have a lot of waste. One car may be driven by a 65-year-old male, while the next car has four teenage girls in it.

Messages on billboards need to be very brief—generally no more than eight words is a good target. The creative image should include large, easy to read font and lower case letters. Also, think about which font colors are easiest to read and use those (in other words, no purple letters on a black background) (http://cdn.creativeguerrillamarketing.com/wp-content/uploads/HLIC/f8c8330ea0d06e7fb9bb4515bee8ac75.jpg).

Direct Mail Advertising

Direct Mail has the advantage of being highly targeted—by zip code or even by a particular block or individual. However, this medium is associated with junk mail so it would be easy for an ad to be overlooked by the recipient. Depending on the type of direct mail utilized, it can also be very expensive.

Mobile Advertising

Mobile is great for "just in time" promos and it's very personal in that most people do read every text message they receive. You might be able to use it for geo-targeting, which is the practice of delivering different content to a website user based on his or her geographic location. Geo-targeting can be used to target local customers through paid (PPC) or organic search, but be careful with that—many users are taking their phone numbers with them when they move away from the city where they zip code is associated. Disadvantages are that you have a very short space in which to place a message and it is really easy to annoy consumers with too many text messages or with content that they don't find useful, either of which can prompt them to unsubscribe (Stuart, 2009).

Online or Digital Advertising

Online or digital advertising, offers the music industry a lot—we can finely target our audience, we can use video and audio, graphics and photos, and we can set up a campaign so that we only pay for people who care enough to actually click on our ad. Disadvantages are that the environment is extremely cluttered and many consumers have trained themselves to ignore most standard ad units. Also, the Internet is vast so finding adequate coverage to the target market may be challenging.

When you think about it, in addition to straight up online advertising opportunities, online marketing is really woven into every other medium. For instance, in our industry we have online radio, radio station websites, local and national TV station websites and online versions of entertainment publications. The opportunities are almost endless and clearly, this form of advertising should be woven into every part of a marketing plan.

Display advertising refers to an image-based way of promoting products and services Another important term is contextual advertising, which is a form of targeted advertising that would allow a user on a rock music website to be served up ads on related content like rock memorabilia or tickets to local rock concerts coming up in the area.

There are three main types of digital ads: There are **standard text ads**, which is what you typically see when you conduct a search for a product on Google. Then we have a standard display ad that has images and basic text. It generally will have only one function though. For instance, if you click on it the ad will take you to a website. **Banner ads** are a good example of this. Then we have **rich media ads**, which can include video and a lot of opportunities to interact with the consumer.

Text Ad	Standard Display Ad	Rich Media Ad
No images	Images & basic interactivity, no expansion	Can include video, high interactivity, expansion and more

IMAGE 19.10

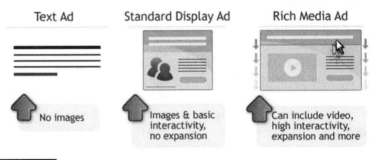

The Grammy Awards
www.**grammy**.com/ ▾ Grammy Awards ▾
Performers announced for GRAMMY holiday special airing Dec. 5 ... Jesse & Joy,
Ricky Martin and Pablo Alborán among first 15th Latin **GRAMMY Awards** performers.
Full Latin ... Set List Bonus: **2014** Made In America Festival, Los Angeles.

Ads ⓘ
Grammy Awards Tickets
www.vividseats.com/**GrammyAwards** ▾
4.6 ★★★★⯪ rating for vividseats.com
Securely Buy **Tickets**:
100% Ticket Guarantee-Official Site

IMAGE 19.11

IMAGE 19.12

This is an example of a standard text ad that sells with words.

This is an example of a Standard display ad. These types of ads may incorporate Flash but again, there is generally one interaction—when you click on it, it takes you to one site.

Rich Media ads offer much more interactivity with the audience. Rich media ads might expand when you click or roll over them or allow videos to be played inside the ad unit itself.

This advertisement allows for interaction with device simulation along with a view of actual "record breaking event."

IMAGE 19.13

Search engine advertising is a core component to every marketing plan. Ads can be on search engines themselves. When searching for "Christian Music" on Google, key word search profiles lift to the top several ad-driven items for searchers to view first.

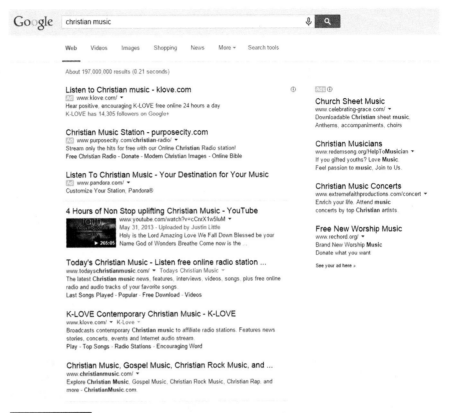

FIGURE 19.14

How does this work? Advertisers pay on a per-click basis; in other words, they pay a few cents for each time a web visitor clicks on their sponsored link. The advertiser enters in a series of keywords (search terms or words their customers are likely to use in a search engine when looking for a particular product or type of web site). Advertisers place bids to have their ad strategically located in the sponsored links category on search engine results. The advertiser can actually create a list of terms and bid independently on each one. The highest bid for that particular set of search terms has the top spot. You may not want to be the top spot for blues music if you depend on live shows for income because blues fans from all over would be likely to click to your site only to learn that you are not performing in their area. But if the term "blues music" was combined with "East Texas" and you are performing in that area, then perhaps you want one of the top advertiser spots. Finding the right keywords and combination of keywords may take a bit of trial and error at first.

Contextual Advertising

Contextual advertising is defined as advertising on a web site that is targeted to the specific individual who is visiting the web site based on the subject matter of the site and then featuring products that relate to that subject matter. For example, if the user is viewing a site about playing music and the site uses contextual advertising, the user might see ads for music-related companies such as music stores. Google has added AdSense as a way for web site owners to feature relevant advertising on their sites and share in the CPC revenue generated by sponsors. The source of these ads comes from the AdWords program, so that those who sign up for AdWords can specify if they want their ad to appear on these related web sites—the *content network*. Google's website states, "A content network page might be a web site that discusses a product you sell, or a blog or news article on a topic related to your business."

Look on other websites with whom search engines partner based on the specific keywords or phrases the marketer is targeting. For example, this contextual ad for Publix Grocery Stores was generated on a website called Everyday Health. The keywords Publix's likely used when they placed the ad on the ad network could have been "healthy eating," "weight loss," "diet," or "menu planning" (http://www.everydayhealth.com/groups/.

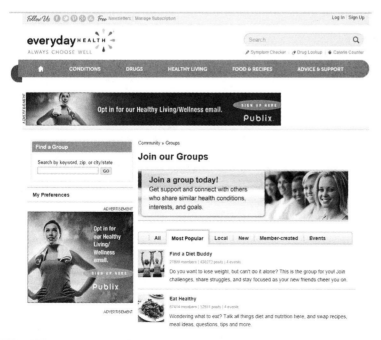

FIGURE 19.15

Digital advertising is typically sold on a **cost per click, or CPC,** basis. This is also referred to as Pay per click, which simply means that an ad might be seen by millions of people but you are only charged for the ones who actually clicked on it.

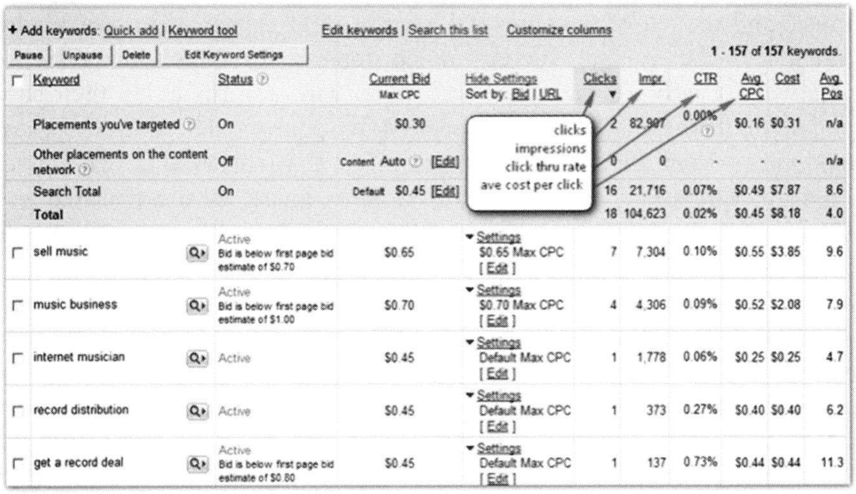

Keyword	Status	Current Bid Max CPC	Hide Settings Sort by: Bid | URL	Clicks	Impr.	CTR	Avg. CPC	Cost	Avg. Pos
Placements you've targeted	On	$0.30		2	82,907	0.00%	$0.16	$0.31	n/a
Other placements on the content network	Off	Content Auto [Edit]		0	0	-	-	-	n/a
Search Total	On	Default $0.45 [Edit]		16	21,716	0.07%	$0.49	$7.87	8.6
Total				18	104,623	0.02%	$0.45	$8.18	4.0
sell music	Active Bid is below first page bid estimate of $0.70	$0.65	Settings $0.65 Max CPC [Edit]	7	7,304	0.10%	$0.55	$3.85	9.6
music business	Active Bid is below first page bid estimate of $1.00	$0.70	Settings $0.70 Max CPC [Edit]	4	4,306	0.09%	$0.52	$2.08	7.9
internet musician	Active	$0.45	Settings Default Max CPC [Edit]	1	1,778	0.06%	$0.25	$0.25	4.7
record distribution	Active	$0.45	Settings Default Max CPC [Edit]	1	373	0.27%	$0.40	$0.40	6.2
get a record deal	Active Bid is below first page bid estimate of $0.80	$0.45	Settings Default Max CPC [Edit]	1	137	0.73%	$0.44	$0.44	11.3

FIGURE 19.16

This is a chart of media ad spending by category for 2011 projected until 2017. You can see that TV is still expected to generate the most in ad

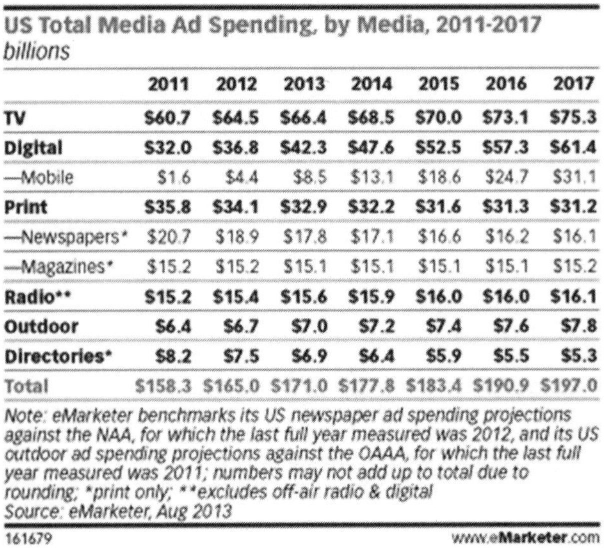

US Total Media Ad Spending, by Media, 2011-2017
billions

	2011	2012	2013	2014	2015	2016	2017
TV	**$60.7**	**$64.5**	**$66.4**	**$68.5**	**$70.0**	**$73.1**	**$75.3**
Digital	**$32.0**	**$36.8**	**$42.3**	**$47.6**	**$52.5**	**$57.3**	**$61.4**
—Mobile	$1.6	$4.4	$8.5	$13.1	$18.6	$24.7	$31.1
Print	**$35.8**	**$34.1**	**$32.9**	**$32.2**	**$31.6**	**$31.3**	**$31.2**
—Newspapers*	$20.7	$18.9	$17.8	$17.1	$16.6	$16.2	$16.1
—Magazines*	$15.2	$15.2	$15.1	$15.1	$15.1	$15.1	$15.2
Radio**	**$15.2**	**$15.4**	**$15.6**	**$15.9**	**$16.0**	**$16.0**	**$16.1**
Outdoor	**$6.4**	**$6.7**	**$7.0**	**$7.2**	**$7.4**	**$7.6**	**$7.8**
Directories*	**$8.2**	**$7.5**	**$6.9**	**$6.4**	**$5.9**	**$5.5**	**$5.3**
Total	$158.3	$165.0	$171.0	$177.8	$183.4	$190.9	$197.0

*Note: eMarketer benchmarks its US newspaper ad spending projections against the NAA, for which the last full year measured was 2012, and its US outdoor ad spending projections against the OAAA, for which the last full year measured was 2011; numbers may not add up to total due to rounding; *print only; **excludes off-air radio & digital*
Source: eMarketer, Aug 2013

161679 www.eMarketer.com

FIGURE 19.17

sales, but look at the projected growth of digital and within digital is mobile advertising—again, huge increases expected. Magazines are expected to hold steady, which newspaper advertising (and this reflects print only) will continue its decline. Radio and outdoor are each expected to increase slightly.

Social Media Advertising

Advertising options on social networks continue to grow. Sites like Twitter, Facebook, and Instagram provide a cost-efficient way to reach a preferred target market.

Facebook

Facebook is able to provide laser-focused targeting by combining its own data from user activity with information provided by third-party providers of syndicated consumer data, giving marketers access to more than 500 partner categories including Demographics, Home, Financials, Vehicles, Travel, Hobbies and Interests, Lifestyles, Charitable and Political Affiliations, and more. For instance, if you want to target ads to people 18–24 years old who like pages related to Beyoncé and are anticipating the purpose of a vehicle in the next six months you can do that through a combination of both Facebook and third-party information. You can also use Facebook's Custom Audiences feature to use your existing database (e.g., your list of email subscribers) to target those people on the platform.

Facebook has made it easy for even novices to implement an ad campaign by first asking advertisers to identify an objective (e.g. engagement, page likes, clicks to a website), then walking you through the options step by step. The most commonly used ad types on Facebook are Page Post Ads and Promoted Posts. Display options include Desktop News Feed, Right Sidebar (the smaller ads to the right of the news feed) and Mobile News Feed.

As mentioned, targeting capabilities in Facebook are vast including: Location, Age, Gender, Relationship Status, Language, Education, Workplaces, Financial, Home, Market Segments (e.g. ethnicity, generation, family size, moms), Parents, Political affiliation, Life events, and more.

There are three pricing options for Facebook ads:

1. CPC or Cost Per Click: the advertiser only pays when someone clicks on the ad

2. CP (Thousand) Impressions—or CPM: the advertiser pays each time an ad is viewed even if there is no interaction with the ad

3. Optimized CPM: the advertiser agrees to let Facebook use its data to serve the ad to the most likely to take the action you want. For example, if your objective is "App Installs" your ad might be shown to the people

in your defined target audience who have a history of downloading apps. With this method you are still paying every time someone sees your ad but because the algorithm is identifying the best targets for you there is less "waste" involved, resulting in a more efficient buy.

Twitter

Twitter follows a similar approach to Facebook in that they ask advertisers to first state their objective (e.g., tweet engagement, app installs, website clicks, video views) which will determine the correct type of ad, of which there are three main types:

1. Promoted Tweets—tweets that show up in the main news feed.

2. Promoted Accounts—shows up in the "who to follow section"

3. Promoted Trends—shows as "promoted" in the "Trends" list. This option allows an advertiser to be at top of the trending list for 24 hours and is only available to large advertisers as the cost can be substantial.

For Promoted Tweets, Twitter considers the quality of tweets, in addition to the budget, in determining with ad units to show. Generally, favored tweets include those from accounts who have demonstrated engagement with consumers (do they re-tweet, favorite or reply often to the tweets?), is the tweet related to subjects the user is interested in, and is the tweet timely. The last one is important as more timely tweets get priority.

Twitter also has robust targeting options including location, gender, language, interests and desires. Ad advertiser may choose to target specific keywords (searches or tweets) or specific television programs watched. An advertiser may also select people with specific interests or people who are similar to followers of specific accounts. As with Facebook, Twitter allows advertisers the ability to reach tailored audiences using data like email addresses.

Advertisers pay for Twitter ads on a CPE (Cost per Engagement) basis.

1. Promoted Tweets—the advertiser only pays when someone clicks, retweets, replies to or favorites an ad.

2. Promoted Accounts—the advertiser only pays when someone follows.

Instagram

Instagram is slowly rolling out ads across its platform, beginning with a few select partners that were already strong players in the Instagram community who have organically built a loyal following. Criteria for ads are more restrictive on this platform than other social networks with a focus on high quality and engaging ads. As such, ads must meet certain criteria such as:

- No logos in ads other than as a natural part of the scene
- No product features in the images
- Images used should be authentic to the partner's brand

Instagram also encourages the design of ads with mobile in mind to better serve that large and growing audience.

Other digital advertising options in the music industry include retargeting ads that places a piece of code on artist and industry websites that allows an advertiser to present an ad to that consumer on social networks or other websites; ads on YouTube such as banner ads, in-video overlay ads and in-stream video ads; and advertising on online radio and streaming services like website placement sold by radio station groups, Spotify and Pandora.

MEDIA PLANNING

Media planning involves the decisions made in determining in which media to place advertisements. It consists of a series of decisions made to answer the following questions:

1. How many potential consumers do I need to reach?

2. In what medium (or media) should I place ads?

3. When and how often should these ads run?

4. What vehicles and in which markets should they run?

5. What choices are most cost-effective?

It also involves analyzing the costs to determine which mediums are most cost effective.

HOW ADVERTISING EFFECTIVENESS IS MEASURED

As we learned in the radio chapter, the total number of unique views of an ad over a period of time is referred to as the **Cume**. A household or person is counted only once even if the ad was viewed by that person multiple times. This is also known as net unduplicated audience or net reach. Cume, or **Net Reach**, may be expressed as an absolute number or as a percentage of the population.

Frequency is the number of times a consumer or population is exposed to the ad message.

Reach X Frequency = Gross Impressions.

Gross Impressions are the total number of times an ad was viewed. The figure counts duplicates or repeated viewings by the same person. For example, one million Gross Impressions could be one million people each exposed once or 10,000 people each exposed 100 times or any other such combination of numbers. So it includes duplicate views by the same person.

Another Example:

Visitor A viewed: Ad1, Ad2, Ad3, Ad2
Visitor B viewed: Ad1, Ad4
To count: Ad1 has 2 unique (views) and 2 gross impressions
Ad2 has 1 unique (views) and 2 gross impressions
Ad3 and Ad4–1 unique and 1 gross impression each

A **rating point** is a value that is equal to one percent of the total population or households that are tuned into a particular program or station at a specific time. For example, a six rating for women 18–49 means that six percent of all women 18–49 in a specified geographic region were viewing that station or program.

Gross Rating Points (or GRPs) quantify impressions as a percentage of the population reached rather than in absolute numbers.

GRPs measure the sum of all Rating Points during an advertising campaign without regard to duplication. Just like in our recent Gross Impressions example, a GRP of 100 could mean that you bought one hundred spots with a one percent reach or that you bought 2 spots with a 50% reach.

GRPs are calculated by determining the reach of your spots (which is a function of the stations or programs you purchase) times the frequency or number of spots.

GRP = Reach X Frequency

Each GRP represents one percent. For example, if 25 percent of all televisions are tuned to a show that contains your spot, you have 25 Rating

Points. If, the next time the show is on the air, 30% are tuned in, you have a total of 25+30 or 55 Rating Points.

Targeted Rating Points or TRPs are similar to GRPs, but they express the reach times frequency of only your most likely prospects. For example, if you were marketing makeup for women and knew that the program in which your message ran consisted of 50% women, your TRP would be half your GRP.

In other words, GRPs relate to the total audience exposure to the message whereas TRPs relate to the target audience exposure.

Cost per Point is used by most media planners as a way to determine the cost for reaching one percent of the target audience. The average cost per spot divided by the average rating point per spot would give you this number.

> *CPP= Avg. cost per spot/Avg. rating point per spot*
>
> *or*
>
> *CPP= Cost of schedule/Gross ratings points*
>
> www.tvb.org

Cost per Thousand or CPM is the cost of reaching 1,000 homes or individuals with your message. By the way, M is the roman number representing 1,000, that's why it's CPM. So in this example, if Option A costs $50,000 and reaches 100,000 people and Option B costs $25,000 but only reaches 40,000 people, how do you determine which option gives you the biggest bank for your buck? Figure out the cost of reaching one thousand people and you can see that Option A is the most efficient buy.

Example:

> *Option A costs $50,000 and reaches 100,000*
> *   CPM = $50,000/100=$500*
> *Option B costs $25,000 and reaches 40,000*
> *   CPM = $25,000/40= $625*

Advertising and its impact is not something that is easy to measure. Sure, we can measure the individual units within a campaign, but for a national effort you might have national and local radio, national and local television spots, online ads, publicity efforts, promotions and other elements that all contribute to the success or failure of a project. It's really a matter of trying to test as many variables as you can to help pinpoint the main drivers of sales.

Types of Trade Advertising

FIGURE 19.18

FIGURE 19.19

Co-op or cooperative advertising generally refers to the joint promotional effort of two or more parties in the selling chain, for example, a release being featured in a Best Buy circular, or premium placement of the product in exchange for ad dollars. This financial implication of this type of advertising is discussed in the Distribution and Retail chapter but here are some examples.

These are all examples of Co-op advertising. There's an end cap display in a music store. Then you have in the lower left, an album Point of Purchase poster that is "available at Target"—in this case, the record label paid for favorable in-store positioning of the album. Then to the right, we have an example outside of the music industry—Amazon Prime and American Express partnering on a banner ad.

Television and Radio Campaign: Here is the schedule of an actual ad buy for the CMA Awards show in Atlanta that included both tv and radio. You can see the Rating Points purchased, the cost $6,000 and the Cost Per Point which is 800 or 6,000 divided by 7.5. One spot was purchased and that generated 75,000 impressions among women 25–54 for a Cost per Thousand or CPM of 80.

FIGURE 19.20

| Table 19.2 | Sample TV Advertising Buy | | | | | | | |

MARKET	PURCHASED			# of SPOTS	WM: 25–54		HOMES	
DAYPART	PNTS	DOLLARS	CPP		IMPS	CPM	IMPS	CPM
ATLANTA								
PRI	7.5	6000	800	1	75	80	115	52.17

	Market:	Atlanta
PRI	Daypart:	Primetime
PNTS	GRP:	7.5
	Cost:	$6,000.00
CPP	Cost/Point:	$800.00
	# of Spots:	1
IMPS	# of Impressions of Target Market—Women 25–54:	75,000
CPM	Cost/Thousand on Target:	$80.00
IMPS	# of Homes:	115,000
CPM	Cost/Thousand Of Homes:	$52.17

This television advertising buy was bolstered by additional ad exposure at radio. These time buys were during the week leading up to the CMA Awards Show.

Table 19.3	Radio Buy							
MARKET	**PURCHASED**			**# of SPOTS**	**WM: 25–54**		**HOMES**	
DAYPART	**PNTS**	**DOLLARS**	**CPP**		**IMPS**	**CPM**	**IMPS**	**CPM**
ATLANTA								
AMD	40	8000	200	27	440	18.18		
PMD	35	7000	200	23	375	18.66		
TOTAL	75	15000	200	50	815	18.4		
	Market:				Atlanta			
AMD	Daypart:				AM Drive			
PMD					PM Drive			
PNTS	GRP:				75			
	Cost:				$15,000.00			
CPP	Cost/Point:				$200.00			
	# of Spots:				50			
IMPS	# of Impressions of Target Market—Women 25–54:				815,000			
CPM	Cost/Thousand				$18.40			

Radio Ad Buy

The combined advertising gross rating point (GRP) for the Atlanta market was 82.5. Although $15,000 was spent at radio, the CPM was dramatically lower than at television. Additionally, many more women aged 25–54 were reached because of the focused advertising that radio and its specific targeted demographics can deliver. Generally, radio advertising does not include gross impressions of homes.

With any budget, the greater the GRP, the better the advertising "bang for the buck." An actual advertising campaign would be comprehensive in attempting to reach the target market while maintaining a budget. To do so, buys at many outlets would be secured. The optimal equation to increase the GRP includes varying the combinations of media, dayparts, and number of spots. Depending on the agenda as well as the budget, all advertising should help increase visibility and sales of a specific artist.

Online Campaign

As mentioned earlier, many labels have opted to use fans as their mouthpiece by giving them tools such as widgets to virally spread news about artists and their new releases. But online advertising campaigns can be directed in a way to maximize dollars while creating exposure and immediate sell-through. Instead of using impressions, as print advertising does, online ad costs are based on PPC or Pay-Per-Click impressions, where sites charge those advertising a fee based on consumers who have "clicked" through the actual banner ad. By researching analytics of website activities, labels can derive best sites on which to advertise, based on all sorts of great data: gender, age, ethnicity, income, education, and lifestyles. By marrying the target market of the artist with website analytics, the probability of click-through traffic increases dramatically—and the impression of hearing and selling the artist increases as well.[1]

FIGURE 19.21

The online music magazine, No Depression targets Americana—roots music. Their advertising fees are strictly impression-based, but the actual ads are interactive, meaning that they have click-through and flash capability.[2]

FIGURE 19.22

---

[1] Jenn Barbin, UMG, Personal Interview, March 5, 2009.
[2] Kyla Fairchild, No Depression, Personal Interview, March 5, 2009.

The Steve Earle "Live At The BBC" release was a click-through banner ad that rotated on No Depression's home page. As an Americana artist, this banner ad placement made perfect sense, notifying the target market of the pending release a week prior to street date. To click through the banner, the reader could instantly become a consumer, landing on Amazon.com's site where a pre-order sale could occur along with the consumer being able to click through to a specified Steve Earle artist store. Here, catalog releases along with select merchandise could be purchased as well.

COORDINATING WITH OTHER DEPARTMENTS

Any media campaign should be designed in conjunction with the publicity department. A coordinated media campaign is necessary because the vehicles targeted for advertising are the same vehicles targeted for publicity and some synergy may occur. For example, placing an ad in a particular publication may increase the likelihood of getting some editorial coverage. Or at least, if it has been determined that the vehicle reaches the targeted market, those consumers will be receptive to the editorial content of the publication, as well as the advertisement content.

The placement of co-op advertising is usually secured through the sales department. Although ad dollars are designated, these funds are usually attached to a "buy" from the account that the co-op ad is being placed. Recognizing and leveraging the value of the ad dollars beyond that of just "impressions at the consumer level" is key for maximizing the advertising exchange.

There are a myriad of choices when deciding how to alert consumers as to a new release from an artist, and as technologies emerge, these options will continue to increase. What is important is to keep the target market top-of-mind with the budget close-to-the-vest. Understanding all the "players" in the drama with their various agendas will help keep the message from getting garbled. For in the end, the label needs to do the "right thing"—which is to sell records.

Remember that with any advertising campaign, you are looking to reach your audience is the most efficient and effective way, or the best bang for the buck. You also want to be sure your marketing efforts are not too focused on one medium. For instance, you wouldn't want to focus on television at the exclusion of any radio or digital ads. The goal is to have a plan that is comprehensive and includes a good mix of mediums to ensure, as much as we can anyway, that we are targeting our best consumers in multiple places with our sales message.

GLOSSARY

Banner Advertising—a type of contextual advertising that includes a designed message that allows for a click-through interface to another site, engaging the consumer with a sell-through missive.

Consumer Advertising—targeted advertising that speaks directly to consumers through mediums such a radio, television and publications that are consumed by consumers.

Contextual Advertising—Advertising on a website that is targeted to the specific individual who is visiting the web site based on the subject matter and then featuring products that relate. A contextual ad system scans the text of a web site for keywords and returns ads to the web page based on what the user is viewing, either through ads placed on the page or pop-up ads.

Cooperative advertising (co-op)—An advertisement where the cost of the advertising is funded by the manufacturer of the product and the retail outlet agrees to feature the product within the store.

Cost per Click (CPC)—Also referred to as Pay per Click, advertisers are charged for an online ad only when consumers actually click on it.

Cost per Point—a way to determine the cost for reaching one percent of the targeted audience.

Cost per thousand (CPM)—A dollar comparison that shows the relative cost of various media or vehicles; the figure indicates the dollar cost of advertising exposure to a thousand households or individuals.

Cume—total number of unique viewers of an advertisement over a period of time.

Dayparts—Specific segments of the broadcast day; for example, midday, morning drive time, afternoon drive time, late night.

DIY or Do-It-Yourself—refers to the independent artist who does not have a record deal and pursues and musical career using social media tools while creating the music that they like.

Frequency—the number of times the target audience will be exposed to a message.

Geotargeting—A method of detecting a website visitor's location to serve location-based content or advertisements.

Gross impressions—The total number of advertising impressions made during a schedule of commercials. GIs are calculated by multiplying the average persons reached in a specific time period by the number of spots in that period of time.

Gross rating point (GRP)—In broadcasting/cable, it means the size of the audience during two or more dayparts. GRPs are determined by multiplying the specific rating by the number of spots in that time period.

Keywords—search terms of words consumers are likely to use in a search engine when looking for a particular product or type of web site.

Media fragmentation—The division of mass media into niche vehicles through specialization of content and segmentation of audiences.

Media planning—Determining the proper use of advertising media to fulfill the marketing and promotional objectives for a specific product or advertiser.

Medium—a class of communication carriers such as television, newspapers, magazines, outdoor, and so on.

Out-of-Home (OOH)—advertising that targets consumers outside of his or home with four main categories: Billboards, Street Advertising, Transit Advertising, and Alternative Advertising such as building wraps.

Rating—In TV, the percentage of households in a market that are viewing a station divided by the total number of households with TV in that market. In radio, the total number of people who are listening to a station divided by the total number of people in the market.

Rating point (a rating of 1%)—1% of the potential audience; the sum of the ratings of multiple advertising insertions; for example, two advertisements with a rating of 10% each will total 20 rating points.

Rich Media Advertising—a type of contextual advertising that can include video and other interactive media with consumer.

ROI (return on investment)—deals with the money you invest in the company and the return you realize on that money based on the net profit of the business.

ROS (Run-of-Schedule)—the scheduling of advertising that the promotional spot will runt run any time within a specified timeframe, usually at radio or television. Typically less expensive than if a specific daypart or placement within particular programming was requested.

Road Blocking—When another channel in the same medium is running the same commercial in the same timeframe. This strategy is to catch channel surfers to expand the reach of the commercial to the same targeted market.

Reach—The total audience that a medium actually reaches; the size of the audience with which a vehicle communicates; the total number of people in an advertising media audience; the total percentage of the target group that is actually covered by an advertising campaign.

Search Engine Advertising—these advertisements are shown on search engine pages when keywords are searched. Used as part of Search Engine Optimization strategies when creating keyword elements in placing ads on search engines.

Targeted Rating Points (TRP)—similar to GRPs, but express the reach X frequency of only the most likely prospects of your advertising.

Trade advertising—Advertising aimed specifically for retailers and media gatekeepers through trade publications and mediums.

Trade publication—A specialized publication for a specific profession, trade, or industry; another term for some business publications.

Vehicle—A particular carrier within the group, such as *Rolling Stone* magazine or MTV network.

REFERENCES

Cano, A. *Advertising and Characteristics*. University of North Carolina at Pembroke, Dr. Anthony Curtis, 10 Accessed Nov. 16, 2014. https://prezi.com/mt05vp3x9hjj/copy-of-advertising-characteristics/.

Stuart, G. Mobile Advertising: Maybe Next Year . . . or the Year after That. (January 12, 2009). AdWeek Online. Accessed Nov. 16, 2014. http://www.adweek.com/aw/content_display/community/columns/other-columns/e3ice058ab1756ad1650d14d274af7e1ec8.

Epilogue

The Music Industry of the Future

"When I grow up I want to invent a technology that will disrupt something."

FIGURE 20.1

The only constant in the entertainment industry is change and much of that change comes from forces outside of the industry. The history of the recording industry is full of technological advances that improved the quality of recordings, made them more accessible to the masses and more convenient to use. The Internet is the first technology adopted by (forced upon) the music industry that has hurt rather than boosted sales. It didn't have to be this way (Knopper, 2009).

Instead of embracing the Internet from the beginning the recording industry resisted adopting the new technology, and although piracy may have declined as a percentage of music acquisition, other ways of consuming music, such as streaming, have replaced the sales of CDs. Shawn Fanning's Napster and Pirate Bay are not the only contributors to the commoditization of music. Chris Anderson, former editor of *Wired* magazine, has written several books and articles suggesting that all music should be made available on the Internet, but he also suggested that it should all be free.

And so it has come to pass that much of the newly recorded music is given away. Indie or major, signed or unsigned, musicians are giving away songs and even entire albums to entice listeners to buy tickets to their concerts (Fard, 2014). In September of 2014, Apple iTunes worked a deal with U2 to give away as many as 500 million copies of their new album, Song of Innocence, during the launch of the iPhone6 and Apple Watch. While U2 was reportedly paid around $100 million by Apple, the recipients of the album paid nothing, further enforcing the notion that music should be free to own (Reilly, 2014).

Exceptions and counter trends do exist. Taylor Swift and others have pushed back against streaming because it cannibalizes sales and pays a fraction of what is earned on a download or video streams (Dickey, 2014). The resurgence of vinyl, and at a premium price, and the proliferation of specially packaged box sets encourage fans to opt for owning music produced by their favorite artists, not just renting it from a streaming service.

Streaming services and live concerts are the growth areas in the music industry. Jim Donio, President of the Music Business Association says, "As we look toward the future of the music industry, the issue that looms largest is clearly the ongoing transition from a commercial marketplace that is predominately unit-based to a predominately access-based consumption model. This transition will affect the industry in any number of ways. We are already seeing that certain artists and genres perform far better in the access-based streaming model than others. There are also shifts in the ways we measure consumption that will allow us to pivot quickly" (Donio, 2015). Streaming services were popularized in Europe as early as 2008, even while most Americans were still resistant to the idea (Gloor and

Rolston, 2010). Wayne Rosso who was president of Grokster and Mash-boxx, in an interview for *The Future of the Music Business* said, "Where once [record labels] spurned the subscription model, they now want to push it as hard as possible. Unfortunately, it has proven a bust with consumers." (Gordon, 2008). My, how things have turned around!

But one thing has not changed in the last 100 years, and that is the fact that the industry was and is built around the recording of the song. The music business systems are built around the monetization, distribution and marketing of the recorded song. Think about it. Unless you are at a live performance, what you are listening to is recorded. And you are probably at that concert because you first heard a recording of the artist. Radio, television, Internet, movies, video games—they all have recorded music. Call them record companies or music companies or anything else you want but their primary function, to record and market music is the same and without that recording much of the rest of the entertainment world goes silent.

Record labels, companies that made and sold recorded music, are quickly becoming extinct. They have been replaced with the music company that still makes records but also oversees merchandising, touring and branding partnerships. The old business model structured around music as a physical product sold in brick and mortar stores as the primary source of income is giving way to the sale of digital product sold and streamed through online retailers. Music companies have moved away from focusing on providing product, selling off their manufacturing capabilities, and focusing on being more of a content and service provider for the artist's entire career—recording, touring, merchandise and even publishing, if they can get it. The role of labels of the future may be limited to developing content, and then marketing that content to consumers while others do the distribution.

This represents a drastic paradigm shift for record labels that resisted moving into the digital age. Not everyone sees this as a bad thing. Donald S. Passman, an entertainment attorney and author of *All You Need to Know About the Music Business* thinks,

> In the next few years, the music business may be larger than it's ever been. As subscription streaming services grow, we will get money from people who would have never gone near a record store (when those existed)—namely people older than their early twenties. And as to people who did buy music, the average consumer used to spend $40 per year on CDs. Subscriptions are $10 per month, or $120 per year. How that translates into money for artists is still to be seen, but the industry itself has enormous potential and artists will in time get their fair share

(Passman, 2015).

Casey Rae, the CEO of the Future of Music Coalition, agrees:

Technological shifts continue to impact the marketplace for music in ways that have made it difficult for many artists and music companies to adapt. Lawmakers are also feeling the pressure, as a great many of the policies that govern what is and isn't possible in the music industry were devised in an analog era. The complexities only increase from there. The Internet has made music a truly global phenomenon, which requires a more intentional approach to how rights are enumerated and revenue collected and distributed. The good news is that these changes have grown the potential market for music. Artists now have the opportunity to be heard in places that in years past would have been unreachable except by the biggest stars. For the industry, this means shifting to more of a service model, in which 'plays' overtake unit sales. To be sure, this impacts the economics for music creators, as well as up-front investment in songs and recordings. To ensure that artists aren't left behind in this transition requires a new commitment to transparency, accuracy and accountability. Any party that distributes music, manages rights or collects money from an expanding universe of access should be looking at not only how to grow the industry, but also how to best serve their artist partners. Technology—which caused so much disruption to traditional business models—may offer a new set of solutions. But at the end of the day, it's not a technical issue; it's a vision thing. To truly achieve a better future for music, artists must articulate theirs.

(Rae, 2015)

DAVE POMEROY PRESIDENT, NASHVILLE MUSICIANS ASSOCIATION, AFM LOCAL 257

We have seen dramatic shifts in the way in which an artists and labels promote their products. As we move towards more independent and artist owned labels competing with the majors, it has become essential to find ways to innovate without breaking the bank. Viral marketing and social media have created new opportunities and markets, but the new paradigm is not without its challenges.

Digital piracy dealt an enormous blow to the music industry as a whole, and while digital sales have finally begun to make up for the huge losses in sales of physical product, the concept of respect for intellectual property is still evolving. As an industry, it is essential that we re-educate the public about the need to compensate artists and copyright owners fairly, and find solutions that work for all shareholders, or generations of future musicians will find it very difficult to make a living.

The American Federation of Musicians is dedicated towards finding positive and constructive ways to solve problems, and we welcome the opportunity to be a part of this important and necessary dialogue as we move forward in a constantly evolving business environment. The future is what we make it.

Dave Pomeroy is a professional bassist, an independent artist, writer, studio musician, and producer. He has served as the president of Nashville Musicians Association, AFM Local 257 since June of 2010.

With that paradigm shift comes a shift in power. Record labels with their big budgets, worldwide distribution networks and marketing machines used to hold all the power. Today, music companies have positioned themselves more as partners with the artist, giving them more control, more favorable contracts (bye-bye controlled composition clauses) and larger royalties in exchange for smaller advances and participation in the revenue from booking and merchandising, services music companies now provide.

A NEW "RENTAL CULTURE"

Young people choosing to stream rather than purchase music may be part of a wider cultural shift away from ownership. Britany Robison wrote, "I'm addicted to the freedom of renting over buying, as are many of my friends and peers. Even those who are making enough to purchase a home are often opting to rent instead. We've seen the demise of the housing market, we've experienced the pain of digging our way out of debt, and we know the freedom of location-independence is available if we're interested" (Robinson, 2014). The rental culture is not exclusive to real estate, either. Auto Rental News reports that more than 50% of millennials choose car sharing over ownership (Zipcar, 2014). This finding is reinforced by the popularity of ride share services like Lyft and Uber.

There are always going to be those artists who break through on an emotional level and end up in people's lives forever. The way I see it, fans view music the way they view their relationships. Some music is just for fun, a passing fling (the ones they dance to at clubs and parties for a month while the song is a huge radio hit, that they will soon forget they ever danced to). Some songs and albums represent seasons of our lives, like relationships that we hold dear in our memories but had their time and place in the past.

> However, some artists will be like finding "the one." We will cherish every album they put out until they retire and we will play their music for our children and grandchildren. As an artist, this is the dream bond we hope to establish with our fans. I think the future still holds the possibility for this kind of bond, the one my father has with the Beach Boys and the one my mother has with Carly Simon.
>
> —Taylor Swift July 7 2014—Wall Street Journal "For Taylor Swift, the Future of Music is a Love Story"

When it comes to entertainment, the same rental culture prevails. Millennials and their younger siblings have all but abandon television for streamed viewing on their schedule. Netflix, Hulu, iTunes, Amazon Prime, and others have catered to mostly young consumers who want to watch on their schedule rather than the networks'. Streaming music rather than buying CDs is just one more piece of the young person's rental culture.

Increased Branding and Partnerships

Next Big Sounds 2014 Industry Report begins with this headline: "Brands. No longer a dirty word in the music industry " (Next Big Sound, 2014). Branding and brand partnerships have taken on tremendous importance under the new model and we believe it will continue to do so in the near future. Labels have reduced the money spent on tour support and it is being replaced with strategic partnerships of all sorts. Universal and their labels account for 57% of branding deals while Sony, Warner, and the indies each account for about 15%. Weird Al Yankovic had his first number one album, *Mandatory Fun*, at age 55 as much because of the partnerships created to market the videos as the content of the music itself. Yankovic partnered with eight different websites giving them each a limited-time exclusive in exchange for funding the production of the video. "Because RCA did not provide any production budget, Mr. Yankovic said, the videos were paid for by various partner sites that brought their own audiences, like Nerdist, Funny or Die, and College Humor" (Sisario, 2014). The partnership is mutually beneficial: "These sites benefit from extra traffic and ad revenue, while Al gets their extra promotional muscle in a different corner of the online world. And getting these third parties to pick up video costs means that Al's label **RCA** isn't paying for it either, meaning the record will potentially recoup faster" (Erikson, 2014). Yankovic's sponsorship, like Lady Gaga and Doritos', is not unique but is interesting because of who the brands are and the possibilities they foretell.

Labels and Radio Doing Direct Licensing

Another trend to watch is radio stations signing direct agreements with record labels to pay performance royalties. Big Machine Label Group entered into agreements with Cox Meida Group, Entercom, and Clear-Channel, who also did a deal with Warner Music, to pay the labels and their artist for the use of their recordings on the radio (Big Machine and Cox Media Seal Direct Licensing Deal). In the U.S. terrestrial radio stations are not required to pay royalties of any sort to artists or their labels (just to songwriters and publishers). But these deals include a direct payment for the use of the songs on the radio groups' Internet properties (e.g., iHeartRadio) as well, circumventing Sound Exchange and their direct payment to the artist. Online streaming service, Pandora also entered direct licensing agreements with Sony/ATV and BMG Publishing and independent label group, Merlin (Arcade Fire, Lenny Kravitz). Although this will probably have little effect on music marketing it may mean that recording artists will have their Internet royalties used to recoup their advances.

More important to music marketers, Edison Research confirmed in the fall of 2014 that American teenagers listened to more streamed music (64 minutes per day) than terrestrial radio (53 minutes per day). UK teenagers still listen to more broadcast radio (124 minutes per day) than streaming (81 minutes per day) (Cridland, 2015). How consumers discover new music has changed with the Internet. The IFPI Investing in Music report says, "Record companies have generally switched their marketing spend towards digital platforms, which enable them to more tightly target individual groups of consumers. Yet the costs of online marketing are increasing, and record companies also have to continue to advertise across television, radio, print and outdoor media to drive awareness of their artists" (IFPI, 2014). Young music consumers are bypassing the radio stations their parents depended on for music discovery and using the Internet to find their music. In the process, they are also bypassing the filtering of music by record labels and radio stations. Instead, they rely heavily on social media and their own research to discover new music. And record labels are too. Martin Mills, chairman of Beggars Group whose artists include Adele, told the IFPI, "We've been discovering artists online for the last decade or so. We're not looking for the metrics, the most hits or the most views, but we're using online platforms as a tool to find artists we are interested in" (IFPI, 2014). We have not yet hit a new equilibrium and the consumers' time and the labels budget will probably continue to shift toward the Internet for the foreseeable future.

CONTINUED FRAGMENTATION OF THE MARKET

Social media allows friends, no matter how separated by time or distance, to share music and make recommendations. The declining costs of music production technology and digital storage make it possible for almost anybody to record a song and make it available on the Internet. This is what Chris Anderson suggested in "The Long Tail"—make everything available (Anderson, 2004). How has this played out in the Internet world? A study conducted by Will Page and Andrew Bud of MCPS-PRS in 2008 of the 13 million songs available for download, more than 10 million had failed to sell a single copy (Michaels, 2008). The number of new releases in the U.S. peaked that year at 105,575 different titles and continued to decline to about 75,000 in 2011 before rebounding slightly in 2012 to 76,875 titles. Every conceivable genre and every niche market is represented on the Internet, but without the distribution and marketing, the kind provided by record label professionals, it appears that most of them will never be discovered, not even by the artist's own mother. Wills and Bud's research bares this out finding the 52,000 of the 13 million tracks accounted for 80% of all revenue. Maybe we need the hits after all.

But a hit record isn't what it used to be. The best selling album of 2014 was Taylor Swift's *1989* with 3.66 million units sold, including 1.4 million digital copies. This is a significant increase over 2013's *The 20/20 Experience* (Justin Timberlake, 2.43 million), but a far cry from NSYNC's *No Strings Attached* which sold 9.936 million copies in 2000 alone. In fact, *1989* would not have made the top 12 in 2000 (Ryneski, 2001). The industry returned to a singles models as evidenced by the over 9 million tracks downloaded from *1989* and other top sellers in 2014, but appears ready to move into a streaming model based on shifting trends in music consumption. Nielsen's end of year report indicates that year over year sales were down in almost every category, including digital downloads of albums and songs, but on demand streams of audio and video increased 54% to 164 billion (Houghton, 2015).

MORE MOBILE THAN EVER

Music is already more mobile than ever and all indications are that it will continue in that direction. Laptops, tablets, and Internet-connected phones were just the beginning. Now automobiles and watches connect to the Web and access our Spotify, iHeartRadio, and iTunes accounts so we can access our playlists and favorite songs. All of this information is stored in the cloud so it can be accessed anywhere there is an Internet connection. According to Apple analyst Gene Munster, neither iTunes nor the iPad makes the top five in the list of Apple's priorities; but the Apple Watch and iPhone do (Yarrow, 2015). iTunes radio has already overtaken Spotify, and is now the third largest music streaming service by market share (Elmer-DeWitt, 2015). Apple now dominates both software and hardware when it comes to music.

WATCHING THE HORIZON

Traditional business management and strategy warns us to keep our eyes not only on our own industry, however broadly we have defined it, but also to keep our eyes on other industries that may impact our own. As we said previously, every major change in the music industry has been a result of some new technology, many of which originated from the outside. A PricewaterhouseCoopers survey of CEOs found that a full 86% of respondents believe that technological advances will most transform their business in the coming years ("CEOs"). Since we often can't predict which technologies will impact our business or how we must remain flexible and open to change when it comes. Embracing and adapting to change is how we survive and those who do it most quickly and adeptly will reap the greatest rewards.

REFERENCES

Anderson, C. "The Long Tail." *Wired*, October 2004. Accessed January 23, 2015.

"Big Machine and Cox Media Seal Direct Licensing Deal." Big Machine Label Group, June 12, 2014. Accessed November 13, 2014.

"CEOs Foresee Changes Resulting from Technology." *emarketer.com*. N.p., February 6, 2014. Accessed January 23, 2015.

Cridland, J. "Streaming Overtakes Radio Use for US Teens: but UK Fares Better." *Media Info*. N.p., January 21, 2015. Accessed January 23, 2015.

Dickey, J. "Taylor Swift on 1989, Spotify, Her Next Tour and Female Role Models." Time.com. Time Inc., November 13, 2014. Accessed November 13, 2014.

Donio, J. Personal interview. January 16, 2015.

Elmer-DeWitt, P. "iTunes Radio overtakes Spotify, gaining on iHeartRadio in U.S." *Fortune*. N.p., March 11, 2014. Accessed January 23, 2015.

Enderle, R. Starbucks and HP: The Future of Digital Music, Ecommerce Times. http://www.ecommercetimes.com/story/33164.html. 2004.

Erikson, K. "5 Lessons for the Music Industry from 'Weird Al' Yankovic." *Hypebot. com*. N.p., July 29, 2014. Accessed January 19, 2015.

Fard, F. J. "4 Reasons Why Giving Away Your Tracks Can Be Priceless." Sonicbids Blog. N.p., July 10, 2014. Accessed January 19, 2015.

Gloor, S. and Rolston, C. P. "Can The Madness Be Monetized? An Exploratory Survey of Music Piracy and Acquisition Behavior." *Journal of the Music & Entertainment Industry Educators Association* Vol. 10, No. 1 (2010).

Gordon, S. *The Future of the Music Business*. 2nd ed. Milwaukee, WI: Hal Leonard, 2008. 317.

Houghton, B. "Music Sales Down In Almost Every Category In 2014." *hypebot. com*. N.p., January 2, 2015. Accessed January 23, 2015.

Investing in Music. Ed. Placido Domingo. IFPI, 2014.

Knopper, S. *Appetite for Self-destruction: The Spectacular Crash of the Record Industry in the Digital Age*. New York: Free Press, 2009.

Michaels, S. "Most music didn't sell a single copy in 2008." *The Guardian*. N.p., December 23, 2008. Accessed January 23, 2015.

"Next Big Sound Presents 2014: State of the Industry." *Next Big Sound*. N.p., January 2015. Accessed January 23, 2015.

Passman, D. S. Personal Interview. January 14, 2015.

Rae, C. Personal Interview. January 13, 2015.

Reilly, D. "U2's iTunes Deal Reportedly Cost Apple $100 Million." *SPIN*. N.p., September 12, 2014. Accessed January 19, 2015.

Robinson, B. *The Cultureist*. N.p., August 18, 2014. Accessed January 19, 2015.

Ryneski, P. *Neo Soul*. N.p., January 19, 2001. Accessed January 23, 2015. http://www.neosoul.com/music/2000/bsalbums.html.

Sandoval, G. "Labels Size up Web 2.0 Music Services." CNet News. http://news.cnet.com/8301-1023_3-10184877-93.html. March 2, 2009.

Sisario, B. "Weird Al Yankovic Scores with 'Mandatory Fun'." *The New York Times*. N.p., July 23, 2014. Accessed January 19 2015.

Yarro, J. "Here are the Top 5 Priorities for Apple, According to Top Apple Analyst Gene Munster." *Business Insider*. N.p., January 22, 2015. Accessed January 23, 2015.

"Zipcar: More than 50% of Millennials Choose Car Sharing Over Driving Own Car." Auto Rental News. N.p., January 28, 2014. Accessed January 19, 2015.

Index